PATTERN RECOGNITION

PATTERN RECOGNITION
A Quality of Data Perspective

WŁADYSŁAW HOMENDA
The Faculty of Mathematics and Information Science, Warsaw University of Technology
Warsaw, Poland
and
The Faculty of Economics and Informatics in Vilnius, University of Białystok
Vilnius, Lithuania

WITOLD PEDRYCZ
The Systems Research Institute, Polish Academy of Sciences
Warsaw, Poland
and
The Department of Electrical and Computer Engineering, University of Alberta
Edmonton, Alberta, Canada

This edition first published 2018
© 2018 John Wiley & Sons, Inc.

The right of Władysław Homenda and Witold Pedrycz to be identified as the authors of this work has been asserted in accordance with law.

Registered Offices
John Wiley & Sons, Inc., 111 River Street, Hoboken, NJ 07030, USA

Editorial Office
111 River Street, Hoboken, NJ 07030, USA

For details of our global editorial offices, customer services, and more information about Wiley products visit us at www.wiley.com.

Wiley also publishes its books in a variety of electronic formats and by print-on-demand. Some content that appears in standard print versions of this book may not be available in other formats.

Library of Congress Cataloging-in-Publication Data

Names: Homenda, Władysław, author. | Pedrycz, Witold, 1953- author.
Title: Pattern recognition : a quality of data perspective / by Władysław
    Homenda, Witold Pedrycz.
Description: Hoboken, NJ : John Wiley & Sons, 2018. | Series: Wiley series on
    methods and applications in data mining | Includes bibliographical
    references and index. |
Identifiers: LCCN 2017045206 (print) | LCCN 2017055746 (ebook) | ISBN
    9781119302834 (pdf) | ISBN 9781119302858 (epub) | ISBN 9781119302827
    (cloth)
Subjects: LCSH: Pattern recognition systems. | Pattern perception. | Data
    mining.
Classification: LCC TK7882.P3 (ebook) | LCC TK7882.P3 H66 2018 (print) | DDC
    006.4–dc23
LC record available at https://lccn.loc.gov/2017045206

Cover design by Wiley
Cover image: © turbodesign777/Gettyimages

Set in 10/12pt Times by Spi Global, Pondicherry, India

Printed in the United States of America

10 9 8 7 6 5 4 3 2 1

CONTENTS

PREFACE

Pattern recognition has established itself as an advanced area with a well-defined methodology, a plethora of algorithms, and well-defined application areas. For decades, pattern recognition has been a subject of intense theoretical and applied research inspired by practical needs. Prudently formulated evaluation strategies and methods of pattern recognition, especially a suite of classification algorithms, constitute the crux of numerous pattern classifiers. There are numerous representative realms of applications including recognizing printed text and manuscripts, identifying musical notation, supporting multimodal biometric systems (voice, iris, signature), classifying medical signals (including ECG, EEG, EMG, etc.), and classifying and interpreting images.

With the abundance of data, their volume, and existing diversity arise evident challenges that need to be carefully addressed to foster further advancements of the area and meet the needs of the ever-growing applications. In a nutshell, they are concerned with the data quality. This term manifests in numerous ways and has to be perceived in a very general sense. Missing data, data affected by noise, foreign patterns, limited precision, information granularity, and imbalanced data are commonly encountered phenomena one has to take into consideration in building pattern classifiers and carrying out comprehensive data analysis. In particular, one has to engage suitable ways of transforming (preprocessing) data (patterns) prior to their analysis, classification, and interpretation.

The quality of data impacts the very essence of pattern recognition and calls for thorough investigations of the principles of the area. Data quality exhibits a direct impact on architectures and the development schemes of the classifiers. This book aims to cover the essentials of pattern recognition by casting it in a new perspective of data quality—in essence we advocate that a new framework of pattern recognition along with its methodology and algorithms has to be established to cope with the challenges of data quality. As a representative example, it is of interest to look at the problem of the so-called foreign (odd) patterns. By foreign patterns we mean patterns not belonging to a family of classes under consideration. The ever-growing presence of pattern recognition technologies increases the importance of identifying foreign patterns. For example, in recognition of printed texts, odd patterns (say, blots, grease, or damaged symbols) appear quite rarely. On the other hand, in recognition problem completed for some other sources such as geodetic maps or musical notation, foreign patterns occur quite often and their presence cannot be ignored. Unlike printed text, such documents contain objects of irregular positioning, differing in size, overlapping, or having complex shape. Thus, too strict segmentation results in the rejection of many recognizable symbols. Due to the weak separability of recognized patterns, segmentation criteria need to be relaxed and foreign patterns similar to recognized symbols have to be carefully inspected and rejected.

The exposure of the overall material is structured into two parts, Part I: Fundamentals and Part II: Advanced Topics: A Framework of Granular Computing. This arrangement reflects the general nature of the main topics being covered.

Part I addresses the principles of pattern recognition with rejection. The task of a rejection of foreign pattern arises as an extension and an enhancement of the standard schemes and practices of pattern recognition. Essential notions of pattern recognition are elaborated on and carefully revisited in order to clarify on how to augment existing classifiers with a new rejection option required to cope with the discussed category of problems. As stressed, this book is self-contained, and this implies that a number well-known methods and algorithms are discussed to offer a complete overview of the area to identify main objectives and to present main phases of pattern recognition. The key topics here involve problem formulation and understanding; feature space formation, selection, transformation, and reduction; pattern classification; and performance evaluation. Analyzed is the evolution of research on pattern recognition with rejection, including historical perspective. Identified are current approaches along with present and forthcoming issues that need to be tackled to ensure further progress in this domain. In particular, new trends are identified and linked with existing challenges. The chapters forming this part revisit the well-known material, as well as elaborate on new approaches to pattern recognition with rejection. Chapter 1 covers fundamental notions of feature space formation. Feature space is of a paramount relevance implying quality of classifiers. The focus of the chapter is on the analysis and comparative assessment of the main categories of methods used in feature construction, transformation, and reduction. In Chapter 2, we cover a variety of design approaches to the design of fundamental classifiers, including such well-known constructs as k-NN (nearest neighbor), naïve Bayesian classifier, decision trees, random forests, and support vector machines (SVMs). Comparative studies are supported by a suite of illustrative examples. Chapter 3 offers a detailed formulation of the problem of recognition with rejection. It delivers a number of motivating examples and elaborates on the existing studies carried out in this domain. Chapter 4 covers a suite of evaluation methods required to realize tasks of pattern recognition with a rejection option. Along with classic performance evaluation approaches, a thorough discussion is presented on a multifaceted nature of pattern recognition evaluation mechanisms. The analysis is extended by dealing with balanced and imbalanced datasets. The discussion commences with an evaluation of a standard pattern recognition problem and then progresses toward pattern recognition with rejection. We tackle an issue of how to evaluate pattern recognition with rejection when the problem is further exacerbated by the presence of imbalanced data. A wide spectrum of measures is discussed and employed in experiments, including those of comparative nature. In Chapter 5, we present an empirical evaluation of different rejecting architectures. An empirical verification is performed using datasets of handwritten digits and symbols of printed music notation. In addition, we propose a rejecting method based on a concept of geometrical regions. This method, unlike rejecting architectures, is a stand-alone approach to support discrimination between native and foreign patterns. We study the usage of elementary geometrical regions, especially hyperrectangles and hyperellipsoids.

Part II focuses on the fundamental concept of information granules and information granularity. Information granules give rise to the area of granular computing—a paradigm of forming, processing, and interpreting information granules. Information granularity comes hand in hand with the key notion of data quality—it helps identify, quantify, and process patterns of a certain quality. The chapters are structured in a way to offer a top-down way of material exposure. Chapter 6 brings the fundamentals of information granules delivering the key motivating factors, elaborating on the underlying formalisms (including sets, fuzzy sets, probabilities) along with the operations and transformation mechanisms as well as the characterization of information granules. The design of information granules is covered in Chapter 7. Chapter 8 positions clustering in a new setting, revealing its role as a mechanism of building information granules. In the same vein, it is shown that the clustering results (predominantly of a numeric nature) are significantly augmented by bringing information granularity to the description of the originally constructed numeric clusters. A question of clustering information granules is posed and translated into some algorithmic augmentations of the existing clustering methods. Further studies on data quality and its quantification and processing are contained in Chapter 9. Here we focus on data (value) imputation and imbalanced data—the two dominant manifestations in which the quality of data plays a pivotal role. In both situations, the problem is captured through information granules that lead to the quantification of the quality of data as well as enrich the ensuing classification schemes.

This book exhibits a number of essential and appealing features:

Systematic exposure of the concepts, design methodology, and detailed algorithms. In the organization of the material, we adhere to the top-down strategy starting with the concepts and motivation and then proceeding with the detailed design materializing in specific algorithms and a slew of representative applications.

A wealth of carefully structured and organized illustrative material. This book includes a series of brief illustrative numeric experiments, detailed schemes, and more advanced problems.

Self-containment. We aimed at the delivery of self-contained material providing with all necessary prerequisites. If required, some parts of the text are augmented with a step-by-step explanation of more advanced concepts supported by carefully selected illustrative material.

Given the central theme of this book, we hope that this volume would appeal to a broad audience of researchers and practitioners in the area of pattern recognition and data analytics. It can serve as a compendium of actual methods in the area and offer a sound algorithmic framework.

This book could not have been possible without support provided by organizations and individuals.

We fully acknowledge the support provided by the National Science Centre, grant No 2012/07/B/ST6/01501, decision no. UMO-2012/07/B/ST6/01501.

Dr Agnieszka Jastrzebska has done a meticulous job by helping in the realization of experiments and producing graphic materials. We are grateful to the team of professionals at John Wiley, Kshitija Iyer, and Grace Paulin Jeeva S for their encouragement from the outset of the project and their continuous support through its realization.

Władysław Homenda and Witold Pedrycz

PART I
FUNDAMENTALS

CHAPTER 1

PATTERN RECOGNITION: FEATURE SPACE CONSTRUCTION

In this chapter, we proceed with a more detailed discussion on the essence and concepts of pattern recognition. We focus on the initial phase of the overall scheme that is focused on feature formation and analysis as well as feature selection. Let us emphasize that, in general, patterns come in various forms: images, voice recordings, text in some natural language, a sequence of structured information (tuples formed according to some key), and so on. A pattern described through a collection of features can be regarded as a generic chunk of information. Features, generally speaking, are descriptors of the patterns. Naturally, the number of features, their nature, and quality influence the quality of ensuing modeling, especially classification. In this chapter, we look at these issues in detail.

The chapter is structured as follows. First, we formulate a theoretical basis of the problems of pattern recognition. We introduce necessary notation and concepts required in the ensuing discussion across the overall book. We formally define features and pattern recognition process. Next, we present practical approaches to feature extraction applied to visual pattern recognition. In particular, we use symbols of printed music notation as examples of patterns. Then, we discuss some elementary feature transformations. Finally, we present various strategies developed for feature selection.

1.1 CONCEPTS

Formally, a standard pattern recognition problem is a task of splitting a set of objects (patterns)

$$O = \{o_1, o_2, \ldots\}$$

Pattern Recognition: A Quality of Data Perspective, First Edition. Władysław Homenda and Witold Pedrycz.

into subsets composed of objects belonging to the same class

$$O_1, O_2, ..., O_C$$

such that

$$O = \bigcup_{l=1}^{C} O_l \text{ and } (\forall l, k \in \{1, 2, ..., C\}, l \neq k) O_l \cap O_k = \varnothing \tag{1.1}$$

A task that results in the formation of subsets $O_1, O_2, ..., O_C$ is defined by a mapping called *a classifier*

$$\Psi: O \to \Theta \tag{1.2}$$

where $\Theta = \{O_1, O_2, ..., O_C\}$ is the set of classes under consideration. For the sake of simplicity, we assume that the mapping Ψ takes values from the set of class indices $\Theta = \{1, 2, ..., C\}$, that is, class labels, instead of classes themselves.

Pattern recognition is usually performed on some observed set of features that characterize objects, rather than on objects directly. Therefore, we formulate and distinguish a mapping from the space of objects O into the space features X:

$$\varphi: O \to X \tag{1.3}$$

The mapping φ is called a features extractor. Subsequently, we consider a mapping from the space of features into the space of classes

$$\psi: X \to \Theta \tag{1.4}$$

Such a mapping is called a classification algorithm or simply a *classifier*. It is important to notice that the term *classifier* is used in different contexts: classification of objects, classification of features characterizing objects, and, more precisely, classification of vectors of features from features space. The meaning of this term can be inferred from the context. Therefore, in the ensuing considerations we will not distinguish explicitly which meaning is being used. A composition of the previously mentioned two mappings constitutes the classifier: $\Psi = \psi \circ \varphi$. In other words, the mapping $O \xrightarrow{\Psi} \Theta$ can be decomposed into the two mappings realized in series $O \xrightarrow{\varphi} X \xrightarrow{\psi} \Theta$.

In general, a classifier Ψ is not known, that is, we do not know the class that a pattern belongs to. However, in pattern recognition problems, it is assumed that the classifier Ψ is known for some subset of the space of patterns called a *learning set*, namely, labels of patterns are known in supervised learning task. A learning set is a subset of the set of objects, $L \subset O$ for which class labels are known, that is, for any pattern from the learning set $o \in L$, the value $\Psi(o)$ is known. Building a classifier $\Psi: O \to \Theta$ with the use of known classes of objects from a learning set, that is, a known mapping $\Psi: L \to \Theta$ is the ultimate goal of pattern recognition. A premise motivating this action is that we hope that having a *good enough* subset of all objects will let us construct a classifier that will be able to successfully assign correct class labels to all patterns. Summarizing, we explore the pattern recognition problem as a design problem aimed at the development of a classifier regarded as the following mapping:

$$\Psi: O \to \Theta$$

assuming that we are provided with $L \subset O$, a subset of all objects with correctly assigned class labels. Such a classifier is decomposed to a feature extractor

$$\varphi : O \to X$$

and a (features) classifier (or, in other words, a classification algorithm)

$$\psi : X \to \Theta$$

as illustrated in Figure 1.1.

Both the feature extractor and the classification algorithm are built based on a learning set L. The classifier ψ divides features space into so-called decision regions,

$$D_X^{(l)} = \psi^{-1}(l) = \{x \in X : \psi(x) = l\} \quad \text{for every } l \in \Theta \qquad (1.5)$$

and then, of course, the features extractor splits the space of objects into classes

$$O_l = \varphi^{-1}\left(D_X^{(l)}\right) = \varphi^{-1}\left(\psi^{-1}(l)\right) \quad \text{for every } l \in \Theta \qquad (1.6)$$

(a)

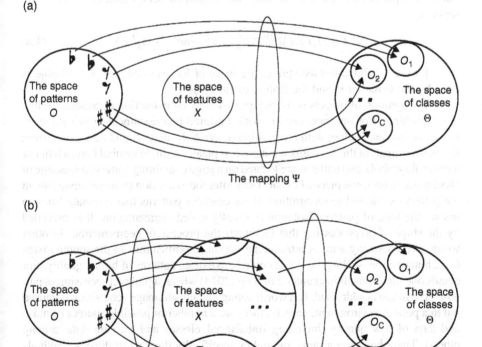

(b)

Figure 1.1 Pattern recognition schemes: direct mapping from the space of patterns into the space of classes (a) and composition of mappings from the space of patterns into the space of features and from the space of features into the space of classes (b).

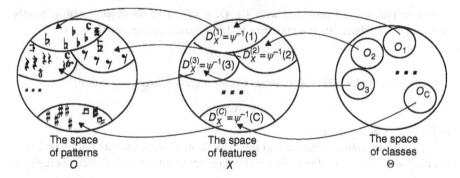

The space of patterns O The space of features X The space of classes Θ

Figure 1.2   A typical pattern recognition scheme.

or equivalently

$$O_l = \Psi^{-1}(l) = (\psi \circ \varphi)^{-1}(l) = \varphi^{-1}\left(\psi^{-1}(l)\right) \quad \text{for every } l \in \Theta \qquad (1.7)$$

We assume that the classification algorithm splits the space of feature values, that is, it separates the space X into pairwise disjoint subsets, which cover the entire space X:

$$(\forall l, k \in M, l \neq k) D_X^{(l)} \cap D_X^{(k)} = \emptyset \quad \text{and} \quad \bigcup_{l \in M} D_X^{(l)} = X \qquad (1.8)$$

Figure 1.2 illustrates the split of the space of features and the space of objects done by the classifier ψ and the feature extractor φ.

Recognition of objects is usually preceded by an extraction of patterns from a given problem. For instance, dealing with a printed text document or printed sheet music, before proceeding with recognition of symbols, they should be isolated from the environment. In this scenario, a pattern is typically a single symbol (say, a letter or a musical symbol), and patterns are located on a page containing some text message or sheet music with some piece of music. Only after the extraction from the environment are patterns subjected to recognition. If we consider patterns that originate from an image, the task of patterns isolation is usually called segmentation. It is preceded by the stage of preprocessing that facilitates the process of segmentation. In other words, preprocessing aims at introducing various modifications to the source image (e.g., binarization, scaling, etc.) that could help extract patterns of higher quality. For details one may consult in chapter 2 of Krig (2014) where signal preprocessing in pattern recognition is addressed. It is worth noting that not all image acquisition is carried out in a perfect environment, namely, there are a number of possible sources of noise and data of low quality (including imbalanced classes and missing data, among others). There has been a range of studies specifically directed to develop methods for image preprocessing for poor quality signals, for instance, with difficult lighting conditions (Tan and Triggs, 2010) or noise (Haris *et al.*, 1998).

Not all pattern recognition tasks have well-defined and clearly delineated preprocessing and symbols extraction (segmentation) stages. Automatic patterns acquisition often produces excessive, undesirable symbols and ordinary garbage. Let us refer to such patterns as *foreign patterns*, in contrast to *native patterns* of proper, recognized classes (cf. Homenda *et al.*, 2014, 2016). In such a case a classification module, which assigns all extracted symbols to designed classes (proper classes of native symbols, labeled and present in the learning set), will produce misclassification for every undesirable symbol and for every garbage symbol. In order to improve the performance of the classification procedure, it is required to construct such classifiers that could assign native symbols to correct class labels and reject undesirable and garbage symbols.

Rejection of symbols can be formally interpreted by considering a new class O_0, to which we classify all undesirable and garbage symbols. In consequence, we can distinguish a classification decision region, which separates foreign symbols from useful ones through the classifier ψ:

$$D_X^0 = \{x \in X : \psi(x) = 0\} \tag{1.9}$$

This new class (decision region) D_X^0 is a distinct subspace of the space X,

$$(\forall l \in C)D_X^{(l)} \cap D_X^{(0)} = \varnothing \quad \text{and} \quad X = D_X^{(0)} \cup \bigcup_{i \in C} D_X^{(l)} \tag{1.10}$$

where, of course, all former classes $D_X^{(l)}, l \in \Theta$ are pairwise disjoint. Rejecting foreign symbols implies a certain problem. Unlike objects of proper classes, foreign symbols are usually not similar to one another and do not create a consistent class. They are not well positioned in the feature space. Moreover, most often they are not available at the stage of classifier construction. Therefore, instead of distinguishing a decision region corresponding to a family of foreign objects, it is reasonable to separate areas outside of decision regions of native objects (cf. Homenda *et al.*, 2014). Of course, in such a case, we assume that decision regions of native symbols cover only their own areas and do not exhaust the whole feature space X. An area outside of decision regions of native objects can be formally defined in the following form:

$$D_X^0 = X - \bigcup_{i \in C} D_X^{(i)} \tag{1.11}$$

This is illustrated in Figure 1.3.

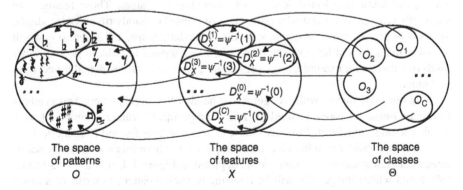

| The space of patterns | The space of features | The space of classes |
| O | X | Θ |

Figure 1.3 Pattern recognition with rejection.

1.2   FROM PATTERNS TO FEATURES

From now on we will use the term *pattern* for objects being recognized as well as for features describing and representing these objects. The exact meaning of this term can be inferred from the context of its usage and, if necessary, it could be explicitly indicated.

Let us distinguish two kinds of features characterizing patterns:

- Numerical features—features whose values form a set of real numbers
- Categorical features—features that are valued in a finite set of values

Values of categorical features can be of any type, for instance, names over some alphabet. Since sets of values of categorical features are finite, they can be enumerated, and values of categorical features can be cast on their indices. Therefore, for the sake of simplicity, categorical features are considered to be numerical ones.

The space of features X is the Cartesian product of individual features X_1, $X_2, \ldots, X_M$, that is, $X = X_1 \times X_2 \times \cdots \times X_M$. Therefore, mappings φ and ψ operate on vectors $(x_1, x_2, \ldots, x_M)^T$. Such vectors are values of the mapping φ and arguments of the mapping ψ. An i-th element of the vector is denoted x_i, $i = 1, 2, \ldots, M$, and it describes the value of the i-th feature value. For the sake of simplicity, a vector of values of features $\mathbf{x} = (x_1, x_2, x_3, \ldots, x_M)$ will be simply called a vector of features or a feature vector.

Now, let us focus on patterns represented as monochrome images, say black and white images, which are in fact rectangular tables with elements called black/white pixels. In this book, we concentrate the discussion on scanned handwritten digits, handwritten letters, and symbols of printed music notation, and we rarely switch to other types of patterns. This choice is motivated by the fact that, in the context of the methods studied here, such patterns have superior illustrative properties. However, it should be clear that the studied methods are of a general nature and could be easily applied to other types of patterns.

As mentioned before, pattern recognition rarely applies to patterns directly, that is, to patterns present in their original form. In almost each case of pattern recognition, features describing patterns are processed. This observation motivates us to discuss in subsequent sections selected features of monochrome images. These features are especially relevant if we consider processing of printed or handwritten letters, digits, symbols of printed music notation, symbols of geodetic maps, and so on. It is worth stressing that different features can be acquired and applied in order to process other kinds of patterns, for example, those present in speech recognition, signal processing, and others.

In our experiments with recognition of handwritten digits, letters, and symbols of printed music notation, we used the following groups of features: numerical, vectorial, vectorial transformed to vectorial, and vectorial transformed to numerical.

Let us consider a treble clef, an example of a pattern taken from a dataset of music notation symbols. A treble clef is depicted in Figure 1.4. It is a monochrome (black and white) image. We will be referring to such a pattern in terms of a raster scan: a rectangular collection of pixels that we locate inside a *bounding box* of width

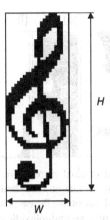

Figure 1.4   A treble clef, a symbol belonging to a data set of printed music notation, taken as an example of a pattern. The pattern is surrounded by a bounding box of width $W = 22$ and height $H = 60$ pixels. The bounding box is not part of the pattern; it has been added only for illustrative purposes.

W and height H (cf. Figure 1.4). In other words, a bounding box is the smallest rectangle part of an image enclosing a pattern. In Figure 1.4, the bounding box is identified as a frame used in order to clearly identify the smallest rectangle of pixels enclosing a pattern.

Specifically, a raster scan pattern is represented as a mapping:

$$I: \langle 1, H \rangle \times \langle 1, W \rangle \rightarrow \{0, 1\} \quad I(i,j) = \begin{cases} 1 & \text{for black pixel} \\ 0 & \text{for white pixel} \end{cases} \quad (1.12)$$

1.2.1   Vectorial Features

Only a very limited number of numerical features, effectively employed in pattern recognition problems, can be derived directly from patterns. These features are discussed later in this chapter. However, many numerical features are derived indirectly from vectorial ones.

Vectorial features are usually created based on a bounding box of a given pattern (cf. Figures 1.5 and 1.6). Now, let us discuss the most prominent examples of vectorial features of monochrome images: projections, margins, and transitions.

1. Horizontal and vertical projections:
   - Horizontal projection is a vector of numbers of black pixels in rows.
   - Vertical projection is a vector of numbers of black pixels in columns.

   Therefore, horizontal projection is a vector (of numbers) of length equal to the height of the bounding box (H), while vertical projection is a vector (of numbers) of the length equal to the width of the bounding box (W):

(a)　　(b)　　(c)　　(d)　　(e)　　(f)　　(g)　　(h)　　(i)

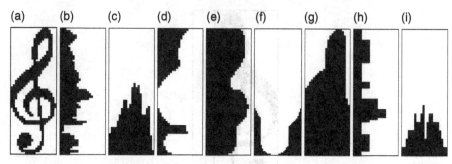

Figure 1.5 Vectorial features: (a) original pattern, (b) horizontal projection, (c) vertical projection, (d) left margin, (e) right margin, (f) bottom margin, (g) top margin, (h) horizontal transition, and (i) vertical transition. Please note that the transition values are very small, so in order to enhance visibility, we multiplied them by 4.

(a)　　(b)　　(c)　　(d)　　(e)　　(f)　　(g)　　(h)　　(i)

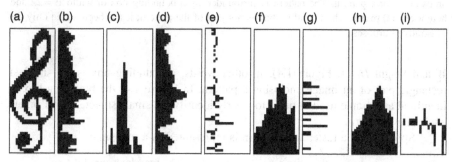

Figure 1.6 Vectorial to vectorial transformations: (a) original pattern, (b) horizontal projection, (c) its histogram, (d) its smoothing, (e) its differentiation, (f) vertical projection, (g) its histogram, (h) its smoothing, and (i) its differentiation. Note: The values of the vertical histogram are multiplied by 4.

$$ProjH(i) = \sum_{j=1}^{W} I(i,j) \quad i = 1, 2, ..., H$$

$$ProjV(j) = \sum_{i=1}^{H} I(i,j) \quad j = 1, 2, ..., W$$

(1.13)

2. Left, right, bottom, and top margins:

- The left margin are indices of the last white pixels preceding the first black ones, pixel indexing starts from 1 from the left side of each row. It is zero if a row begins with a black pixel; it is W (the width of the bounding box) if no black pixel is in the row.

- The right margin are indices of the last black pixels, pixel indexing starts from 1 from the left side in each row; it is 0 if no black pixel is in the row.

- The bottom and top margins are defined like left and right margins, indexing starts from 1 and goes from bottom to top in each column.

Hence, left and right margins are vectors (of numbers) of lengths equal to the height of the bounding box, while bottom and top margins are vectors (of numbers) of length equal to its width; the detailed formulas for margins are as follows:

$$MargL(i) = \begin{cases} W & \text{if } \sum_{j=1}^{W} I(i,j) = 0 \\ \arg\min_{1 \le j \le W} \{I(i,j) = 1\} - 1 & \text{otherwise} \end{cases} \quad i = 1, 2, \ldots, H$$

$$MargR(i) = \begin{cases} 0 & \text{if } \sum_{j=1}^{W} I(i,j) = 0 \\ \arg\max_{1 \le j \le W} \{I(i,j) = 1\} & \text{otherwise} \end{cases} \quad i = 1, 2, \ldots, H$$

$$(1.14)$$

$$MargB(j) = \begin{cases} H & \text{if } \sum_{i=1}^{H} I(i,j) = 0 \\ \arg\min_{1 \le i \le H} \{I(i,j) = 1\} - 1 & \text{otherwise} \end{cases} \quad j = 1, 2, \ldots, W$$

$$MargT(j) = \begin{cases} 0 & \text{if } \sum_{i=1}^{H} I(i,j) = 0 \\ \arg\max_{1 \le i \le H} \{I(i,j) = 1\} & \text{otherwise} \end{cases} \quad j = 1, 2, \ldots, W$$

3. Horizontal and vertical transitions:

- The horizontal transition is the number of pairs of two consecutive white and black pixels in given rows.
- The vertical transition is the number of such pairs in columns.

Like in the case of projections, transitions are vectors of numbers of respective length.

$$TranH : \langle 1, H \rangle \rightarrow \langle 1, W \rangle^H \quad TranH(j) = \sum_{j=2}^{W} \max\{0, I(i,j) - I(i,j-1)\}$$

$$(1.15)$$

$$TranV : \langle 1, W \rangle \rightarrow \langle 1, H \rangle^W \quad TranV(i) = \sum_{i=2}^{H} \max\{0, I(i,j) - I(i-1,j)\}$$

1.2.2 Transformations of Features: From Vectorial to Vectorial

An interesting collection of numerical features can be derived from vectorial features transformed to other vectorial features. Let us present several important vectorial to vectorial mappings: histograms, smoothings, and differentiations. Illustrative examples of such transformations are presented in Figure 1.6:

1. A histogram and a cumulative histogram are defined on a vector V of length L. Let us assume that elements of vector V are integers located in an interval $\langle 1, L_h \rangle$, that is, $V(i) \in \langle 1, L_h \rangle$. Let us also consider a vector V_h of length L_h. The histogram is a mapping from the vector V to the vector V_h that assigns the number of elements that have value i in the vector V to the i-th element of V_h, that is, assigns this number to the $V_h(i)$, $i = 1, 2, ..., L_h$. Given these assumptions we define histogram *Hist* and cumulative histogram *HistC* as follows:

$$V_h(j) = \sum_{i=1}^{L} \begin{cases} 1 & V(i) = j \\ 0 & V(i) \neq j \end{cases} \quad \text{for } j = 1, 2, ..., L_h$$

$$V_h(j) = \sum_{i=1}^{L} \begin{cases} 1 & V(i) \leq j \\ 0 & V(i) > j \end{cases} \quad \text{for } j = 1, 2, ..., L_h$$

(1.16)

For instance, a histogram of a vertical projection is defined for each integer number i between 0 and the number of rows (H). It counts the number of columns, in which the number of black pixels is equal to i.

2. Smoothing is a mapping defined on a vector that replaces a given value by the mean of this value and its p left and p right neighbors. Both original and result vectors have the same length L. For instance, for $p = 1$ the value is replaced by the mean of this value and its left and right neighbors in the vector. Note that, for $p = 1$, the first and the last elements of the vectors do not have their left and right neighbors, respectively. By analogy, for values of p greater than one, some neighbors of p left and p right elements are missing for leftmost and rightmost elements. The following formulas define smoothing mapping $Smth_p$ for any $p < L/2$:

$$V_{\text{Smth}}(i) = \frac{1}{r - l + 1} \sum_{j=i-l}^{i+r} V(j) \quad i = 1, 2, ..., L$$

$$l = \max\{1, i-p\}, \quad r = \min\{L, i+p\}$$

(1.17)

3. Differentiation assigns a difference between current and previous elements of vector V to the second and next elements of the result vector V_d:

$$\text{Diff}: V \to V_d \quad V_d(i) = V(i) - V(i-1) \quad \text{for } i = 2, 3, ..., L \text{ and } V_d(1) = 0 \quad (1.18)$$

Notice that differential values may be negative, positive, or equal to 0. The first element of the result differential vector is arbitrarily set to 0.

1.2.3   Transformations of Features: Vectorial to Numerical

As we have already mentioned, a pattern recognition task usually employs numerical features. We have also shown that quite a few interesting characteristics describing

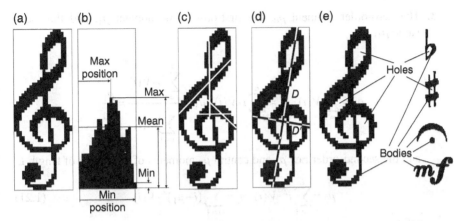

Figure 1.7   Vectorial to numerical transformations: (a) original pattern, (b) numerical features of vertical projection (min = 2, mean = 23, max = 34, min position = 22, max position = 13), (c) directions—white lines on the black pattern (W–E = 13, N–S = 28, NE–SW = 20, NW–SE = 11), (d) eccentricity, and (e) Euler numbers (treble clef: –2, flat: 0, sharp: 0, fermata: 2, mezzo forte: 2).

images could be gathered in the corresponding vectors. Therefore, it becomes imperative to derive numerical characteristics from vectorial features. In this section we discuss principal numerical characteristics of vectorial features. These characteristics can be applied to vectors discussed in the previous sections: projections, margins, and transitions and then histograms, smoothings, and differentiations of projections, margins, and transitions. Transformations from vectorial to numerical features are outlined in the following text and illustrated in Figure 1.7.

1. Minimum, mean, and maximum values of a vector. These transformations can be applied to projections, margins, and transitions. Let V be a vector of length L. Then the following obvious formulas define these concepts:

$$\text{Min value} = \min_{1 \le i \le L} \{V(i)\} \quad \text{Mean value} = \frac{1}{L}\sum_{i=1}^{L} V(i) \quad \text{Max value} = \max_{1 \le i \le L} \{V(i)\}$$

$$(1.19)$$

2. Positions of minimum and maximum values are just indices of vector's elements with the minimal and maximal values, respectively. If the minimal or maximal value appears more than once in a vector, then the position can be chosen arbitrarily. In the following formulas, the first occurrence is taken as the position. Let V be a vector of length L, and then the following formulas define these features:

$$\text{Position of min value} = \arg \min_{1 \le i \le L} \{V(i) = \min \text{ value}\}$$

$$\text{Position of max value} = \arg \min_{1 \le i \le L} \{V(i) = \max \text{ value}\}$$

$$(1.20)$$

where *min value* and *max value* are defined in (1.19).

3. The zero-order moment ρ_0, the first-order raw moment ρ_1, and the mean value μ_1,

$$\rho_0 = \sum_{i=1}^{L} V(i) \quad \rho_1 = \sum_{i=1}^{L} i \cdot V(i) \quad \mu_1 = \frac{\sum_{i=1}^{L} i \cdot V(i)}{\sum_{i=1}^{L} V(i)} = \frac{\rho_1}{\rho_0}$$

and the second-order raw ρ_2 and central μ_2 moments of a vector V of length L,

$$\rho_2 = \sum_{i=1}^{L} i^2 \cdot V(i) \quad \eta_2 = \sum_{i=1}^{L} (i-\mu_1)^2 \cdot V(i) \tag{1.21}$$

1.2.4 Numerical Features

Several important features can be extracted directly from an image. We discuss here the following features: shape of the bounding box (height to width proportion), blackness, raw and central moments, eccentricity, and Euler numbers. In the following text we present descriptions of the listed features and illustrate the discussion with Figures 1.4 and 1.7.

1. Proportion of the bounding box is just the proportion of its height H to width W:

$$\frac{H}{W} \tag{1.22}$$

2. Blackness is the proportion of the number of black pixels to all pixels in the bounding box:

$$\frac{\sum_{i=1}^{H} \sum_{j=1}^{W} I(i,j)}{H \cdot W} \tag{1.23}$$

3. Raw and central moments. Raw moments of an image are defined as follows:

$$\rho_{kl} = \sum_{i=1}^{H} \sum_{j=1}^{W} i^k j^l \cdot I(i,j) \tag{1.24}$$

where $k+l$ is an order of a moment. Please notice that the moment of order zero is equal to the area of the image (it is the number of black pixels) and the first-order moments ρ_{10} and ρ_{01} define the center of the image (which can be interpreted as the mean value or the center of gravity).

Central moments are defined by the formula

$$\mu_{kl} = \sum_{i=1}^{W} \sum_{j=1}^{H} (i-\rho_{10})^k (j-\rho_{01})^l \cdot I(i,j) \tag{1.25}$$

Notice that $\mu_{00} = \rho_{00}$ and $\mu_{10} = \mu_{01} = 0$. If we compare moments of an image with moments of horizontal and vertical projections of this image, we come to a conclusion that they are identical, that is, the first-order moment ρ_{10} of an image and the first-order moment of its horizontal projection ρ_1 are equal:

$$\rho_{10} = \sum_{i=1}^{H}\sum_{j=1}^{W} i^1 j^0 \cdot I(i,j) = \sum_{i=1}^{H}\left(i \cdot \sum_{j=1}^{W} I(i,j) \right) = \sum_{i=1}^{H}(i \cdot ProjH(i)) = \rho_1 \qquad (1.26)$$

Alike, the first-order moment ρ_{01} is equal to the first-order moment ρ_1 of its vertical projection. Analogously, the second-order raw moments ρ_{20} and ρ_{02} and the second-order moments of its vertical and horizontal projections are equal. The same correspondence concerns central moments μ_{20} and μ_{02} and the respective moments of vertical and horizontal projection μ_2.

4. Eccentricity E is defined as the proportion of the length of diameter D to the length of diameter D' perpendicular to D. Diameter D of a pattern is an interval of the maximal length connecting two black pixels of the pattern

$$\text{Length}(D) = \max_{\substack{1 \leq i,k \leq H \\ 1 \leq j, l \leq W}} \{d(I(i,j), I(k,l)) : I(i,j) = 1 = I(k,l)\} \qquad (1.27)$$

The following formula allows a simple computation of this feature:

$$E = \frac{(\mu_{20} - \mu_{02}) + 4\,\mu_{11}^2}{\mu_{00}} \qquad (1.28)$$

where $\mu_{20}, \mu_{02}, \mu_{11}$ are central moments of the second order and μ_{00} is the area of the pattern (equal to the number of black pixels) (cf. Hu, 1962; Sonka *et al.*, 1998).

5. Euler numbers 4, 6, and 9. Euler numbers of a pattern represented as a monochrome image describe topological properties of the pattern regardless of its geometrical shape. The Euler number of a binary image is the difference between the number of connected components (*NCC*) and the number of holes (*NH*) (Sossa-Azuela *et al.*, 2013):

$$EN = NCC - NH \qquad (1.29)$$

A connected component is a region of black (foreground) pixels, which are connected. A hole is a region of connected white (background) pixels enclosed by black pixels. For instance:

- The treble clef has one connected component and three holes, $EN = 1 - 3 = -2$.
- The mezzo forte pattern has two connected components and no holes, $EN = 2 - 0 = 2$.
- The sharp has one connected component and one hole, $EN = 1 - 1 = 0$.

Connectivity depends on the definition of connected components. We consider three types of connectivity:

- 4-Connectivity is computed in a 4-pixel neighborhood, that is, to a given pixel $I(i,j)$, connected are its horizontal $I(i \pm 1, j)$ and vertical $I(i, j \pm 1)$ neighbors.

- 8-Connectivity is calculated in an 8-pixel neighborhood, that is, to a given pixel $I(i,j)$, connected are its horizontal, vertical, and diagonal $I(i \pm 1, j \pm 1)$, $I(i \pm 1, j \mp 1)$ neighbors.

- 6-Connectivity is based in a 6-pixel neighborhood, that is, to a given pixel $I(i,j)$, connected are its horizontal and vertical neighbors and two neighbors on an arbitrarily chosen diagonal, and in this study we consider right-to-left diagonal $I(i \pm 1, j \pm 1)$ neighbors.

6. Directions: vertical (N–S, short for North–South), horizontal (W–E, short for West–East), and diagonal (NW–SE and NE–SW). In brief, direction is the length of the longest run of black pixels of a line in a given direction. For instance, the formulas for vertical and NW–SE diagonal directions read as follows:

$$\max_{1 \leq i \leq H, 1 \leq j \leq W} \max_{l \geq 0, r \geq 0} \left\{ l + r + 1 = \sum_{k=-l}^{r} I(i+k, j) \right\}$$

$$\max_{1 \leq i \leq H, 1 \leq j \leq W} \max_{l \geq 0, r \geq 0} \left\{ l + r + 1 = \sum_{i=-l}^{r} I(i+k, j-k) \right\}$$

(1.30)

with an assumption that for a given i and j, values l and r are such that the following obvious inequalities hold: $1 \leq i-l \leq i+r \leq H$ and $1 \leq j-l \leq j+r \leq W$.

7. A peak in a vector (e.g., in a horizontal/vertical projection, margin, etc.) is an element of this vector that is not smaller than its left and right neighbors, and at the same time it is either not smaller than the 3/4 of the maximum in this vector, or it is not smaller than half of the maximum in this vector, and it exceeds its left and right neighbors by a quarter of the maximum. The following formula defines the number of peaks in a vector V of length L assuming that $MAX = \max_{1 \leq i \leq L}\{V(i)\}$ is the maximal element in the vector V:

$$\sum_{i=2}^{L-1} \begin{cases} 1 & V(i) > 3/4 \cdot MAX \wedge V(i) - \max(V(i-1), V(i+1)) \geq 0 \\ 1 & 3/4 \cdot MAX \geq V(i) > 1/2 \cdot MAX \wedge V(i) - \max(V(i-1), V(i+1)) \geq 1/4 \cdot MAX \\ 0 & \text{otherwise} \end{cases}$$

(1.31)

To sum up, a primary step of pattern recognition described in this section aims at the generation of numerical features from images. The proposed mechanisms generate, in total, 171 features (viz., the feature vector contains 171 elements). It turned out that for the two datasets that we have tackled in the detailed experiments (viz., handwritten digits and printed music notation), some of the computed features were constant, so we removed them. As a result, we obtained a list of 159 features; they are listed in Appendix 1.A.

There are numerous papers on feature extraction in pattern recognition. Many of them discuss very specific cases of image analysis, for instance, involving face recognition (Turk and Pentland, 1991), high-speed methods for feature extraction (Bay et al., 2008), texture analysis (Manjunath and Ma, 1996), and real-world applications in which we aim at matching between different views of an object or a scene (Lowe, 2004).

1.3 FEATURES SCALING

The values assumed by different features may vary, for instance, comparison of salaries and age. Features in their *natural* form are called *raw features* or we may just say *raw values*. Some raw features may *weigh* much more than the other features. As a consequence, when processing raw features with certain algorithms, we risk that *heavy* features will overshadow the *light* ones. Therefore, there is a need to *scale* raw features, and this process aims at unifying ranges of their values. We consider two types of unification: normalization to the unit interval and standardization based on unification of mean and standard deviation. In both types of scaling, a linear transformation is applied, so characteristics of patterns are preserved.

1.3.1 Features Normalization

A typical normalization linearly transforms raw values of features to the unipolar unit interval $[0, 1]$ or to the bipolar unit interval $[-1, 1]$. In order to give details of such transformation, let us assume that patterns are described by features $X_1, X_2, ..., X_M$ and that $x_{i,min}$ and $x_{i,max}$ are the minimal and the maximal values of a feature X_i for all patterns located in a learning set. Hence, for a value $x_{i,j}$ of this feature for a given pattern o_j represented as a vector of features $\mathbf{x}_j = \left(x_{1,j}, x_{2,j}, ..., x_{M,j} \right)^T$, the corresponding unipolar value is computed as follows:

$$a_{i,j} = \frac{x_{i,j} - x_{i,\min}}{x_{i,\max} - x_{i,\min}} \tag{1.32}$$

The corresponding bipolar value is given by

$$b_{i,j} = 2 \cdot \frac{x_{i,j} - x_{i,\min}}{x_{i,\max} - x_{i,\min}} - 1 \tag{1.33}$$

In Table 1.1, we present example values of features. Numerical features (minimal, maximal, and mean values, positions of minimal and maximal values) are derived from two vectorial features: vertical projection and differential of vertical projections. We outline features of printed music notation symbols. In this example, we consider only eight classes and take a single pattern (symbol) coming from each class. In the consecutive segments of this table, we present raw values, parameters of raw values of the entire learning set (minimal, maximal, mean values, and standard deviation), values normalized to the unipolar and bipolar unit intervals, and standardized values. It is worth noting that normalization and standardization may give undefined values such as mean of differential of vertical projection. This feature is constant; therefore, the denominators in formulas (1.32), (1.33), and (1.34) are 0.

In the normalized learning set of patterns, values fall into the unit interval. However, when we consider patterns not belonging to the learning set, for example, patterns subjected to recognition after the original normalization has been performed, their feature values may be less than the minimum or greater than the maximum of a given feature. We can easily normalize new, incoming patterns, but it may happen that the result values will fall out of the desired unit interval. Hence, such irregularity

TABLE 1.1 Numerical features derived from two vectorial features: vertical projection and differential of vertical projections

Class Name	Vertical Projection						Differential of Vertical Projection					
	Min	Min Position	Max	Max Position	Mean	ρ_1	Min	Min Position	Max	Max Position	Mean	ρ_1
					Raw values							
Mezzo forte	3	0	22	13	10	15.07	−9	15	11	11	0	11.81
Sharp	12	26	30	7	13	13.93	−7	9	10	6	0	12.40
Flat	9	31	32	1	17	12.24	−8	6	6	0	0	10.80
Clef G	3	0	17	11	11	15.69	−6	24	6	7	0	15.07
Clef C	9	0	32	4	23	14.88	−14	10	18	1	0	14.79
Quarter rest	4	0	25	16	12	14.68	−4	20	8	3	0	14.02
Eighth rest	3	3	14	15	9	15.88	−5	17	3	3	0	15.26
1/16 rest	2	30	22	15	11	14.04	−3	16	3	1	0	14.64
⋮												
Mean value	4.73	15.47	23.27	11.63	12.53	14.67	−8.67	15.50	8.10	7.27	0.00	14.17
Standard deviation	2.83	13.68	6.76	7.55	4.26	1.23	6.20	8.68	5.62	7.12	0.00	2.62
Min	1.00	0.00	11.00	0.00	7.00	11.23	−25.00	1.00	2.00	0.00	0.00	8.38
Max	12.00	31.00	32.00	23.00	23.00	16.63	6.20	30.00	21.00	27.00	0.00	19.54
					Values normalized to the unit interval							
Mezzo forte	0.18	0.00	0.52	0.57	0.19	0.71	0.70	0.48	0.47	0.41	NA	0.31
Sharp	1.00	0.84	0.90	0.30	0.38	0.50	0.58	0.28	0.42	0.22	NA	0.36
Flat	0.73	1.00	1.00	0.04	0.63	0.19	0.54	0.17	0.21	0.00	NA	0.22
Clef G	0.18	0.00	0.29	0.48	0.25	0.83	0.83	0.79	0.21	0.26	NA	0.60
Clef C	0.73	0.00	1.00	0.17	1.00	0.68	0.48	0.31	0.84	0.04	NA	0.57
Quarter rest	0.27	0.00	0.67	0.70	0.31	0.64	0.91	0.66	0.32	0.11	NA	0.51
Eighth rest	0.18	0.10	0.14	0.65	0.13	0.86	0.87	0.55	0.05	0.11	NA	0.62
1/16 rest	0.09	0.97	0.52	0.65	0.25	0.52	0.96	0.52	0.05	0.04	NA	0.56

Values normalized to the bipolar unit interval

Mezzo forte	-0.64	-1.00	0.05	0.13	-0.63	0.42	0.39	-0.03	-0.05	-0.19	NA	-0.38
Sharp	1.00	0.68	0.81	-0.39	-0.25	0.00	0.15	-0.45	-0.16	-0.56	NA	-0.28
Flat	0.45	1.00	1.00	-0.91	0.25	-0.63	0.09	-0.66	-0.58	-1.00	NA	-0.57
Clef G	-0.64	-1.00	-0.43	-0.04	-0.50	0.65	0.65	0.59	-0.58	-0.48	NA	0.20
Clef C	0.45	-1.00	1.00	-0.65	1.00	0.35	-0.04	-0.38	0.68	-0.93	NA	0.15
Quarter rest	-0.45	-1.00	0.33	0.39	-0.38	0.28	0.83	0.31	-0.37	-0.78	NA	0.01
Eighth rest	-0.64	-0.81	-0.71	0.30	-0.75	0.72	0.74	0.10	-0.89	-0.78	NA	0.23
1/16 rest	-0.82	0.94	0.05	0.30	-0.50	0.04	0.91	0.03	-0.89	-0.93	NA	0.12

Standardized values

Mezzo forte	-0.61	-1.13	-0.19	0.18	-0.60	0.33	-0.05	-0.06	0.52	0.52	NA	-0.90
Sharp	2.57	0.77	1.00	-0.61	0.11	-0.60	0.27	-0.75	0.34	-0.18	NA	-0.68
Flat	1.51	1.14	1.29	-1.41	1.05	-1.97	0.11	-1.09	-0.37	-1.02	NA	-1.29
Clef G	-0.61	-1.13	-0.93	-0.08	-0.36	0.83	0.43	0.98	-0.37	-0.04	NA	0.34
Clef C	1.51	-1.13	1.29	-1.01	2.46	0.17	-0.86	-0.63	1.76	-0.88	NA	0.24
Quarter rest	-0.26	-1.13	0.26	0.58	-0.13	0.01	0.75	0.52	-0.02	-0.60	NA	-0.06
Eighth rest	-0.61	-0.91	-1.37	0.45	-0.83	0.99	0.59	0.17	-0.91	-0.60	NA	0.41
1/16 rest	-0.97	1.06	-0.19	0.45	-0.36	-0.51	0.91	0.06	-0.91	-0.88	NA	0.18

Outlined are features of single patterns (symbols) of a printed music notation.

must be considered in some way. In order to deal with such case, we recommend selecting processing algorithms prone to slight variations in feature values, which with the current advancements in the machine learning area is not a problem.

Alternatively, if there is no other option, one may truncate values located outside the unit interval.

1.3.2 Standardization

Standardization is another method of features unification. Standardization considers not only raw values themselves but also dispersion of values, that is, we employ the mean value and standard deviation of a given feature. Let the following vector $\mathbf{x}_j = \left(x_{1,j}, x_{2,j}, \ldots, x_{M,j}\right)^T$ represent a j-th pattern. The following formula realizes a standardization procedure,

$$u_{i,j} = \frac{x_{i,j} - \bar{x}_i}{\sigma_i} \tag{1.34}$$

where $\bar{x}_i$ is the mean of the feature X_i and σ_i is the standard deviation of this feature:

$$\bar{x}_i = \frac{1}{N} \sum_{j=1}^{N} x_{i,j}, \quad \sigma_i = \sqrt{\frac{1}{N} \sum_{j=1}^{N} \left(x_{i,j} - \bar{x}_i\right)^2} \tag{1.35}$$

N is the number of patterns in the learning set.

Standardized values of selected features are displayed in the bottom segment of Table 1.1. As mentioned earlier, we outline features of one pattern (symbol) from each of 8 classes of printed music notation. Unlike in the case of normalization, there is no fixed interval that includes the values of a feature, so in order to process a standardized dataset, one cannot choose a classifier that requires all features to fall into some arbitrarily predefined interval.

1.3.3 Empirical Evaluation of Features Scaling

In this section, we discuss the influence of feature values scaling on classification quality. We tested two datasets: musical symbols and handwritten digits. Let us recall that handwritten digits are examples of a well-balanced dataset. In contrast, patterns in the dataset with musical symbols are imbalanced with regard to size, shape, and cardinality.

There are some measures that can be taken to mitigate problems of class imbalance. Even though pattern shape is a property that cannot be modified, the other two aspects—samples cardinality and size—can be adjusted to *balance* an imbalanced dataset. Therefore, we *balanced* values of features, that is, standardized and normalized values. Also, we *balanced* cardinalities of classes.

The dataset of musical symbols consists of 20 classes; see Section 3.4 for details. The majority of classes contain from 200 to 3000 samples. One class is extremely small; it includes only 26 patterns. Based on the original dataset, we built two balanced datasets in which all classes contain exactly 500 patterns, that is, we obtained two balanced datasets with 10,000 patterns in each of them. In order to construct these datasets, we had to oversample less frequent classes and undersample the frequent ones. For those classes in which there were more than 500 samples, we randomly selected 500 patterns. With regard to rare classes, new patterns were generated in order to reach the total of 500 patterns in a given class. We applied two distinct methods for generation of new patterns, and hence we obtained two balanced datasets.

The essence of the first method for samples generation is defined as follows:

Algorithm 1.1
On intervals oversampling rare class
Data: class O of patterns of cardinality N
     Set of Features
     N_B the assumed cardinality of the balanced class
Algorithm: initialize the balanced class $O_B = O$
     repeat
       pick up randomly two native patterns from O:
        $\mathbf{X} = (x_1, x_2, \ldots, x_M)$ and $\mathbf{Y} = (y_1, y_2, \ldots, y_M)$
       create new pattern $\mathbf{Z} = (z_1, z_2, \ldots, z_M)$ such that for $l = 1, 2, \ldots, M$
        z_l is a random value from the interval $[\min(x_l, y_l), \max(x_l, y_l)]$
       add $\mathbf{Z}$ to O_B
      until number of patterns in O_B is not less than N_B
Results: the balanced class O_B of patterns

In the procedure shown earlier, we generate a number of patterns so that the total number of patterns (original patterns plus the ones that are generated) is equal to 500. On the input to this procedure, we pass original samples from a given rare class. Technically, we operate on a data frame of M features describing some patterns. We call this method *on intervals* as new patterns are generated with the use of random intervals formed between feature values of existing patterns.

The alternative method for samples generation is realized as follows:

Algorithm 1.2
Gaussian oversampling rare class
Data: class O of patterns of cardinality N
     Set of Features
     N_B—assumed cardinality of the balanced class
Algorithm: initialize the balanced class $O_B = O$
     compute center $\bar{\mathbf{X}} = (x_1, x_2, \ldots, x_M)$ of the class O
     repeat
       create new pattern $\mathbf{Z} = (z_1, z_2, \ldots, z_M)$ such that for $l = 1, 2, \ldots, M$
        use Gaussian probability distribution $N(\mu, \sigma)$ to sample z_l,
         where μ and σ are computed according to (1.35)
       add $\mathbf{Z}$ to O_B
      until number of patterns in O_B is not less than N_B
Results: the balanced class O_B of patterns

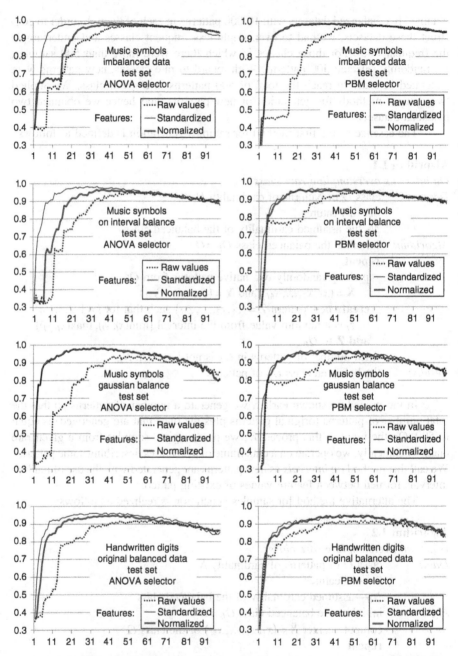

Figure 1.8   Quality of different sets of features selected with the greedy search method. The procedure was adding features one by one: in each iteration one best feature was added. Feature evaluation was performed using the ANOVA F-test and PBM index. We display accuracy (vertical axis) versus feature sets cardinality (horizontal axis). Results concern sets of features ranging from 1 to 100. Plots present accuracy measured on test sets. Plots concern different sets: original, normalized, standardized digits, and musical symbols for the dataset (Homenda *et al.*, 2017). Information about the kind of data is presented in each individual plot.

In this procedure, we use normal (Gaussian) probability distribution to approximate distribution of patterns in a given class. We call this data generation procedure *Gaussian* generation.

The dataset of handwritten digits consisted of 10 classes with around 1000 patterns in each class. The total number of patterns is exactly 10,000. This dataset is balanced, and there is no need to undersample the patterns coming from the dominant class or oversample the patterns that form a minority class.

In Figure 1.8, we present results of empirical tests done for the four datasets described earlier: original dataset of handwritten digits, original (unmodified) dataset of music notation, music notation dataset balanced with the *on intervals* method, and music notation dataset balanced with the *Gaussian* method. In addition, in each individual plot in Figure 1.8, we present (for comparative purposes) results achieved for raw data versus normalized data versus standardized data. We display the accuracy of SVM classifiers, which were built using sets of features of various sizes. We checked sets of features of all cardinalities between 1 and 100. In order to select a particular set of features, we ran a greedy search algorithm employing the ANOVA F-test and PBM index for evaluation of features quality (cf. Section 1.4 for the details of the features selection process).

In the experiment with the results plotted in Figure 1.8, the classes forming learning sets were split randomly into the training and test sets in proportion 70–30%. Training sets were used to construct classifiers, while the accuracy was evaluated on the test sets.

The results displayed in Figure 1.8 indicate that the standardization of a dataset helps achieve better numerical accuracy. The results for raw data are worse than for the standardized one. The results for normalized data are worse or in a few cases similar to the results achieved with standardized data. The differences are clearly visible for feature sets of size ranging from 10 to 50. For very large feature sets, the classification results tend to be very similar, no matter which dataset was used. Naturally, such large feature sets are not recommended as we see a clear effect of overfitting that makes test accuracy drop after achieving some peak. Peaks in the accuracy test occur for feature sets consisting of more than 20 but less than 30 features.

It is important to notice that balancing raw (not normalized and not standardized) musical symbols dataset improves accuracy a lot, especially for the PBM index. On the other hand, accuracy obtained on original (imbalanced) musical symbols dataset for standardized features (both ANOVA F-measure and PBM index) and for normalized features (PBM index) is slightly better than for balanced datasets.

1.4   EVALUATION AND SELECTION OF FEATURES

In the previous sections of this chapter, we have presented the general idea of how to represent patterns using features. The discussion was focused on extracting various features from monochrome, segmented images of patterns. Up to this point we have avoided a very crucial topic: quality of features. In this section, we take a closer look at this issue.

Taking up a pattern recognition task, we have to be aware of the fact that extracted features may be of poor quality. This problem manifests itself in a few aspects. First, we may be in possession of too many features and our processing algorithm may not be able to handle them efficiently. The second apparent flaw is that features may be redundant (duplicated), constant, or may carry utterly useless information (such as the National Insurance Number that is unique for each person and does not carry any predictive strength). Last, we may have heavily correlated features or features of general poor quality (e.g., with missing values). In any case, before actual pattern recognition it is obligatory to get to know the data we process. When it comes to features, we need to evaluate them. The objective of this task is to select such a subset that would be valuable for classification. In the following section, we discuss problems of features evaluation and selection of an optimal feature subset.

1.4.1 Correlation

It may happen that some features are very similar or even identical in the sense that, for instance, their values are linearly dependent for the considered learning set of patterns. A few linearly dependent features do not carry more information than only one of them. If we have a pair of correlated features, only one of them is needed in the process of classifier construction. Redundant features should be dropped because they do not carry new information. Moreover, they complicate the construction of the model, and in this way we risk constructing a poor model.

To illustrate dependence between features, let us consider mean values of vertical and horizontal projections and blackness. These features are defined by formulas (1.13) and (1.23). It is obvious that they are proportional: their values are equal to the number of black pixels in a bounding box of a pattern divided by width, height, and the product of width and height of the box, respectively:

$$\text{Mean}_{\text{vert,proj}} = \frac{\sum_{i=1}^{W}\sum_{j=1}^{H} I(i,j)}{W} \quad \text{Mean}_{\text{hor,proj}} = \frac{\sum_{i=1}^{W}\sum_{j=1}^{H} I(i,j)}{H} \quad \text{Blackness} = \frac{\sum_{i=1}^{W}\sum_{j=1}^{H} I(i,j)}{W \cdot H}$$

$$(1.36)$$

Hence, these three features are strictly correlated. It is sufficient to multiply values of one feature by a constant in order to obtain the values of some other features:

$$\text{Mean}_{\text{hor,proj}} = \frac{W}{H} \cdot \text{Mean}_{\text{vert,proj}} \quad \text{Blackness} = \frac{1}{H} \cdot \text{Mean}_{\text{vert,proj}} \quad (1.37)$$

and, of course, two out of these three features should be eliminated.

The strength of correlation between two features is expressed by the Pearson correlation coefficient. To compute this coefficient, let us consider two numerical features X_k and X_l and their values for all N patterns from the learning set: $x_{1,k}, x_{2,k}, ..., x_{N,k}$

and $x_{1,l}, x_{2,l}, \ldots, x_{N,l}$, respectively. The Pearson correlation coefficient is defined as follows:

$$r_{k,l} = \frac{\sum_{i=1}^{N}(x_{i,k}-\bar{x}_k)(x_{i,l}-\bar{x}_l)}{\sqrt{\sum_{i=1}^{N}(x_{i,k}-\bar{x}_k)^2}\sqrt{\sum_{i=1}^{N}(x_{i,l}-\bar{x}_l)^2}} \tag{1.38}$$

where the mean values of these features are computed according to the formulas as

$$\bar{x}_k = \frac{1}{N}\sum_{i=1}^{N}x_{i,k}, \quad \bar{x}_l = \frac{1}{N}\sum_{i=1}^{N}x_{i,l} \tag{1.39}$$

The Pearson correlation coefficient is a number from the interval $[-1, 1]$. It assumes the value equal to 1 for a pair of strictly correlated features, that is, a pair of features that are in a linear relationship such that when one feature increases, the second one increases as well. Such behavior is exhibited by any two features outlined in (1.36). This conclusion is trivially obtained by replacing feature values in (1.38) with the product of the values of another feature and respective coefficient. On the other hand, two linearly dependent features with a negative linear coefficient produce the correlation value equal to -1. As an example of such two features, we may give blackness and whiteness. Another example of a correlation with a negative coefficient is the mean value of horizontal projection and the mean value of a cumulative histogram of vertical projection. This pair of features has a correlation coefficient very close to -1.

For a given set of features, we compute a correlation matrix, identify groups of correlated features, and then eliminate all but one feature from every such group. It should be mentioned that, typically, we do not see strictly correlated features, that is, pairs with the absolute value of coefficients equal to 1, as in the cases outlined earlier. Usually, we have strongly correlated features, that is, pairs of features such that their absolute correlation values are high. We note that it is up to the model designer to determine a cut threshold that allows making a decision on which pair of features is correlated. Assuming that we have selected such threshold, say, the absolute value of 0.6, we shall investigate pairs of features correlated to a degree exceeding 0.6, and we should drop redundant features.

Table 1.2 presents correlation coefficients between features listed before in Table 1.1. Of course, the matrix is symmetrical and has 1s at the main diagonal. In this matrix each pair of features with the absolute value of the correlation coefficient greater than 0.6 is highlighted in boldface. For example, the following pairs of features are correlated: the raw moment of the differential of vertical projection is correlated with the position of the maximal value of vertical projection (0.64) and the first raw moment of vertical projection with the values of vertical projection (0.65) and with the position of the minimum value of differential of this projection (0.64). If we want to eliminate these correlations, we should either remove the raw moment of the differential of vertical projection or two other correlated features. There is no general rule on which one of them should be removed. If we have, for instance, evaluated the quality

TABLE 1.2 Matrix of the Pearson correlation coefficients for features outlined in Table 1.1

		Vertical Projection						Differential of Vertical Projection					
		Min	Min Position	Max	Max Position	Mean	ρ_1	Min	Min Position	Max	Max Position	Mean	ρ_1
Vertical Projection	Min	1	0.17	0.32	−0.04	**0.72**	−0.09	0.01	0.04	−0.11	0.10	0.02	0.03
	Min Position	0.17	1	0.07	−0.07	0.19	−0.38	0.05	0.02	−0.13	0.04	−0.02	0.06
	Max	0.32	0.07	1	−0.39	**0.60**	−0.39	**−0.63**	−0.24	0.54	−0.04	−0.02	−0.41
	Max Position	−0.04	−0.07	−0.39	1	−0.03	0.57	0.45	0.53	−0.38	0.20	0.03	**0.64**
	Mean	**0.72**	0.19	**0.60**	−0.03	1	−0.06	−0.09	0.16	0.02	0.02	0.02	0.10
	ρ_1	−0.09	−0.38	−0.39	0.57	−0.06	1	0.35	0.41	−0.22	0.24	0.07	**0.65**
Differential Vertical Projection	Min	0.01	0.05	**−0.63**	0.45	−0.09	0.35	1	0.29	**−0.71**	0.02	0.04	0.43
	Min Position	0.04	0.02	−0.24	0.53	0.16	0.41	0.29	1	−0.29	0.11	0.03	**0.64**
	Max	−0.11	−0.13	0.54	−0.38	0.02	−0.22	**−0.71**	−0.29	1	−0.15	0.01	−0.48
	Max Position	0.10	0.04	−0.04	0.20	0.02	0.24	0.02	0.11	−0.15	1	−0.04	0.39
	Mean	0.02	−0.02	−0.02	0.03	0.02	0.07	0.04	0.03	0.01	−0.04	1	0.01
	ρ_1	0.03	0.06	−0.41	**0.64**	0.10	**0.65**	0.43	**0.64**	−0.48	0.39	0.01	1

The bold font is used to highlight coefficients of the absolute value greater than 0.6. Coefficients at the main diagonal are obviously equal to 1.

of single features (cf. Section 1.4.2 for such evaluations), then the weaker and not the stronger one(s) may be removed.

1.4.2   Evaluation of Features: Two Approaches

Features correlation allows to identify similar features and then eliminate dependent ones. Equally important is features evaluation with regard to their usefulness in classification. The challenge is to find a possibly small set of features, which would guarantee construction of a high quality classifier. Unfortunately, there is no single suitable method of low computational cost to select the best subset of features for all data. In practice, we distinguish two kinds of approaches for feature selection:

- Index-based methods
- Wrapper-based methods

The first family of methods relies on relatively simple indices that evaluate features. They rely on dependencies such as relationships between the feature and dependent variables and relationships between features themselves. The upside of applying index-based approaches is that they require moderate computational effort.

The so-called wrapper-based methods rely on constructing multiple models based on various features subsets. After forming collection of classification models, we compare their efficiency and select the best one. In order to be sure that for a given classification method, we selected the best subset of features, and we need to build classifiers for all possible subsets of features. In practice, however, this is not a feasible option, especially when the full feature set is large. To limit computational overhead induced by a brute force feature search, we can apply several greedy algorithms that limit the number of checked feature subsets in some way. Still, this method is computationally costly and time consuming. In addition, if we switch to another classification algorithm, then in order to obtain the same high quality model, it will be desirable to repeat the whole procedure. Even though the mentioned negative aspects are hard to overlook, it shall be mentioned that wrapper-based methods provide models of superior numerical quality. For this reason in further parts of this chapter, we take a closer look at wrapper-based feature search methods.

The literature of the topic offers a wide range of papers in which we find quite elaborate examples of how appropriate feature selection ensures proper processing abilities. One may read on unsupervised similarity-based feature selection in Mitra *et al.* (2002), mutual information-based feature selection in Peng *et al.* (2005), feature selection for bioinformatics data in Saeys *et al.* (2007) and Guyon *et al.* (2002), rough set-based feature selection in Swiniarski and Skowron (2003), and more. There are also papers in which we find elaborations on feature selection techniques directed to be used with some particular classifier, for instance, SVM in Huang and Wang (2006). It is also worth to consult a general-purpose surveys on feature selection, in, for instance, Trier *et al.* (1996) and Kudo and Sklansky (2000).

1.4.3   Index-Based Feature Evaluation: Single Feature Versus Feature Set Evaluation

Let us now address an important distinction arising in a scenario when we apply index-based methods for feature selection. Let us stress that index-based feature evaluation could be applied in two variants:

- To evaluate the quality of a single feature
- To evaluate the quality of a subset of k features

The first strategy, outlined in Algorithm 1.3, simply requires evaluating features one by one.

Algorithm 1.3
Index-based evaluation of features: one-by-one scheme
Data: Set of Features for Evaluation
               Learning Set of Patterns
               Index for Feature Evaluation
Algorithm: **for each** feature in the Set of Features
               evaluate feature **using** Index for Feature Evaluation
Results: vector with quality score for each feature in the Set of Features

If we sort the output of Algorithm 1.3, we construct feature ranking. Subsequently, the model designer will be able to select a subset of features individually evaluated as the best, in the hope that such features together will provide us with a high quality model. Selection of k (the number of features) could be performed with a plot of index values. Looking for a knee-point present in the plot is a standard way when making an empirically guided selection of parameters.

The second strategy evaluates features as a group. This, however, entails a certain problem as to how to select a features subset for evaluation. This issue will be tackled later in this chapter. At this point let us focus on a description of indices for feature evaluation in the context of single feature evaluation. Later on, this discussion will be extended into the second scenario in which we evaluate subsets of features.

1.4.4   Indices for Feature Evaluation

We turn attention to two kinds of methods of low computational complexity: statistical indices for feature evaluation and indices used in clustering quality verification.

Even though clustering is an assignment with a motivation different than pattern recognition, we see a lot of similarities between those two. Intuitively, we see an analogy between clusters and classes: we assume that classes, like clusters, gather similar objects or, in other words, they reside in some designed subspace of the feature space. Therefore, we perceive cluster validity indices as feasible candidates for feature evaluation indices.

In the following sections we describe four indices: ANOVA F-test (statistical index) and three clustering indices.

ANOVA F-Test

The ANOVA F-test (*analysis of variance*) is a statistical test used to assess whether the expected values of a quantitative variable within several predefined groups differ from each other. This test is used to evaluate the ability of a feature to differentiate classes of native patterns. Roughly speaking, the evaluation is described by the following proportion:

$$F = \frac{\text{between} - \text{class variability}}{\text{within} - \text{class variability}} \tag{1.40}$$

Let N_i, $i = 1, 2, ..., C$ be the cardinalities of classes of native patterns, where of course $N_1 + N_1 + \cdots + N_C = N$ and $x_{i,j}$ stands for the value of the (considered) feature of the j-th pattern from the i-th class, $j = 1, 2, ..., N_i$, $i = 1, 2, ..., C$. Let $\bar{x}_i$ $i = 1, 2, ..., C$ and $\bar{x}$ be the mean values of the feature in respective classes and in the whole learning set:

$$\bar{x}_i = \frac{1}{N_i} \sum_{j=1}^{N_i} x_{i,j}, \quad i = 1, 2, ...C, \quad \bar{x} = \frac{1}{N} \sum_{i=1}^{C} \sum_{j=1}^{N_i} x_{i,j} \tag{1.41}$$

Then the ANOVA F-test for a given feature is defined by the following formula:

$$F = \frac{\dfrac{1}{C-1} \sum_{i=1}^{C} N_i (\bar{x}_i - \bar{x})^2}{\dfrac{1}{N-C} \sum_{i=1}^{C} \sum_{j=1}^{N_i} (x_{i,j} - \bar{x}_i)^2} \tag{1.42}$$

It is clear that the more dispersed centers of classes and the more compact classes inside are, the greater the value of the ANOVA F-test becomes. This observation implies that the greater the ANOVA F-test is, the easier it is to separate classes from each other. Finally, the quality of features is consistent with the values of the ANOVA F-test: the greater the value of this test is, the better class separation the feature provides. Interestingly, ANOVA F-test turned out to be a great match for feature selection in natural language processing (Elssied *et al.*, 2014).

Clustering Indices

Among a multitude of cluster validity indices, let us select a few that turned out, after a series of empirical experiments, to be well suited to express the quality of features. These are the McClain–Rao (MCR) index, the generalized Dunn index (GDI), and the PBM index.

The McClain–Rao Index (MCR)

The first index considered here was proposed in McClain and Rao (1975) and Charrad (2014). This index expresses the ratio of two terms that is the average distance between pairs of points of the same cluster and pairs of points of different clusters. The following formula defines this index:

$$\text{MCR} = \frac{P_2(N) - \Sigma P_2(N_i)}{\Sigma P_2(N_i)} \cdot \frac{\displaystyle\sum_{i=1}^{C} \sum_{1 \le k < l \le N_i} |x_{i,k} - x_{i,l}|}{\displaystyle\sum_{1 \le i < j \le C} \sum_{1 \le k \le N_i} \sum_{1 \le l \le N_j} |x_{i,k} - x_{i,l}|}$$

where

$$P_2(N) = N(N-1)/2 \quad \text{and} \quad \Sigma P_2(N_i) = \sum_{i=1}^{C} N_i(N_i-1)/2 \tag{1.43}$$

and $P_2(N)$ is the number of all distances computed between points of all clusters (more precisely, the number of all sets consisting of two points from all clusters) and $\Sigma P_2(N_i)$ is the number of all distances between points in the same cluster (the number of all sets consisting of two points from the same cluster). The meaning of other symbols is analogous to the notation being used in the section devoted to the ANOVA F-test.

Considering (1.42) to be the product of two terms, we may interpret the second term as the sum of within-cluster distances divided by the sum of between-cluster distances. The first term counts the number of the between-cluster distances divided by the number of the within-cluster distances. The product of both terms is just a proportion of the average within-cluster distance and the average between-cluster distance.

The minimum value of the index is used to indicate optimal clustering. Therefore, the smaller the value of this index is, the better the feature quality is. Finally, for the sake of consistency with the other indices, we reverse the rank of features formed with the MCR.

The Generalized Dunn Index (GDI)

The Dunn index (Dunn, 1973) defines the ratio between the minimal between-cluster distance to the maximal within-cluster distance. This index was generalized to the GDI by using different definitions of the minimal between-cluster distance and the maximal within-cluster distance (cf. Desgraupes, 2013). We use the version GDI_{41} given by the following formula (cf. Desgraupes, 2016). Of course, when applying clustering indices for feature set evaluation in the context of classification, we replace cluster belongingness with class memberships:

$$GDI_{41} = \frac{\min_{1 \le k < l \le C} |\bar{x}_k - \bar{x}_l|}{\max_{1 \le k \le C} \max_{1 \le i < j \le N_k} |x_{k,i} - x_{k,j}|} \tag{1.44}$$

If the dataset contains compact and well-separated clusters, the diameters of clusters are expected to be small, and the distances between clusters are expected to be large. Thus, the Dunn index should be maximized. Therefore, the higher the value of this index is, the better the quality of the feature is.

The PBM Clustering Index

The third cluster validity index considered here was proposed in Bandyopadhyay *et al.* (2004), and it is called PBM (after the names of the authors, Pakhira, Bandyopadhyay, and Maulik). Adapting the notation used in the ANOVA F-measure, the following the PBM index is proposed:

$$PBM = \left(\frac{D_B}{C} \cdot \frac{\sum_{i=1}^{C} \sum_{j=1}^{N_i} |x_{i,j} - \bar{x}|}{\sum_{i=1}^{C} \sum_{j=1}^{N_i} |x_{i,j} - \bar{x}_i|} \right)^2 \tag{1.45}$$

where $D_B = \max\limits_{1 \le i < j \le C} |\bar{x}_i - \bar{x}_j|$ is the largest distance between mean values in the set of all clusters (centers of clusters). As in the case of the ANOVA F-test, the greater the dispersion between clusters and the more compact each cluster is, the greater the PBM index is. In conclusion, the greater the PBM index is, the better class separation the feature provides and the better the feature quality is.

1.4.5   Selection between Index-Based Methods and Wrapper-Based Methods

Let us recall that classifiers can serve as a feature evaluation method—following the so-called wrapper-based approach to feature selection. Applying classifiers to feature evaluation is in primal conflict with index-based feature search as it drastically increases computational complexity. Wrappers evaluate the *final product* (classification accuracy), while index-based methods are more sublime: they evaluate *constituents*, features making the final model.

At the same time, we have to face a realistic expectation that a computationally inexpensive index-based method would be only valuable if its results do not fall far behind a superior model constructed at a higher computational cost.

In the next section, we aim at a fair comparison of index-based methods for single feature evaluation with classifier-based single feature evaluation. By analogy, in later parts of this chapter, we aim at comparing index-based methods for multiple features selection with classifier-based multiple features selection. In experimental tests we analyze the consistency of a given index-based evaluation method with the accuracy of a trained model.

In this study we use k-NN and SVM classifiers. An extended description of these classifiers is provided in Chapter 2.

1.4.6   Single Feature Evaluation Scheme Using Indices and Classifiers

There are many indices used for evaluating clustering quality. We investigated more than 20 indices in order to choose a few, which would evaluate features consistently with classifiers; that is, we look for indices that rank features in a way similar to classifiers. Moreover, we discuss selected features in order to illustrate some properties such as the correlation of features or the scoring results. Outlined in Table 1.3 are results produced by classifiers SVM and k-NN ($k = 1$) and the ANOVA F-test and several clustering indices: the Calinski–Harabasz index, the Baker–Hubert gamma index, the G+ index, the GDI with parameters δ_4 and Δ_1 (cf. Dunn, 1973; Desgraupes, 2013), the MCR index, the PBM index, and the point-biserial index (cf. Desgraupes, 2013, 2016). For each index listed in the first column of Table 1.3, given are 15 features (their numbers) with the best rank, in descending order of this rank. The full names of these features and assigned numbers are given in Appendix 1.A. In Table 1.3, we would like to draw attention to rows corresponding to the ANOVA F-test and the Calinski–Harabasz, which are identical, which means that these two indices are fully

TABLE 1.3 Features ranking with two classifiers and different indices applied: classifiers SVM and k-NN (k=1), ANOVA index, and several clustering indices

SVM	154	120	152	89	116	3	134	67	91	106	130	148	61	150	20
k-NN (k=1)	154	134	120	152	116	81	106	130	95	156	70	89	97	92	148
ANOVA	154	3	145	61	8	67	80	64	10	143	17	32	7	22	69
Calinski–Harabasz	154	3	145	61	8	67	80	64	10	143	17	32	7	22	69
Gamma	12	40	126	34	140	26	87	101	52	98	112	46	59	41	115
G+	12	40	126	34	140	26	87	101	52	98	112	46	59	41	115
GDI-41	6	149	95	144	57	96	146	154	61	72	25	33	45	65	116
McClain–Rao	154	32	80	3	145	44	76	17	143	67	61	69	116	8	78
PBM	154	3	61	80	67	76	145	69	156	64	17	8	32	135	152
Point-biserial	40	12	34	126	26	140	87	101	59	41	98	52	115	112	129

This rank was performed for 159 features out of 171 ones in total (prior to the evaluation we removed 12 features that were constant), cf. Appendix 1.A.

correlated (perfectly consistent) and one should be dropped. Further properties of indices are outlined later on in this chapter.

Feature ranks are investigated to evaluate the usefulness of indices. We applied two schemes to compute the score for pairs of indices/classifiers, both based on computing a *distance* between positions of features. Then we look for indices consistent with SVM and k-NN classifiers.

The Distance by Rank (DR)

The first criterion used for measuring similarity is the *distance by rank* (DR) *score*. This score computes the sum of differences between positions of features in ranks. Let us assume that R_1 and R_2 are ranks based on two indices and that $R_1(X_i)$ and $R_2(X_i)$ are indices of the feature X_i in these ranks. This score is defined as follows:

$$DR(R_1, R_2) = \sum_{i=1}^{M} |R_1(X_i) - R_2(X_i)| \qquad (1.46)$$

The results of the DR are shown in Table 1.4. In this table, we display the score for these indices that were used in Table 1.3. The smaller the value of this score is, the more similar indices are. It is equal to 0 for the two perfectly consistent indices, that is, such indices that assign the same rank to each feature. This is the aforementioned case of the ANOVA F-test and the Calinski–Harabasz index and the pair Gamma and G(amma)+ indices. Apart from these two pairs, the highest similarity is encountered for the pair of two classifiers: SVM and k-NN. With regard to similarity of a classifier and an index, relatively low scores, less than 6000 for the SVM classifier and less than 7000 for the k-NN classifier, are encountered for the pair of perfectly consistent ANOVA F-measure and Calinski–Harabasz index and then for the GDI-41, the MCR, and the PBM indices. The point-biserial index is an example of an index that exhibits high DR score values. This index (among a few others) is very inconsistent with other indices.

The Distance by Segments Cardinality (DSC)

The second criterion used for measuring similarity is called the *distance by segments cardinality* (DSC). This score is based on consecutive groups of top features in ranks created by two compared indices. Then, cardinalities of an intersection of corresponding sets of features are summed. The following formula defines the DSC:

$$DSC(R_1, R_2) = \sum_{i=1}^{\lfloor M/r \rfloor} \text{card}\left(\bigcup_{j=1}^{r \cdot i} \{R_1^{-1}(j)\} \cap \bigcup_{j=1}^{r \cdot i} \{R_2^{-1}(j)\} \right) \qquad (1.47)$$

where R_1 and R_2 are ranks based on two indices, R_1^{-1} and R_2^{-1} are inverse mappings of ranks, and r is the segment length parameter. Explicitly, $R_1^{-1}(j)$ is the feature X_i if and only if $R_1(X_i) = j$, and card$\left(\bigcup_{j=1}^{r \cdot i} \{R_1^{-1}(j)\} \cap \bigcup_{j=1}^{r \cdot i} \{R_2^{-1}(j)\} \right)$ is the number of common features in initial segments of length $r \cdot i$ in both ranks. Hence, this criterion operates on initial segments of both ranks, namely, the segments of length $r, 2r, 3r, \ldots, \lfloor M/r \rfloor \cdot r$. From (1.47) we can easily conclude that the first segments

TABLE 1.4 **Features ranking with *distance by rank*: compared are all pairs of classifiers and indices, the scores less than 6000 for the SVM index and less than 7000 for the k-NN index are bolded**

	SVM	k-NN (k=1)	ANOVA F-meas.	Calinski-Harabasz	Gamma	Gamma+	GDI-41	McClain–Rao	PBM	Point biserial
SVM	**0**	**2783**	**5,283**	**5,283**	8862	8862	5214	**4,723**	**5408**	10,560
k-NN (k=1)	**2,783**	**0**	**6,550**	**6,550**	9437	9437	**5341**	**6,042**	**6749**	9,727
ANOVA F-meas.	**5,283**	**6550**	0	0	5913	5913	6135	1,638	1923	11,551
Calinski-Harabasz	**5,283**	**6550**	0	0	5913	5913	6135	1,638	1923	11,551
Gamma	8,862	9437	5,913	5,913	0	0	9698	5,327	5418	8,888
Gamma+	8,862	9437	5,913	5,913	0	0	9698	5,327	5418	8,888
GDI-41	**5,214**	**5341**	6,135	6,135	9698	9698	0	6,607	6792	9,340
McClain–Rao	**4,723**	**6042**	1,638	1,638	5327	5327	6607	0	2199	11,721
PBM	**5,408**	**6749**	1,923	1,923	5418	5418	6792	2,199	0	11,274
Point biserial	10,560	9727	11,551	11,551	8888	8888	9340	11,721	11274	0

Note that the SVM-k-NN score is also low. The score was computed according to (1.46) for M=159 features.

(of length r) are counted $\lfloor M/r \rfloor$ times, the second segments (of length i) are counted $\lfloor M/r \rfloor - 1$, and so on. In this way, DSC score gives priority to shorter segments. Finally, the higher the score for a pair of indices is, the better the consistency of indices is.

The results of the DSC applied to selected indices are outlined in Table 1.5. The SVM classifier is compared with the k-NN classifier and with selected indices. The parameter r is set to 10, so then segments of length tens are considered, that is, in the first row, we have cardinalities of intersection of top 10 features in the SVM rank and in ranks of k-NN and consecutive indices. In the second row, we have cardinalities of top 20 features in the SVM rank, in the ranks of k-NN and consecutive indices, and so on. Specifically, top 10 features in SVM and k-NN ranks have 6 common features, top 10 features in SVM and ANOVA ranks have 3 common features, and so on. Let us notice that two pairs of indices (the pair ANOVA–Calinski–Harabasz and the pair Gamma – Gamma+) give the same results due to their perfect consistency being observed earlier.

The final comparison to the SVM score is shown in Table 1.5 in the row labeled "SVM DSC". In the row labeled "k-NN DSC", we give the final comparison to the k-NN ($k = 1$) score, detailed data are dropped. Notice that the value in the column labeled "k-NN" concerns a comparison for the k-NN rank with itself, and hence it is maximal. In both rows with the final score, we highlight with the bold font scores greater than an (arbitrarily) set threshold 800.

TABLE 1.5   Features ranking with *distance by segments cardinality*: compared are initial segments of ranks created by classifiers and an index

Segment—Index	k-NN	ANOVA	C-H	G	G+	GDI	M-R	PBM	P-B
SVM—[1–10]	6	3	3	0	0	1	3	3	0
SVM—[1–20]	11	8	8	0	0	4	7	6	0
SVM—[1–30]	15	13	13	3	3	8	15	14	0
SVM—[1–40]	24	17	17	6	6	12	20	18	0
SVM—[1–50]	34	24	24	10	10	19	27	26	2
SVM—[1–60]	44	33	33	17	17	30	37	36	3
SVM—[1–70]	53	41	41	27	27	41	46	44	7
SVM—[1–80]	65	51	51	37	37	51	54	53	22
SVM—[1–90]	77	63	63	49	49	64	65	61	37
SVM—[1–100]	88	77	77	59	59	81	79	75	52
SVM—[1–110]	102	89	89	74	74	95	92	87	65
SVM—[1–120]	114	105	105	90	90	111	105	100	83
SVM—[1–130]	126	120	120	106	106	126	122	117	102
SVM—[1–140]	138	137	137	121	121	135	136	133	121
SVM—[1–150]	149	149	149	141	141	143	150	150	141
SVM DSC	**1046**	**930**	**930**	740	740	**921**	**958**	**923**	635
k-NN DSC	**1200**	**860**	**860**	715	715	**917**	**886**	**849**	687

SVM DSC denotes scores of distance by segments cardinality for the SVM classifier and given indices. k-NN DSC denotes scores for k-NN ($k = 1$) and given indices. Indices are taken from Table 1.4, and their abbreviations are in the first row. The bold font is used to highlight relatively high scores.

Finally, the analysis of the DR and the DSC scores allows to recommend five indices for dealing with features selection: the ANOVA F-measure, the Calinski–Harabasz, GDI-41, the MCR, and the PBM indices. Notice that both the DR and the DSC scores recommend the same indices to be consistent with classifiers. Since the ANOVA F-measure and the Calinski–Harabasz index are perfectly consistent, we decided to drop the Calinski–Harabasz index and leave four others.

1.4.7   Selection of Subsets of Features

Needless to say, a successful classification process in most cases requires many features, not only a single one. Therefore, we need to select a number of them from a wider spectrum. One may be tempted to expect that selecting a number of features with the highest individual evaluation will guarantee the best choice. Unfortunately, evaluation of single features is not the best indicator of the predictive power of a model based on more than one feature. The aforementioned feature interactions influence a design of the classifier.

Let us reiterate that a subset of features, say, k features, with the best individual evaluation does not guarantee the best evaluation within other subsets with the same cardinality. Namely, it is necessary to test all subsets including k features out of M in order to have the best set of k features. Since the number of all k subsets of cardinality k out of M features, given in (1.48), is factorial, roughly estimating the complexity of such a method is exponential, so then such method is useless in practice for problems of higher dimensionality:

$$\binom{M}{k} = \frac{M!}{(M-k)!k!} \tag{1.48}$$

In the next section, we discuss some approximation methods that can be used in selecting the best subset of features out of a wider range of them. Such approximation methods, in fact, do not guarantee the best set to be identified, but we can expect that a set close to the optimal one could be selected.

Index-Based Feature Selection Methods
The most straightforward method for feature selection, conceptually speaking, is to generate subsets of features based on a full set of features, evaluate all generated subsets using some index, and select a subset with the highest score. As a quality evaluation index, we can use any of the clustering indices discussed before. Naturally, they are fit to evaluate not only a single feature but also a set of features. This straightforward procedure is outlined in Algorithm 1.4.

Algorithm 1.4
Index-based evaluation of a given set of features
Data: Set of Features for Evaluation
              Learning Set of Patterns
              Index for Feature Set Evaluation
Algorithm: **evaluate** the Set of Features **using** Index for Feature Set Evaluation
Results: quality score for the given Set of Features

Wrapper-Based Feature Selection Methods
Wrapper-based methods rely on comparing the quality of multiple classifiers built on different sets of features. In the following text we formulate two algorithms to be used in the wrapper-based evaluation of sets of features. These algorithms are executed multiple times to form a wide collection of models, out of which the best one is selected.

Algorithm 1.5
Classifier-based evaluation of a given set of features
Data: Set of Features
           Learning Set of Patterns
           Classification Method
           Classifier Evaluation Method
Algorithm: split Learning Set of Patterns into Training and Test Sets
           build classifier for given
                    Classification Method **and** Training Set
           for Training and Test Sets **do**
               evaluate constructed **classifier**
                    using given Classifier Evaluation Method
           use evaluation of **classifier** as evaluation of features
Results: quality score of the given Set of Features

This algorithm works as follows. First, we construct a classifier using a given classification method and the training set of patterns. Once the classifier has been built, the classifier evaluation method produces scores on the training set and the test set. Finally, both scores are involved in evaluation of the set of features.

This algorithm is usually extended with a cross-validation technique, which allows to average classifier construction. We rewrite Algorithm 1.5 with a small update in order to underline that the cross-validation concerns the training set and that the test set is used only for classifier evaluation. Again, we skip details of cross-validation techniques in order to keep clarity of narration. It is worth noting that Algorithm 1.5 is a special case of Algorithm 1.6 with the number of cross-validation folds equal to 1.

Algorithm 1.6
Classifier-based evaluation of a set of features with cross-validation
Data: Set of Features
           Learning Set of Patterns
            Classification Method
            Classifier Evaluation Method
            r—number of cross-validation folds
Algorithm: split Learning Set of Patterns into Training and Test Set
           repeat r **times**
           begin
               use Training Set to build a fold
               build a classifier **using** given Classification Method
                   and the current fold
           end

build the final classifier **based on** obtained r results
evaluate constructed **classifier**
    using given Classifier Evaluation Method
      and Training and Test Sets
    use evaluation of **classifier** as evaluation of features
Results: quality score of the given Set of Features

Algorithms 1.5 and 1.6 provide methods for evaluation of feature sets using a classifier of choice. Typically, they could be employed as a component of the wrapper-based feature selection method that generates sets of features and then employs evaluation algorithms for feature sets to select the best one. Of course, saying *the best set of features*, we mean that such a set of features gets the best evaluation on the given learning set of patterns. We also hope that this set of features will allow to construct such a classifier that will occur to be the best one among the others in future applications.

In the addressed schemes the classifier is built based on a training set (a subset of the learning set used for model construction). Next, we evaluate classifier performance on a test set. The test set, in other words, is a holdout set of patterns unseen at the stage of classifier construction. Finally, the scores obtained on train and test sets can be combined in any way to get an evaluation of the set of features. For the sake of clarity, we do not discuss details such as proportion between cardinalities of the training and test sets, quality evaluation method, relation between evaluation of the training set and the test set of patterns, and so on. Among early researches on wrapper-based feature selection, we may refer to Kohavi and John (1997).

1.4.8 Feature Subsets Generation

As mentioned before, building classifiers for multiple sets of features requires long computation time, and, of course, it is definitely useless for systematic searching in large spaces of features. Therefore, instead of classifiers, we discuss employment of indices to evaluate the quality of sets of features. We collate the results with *classical* wrapper-based methods relying on classifiers only. We place a strong emphasis on comparison between wrapper-based methods, and therefore this discussion is placed in the section devoted to wrapper-based feature search methods.

However, no matter if we use a classifier or an index, still one problem remains unsolved: how to generate subsets of features that need to be checked.

In subsequent sections let us discuss several methods that could be used in generating an optimal set of features. Saying *an optimal set of features*, we mean that performance of such set will be as good as possible. It is important to underline that such a set of features may not guarantee the best performance among other sets. This is for several reasons such as cost reduction of the features selection process. There is no one universal and objective method that could be used to find the best set of features.

Naïve (Brute Force) Selection

We begin this discussion with recalling the naïve (brute force) method. More specifically, we may use the naïve (brute force) selection, that is, investigation of all nonempty subsets of the set of features in order to select the best one. Of course, this method, formulated in Algorithm 1.7, guarantees selection of the best set in terms of the learning set of patterns. But such a set of features may be less successful when it is used to process patterns not belonging to the learning set.

Algorithm 1.7
Naive (brute force) selection of the best subset of features
Data:	Set of Features
	Learning Set of Patterns
	either Index for Feature Set evaluation
	or Classification Method and Classifier Evaluation Method
Algorithm:	**for** each nonempty subset of the Set of Features **do**
	call Algorithm 4 or 5 or 6 to evaluate this subset of features
	choose the subset with the best evaluation
Results:	the subset of the Set of Features with the best evaluation

However, computational complexity of the naïve selection is exponential, and therefore it can be used only for a small set of features, that is, for a small M. Therefore, in practice, it is useless for more than 10 features. Saying it useless, we mean that although the running time of such a method is finite, it is so long that we will not get results in a reasonable time. In such cases, instead of the exact naïve method, different approximations are applied. In the following text we discuss four methods, three greedy ones and a method with a limited expansion.

Greedy Selection by Rank of Single Features
The first attempt to select an approximated best set of k features is just using the top k features from the rank provided by any quality measure of single features. We may use any measure to evaluate single features, for instance, those discussed in Section 1.4, that is, a classifier, the ANOVA F-test, the GDI-41, the MCR, and the PBM indices. Using single features ranking would be appropriate for a raw classification task, where the highest quality result is not the prime issue. Unfortunately, if high classification accuracy is a prime objective, this simple method is not sufficient. Therefore, we need to utilize more sophisticated selection in order to produce a set of features of better quality.

When we apply the single feature rank selection procedure and our objective is to choose k features; we select top k features. They were evaluated individually as the best, but we do not know how they *cooperate* with one another. In contrast, when we use greedy forward/backward selection (that will be described in the following two sections), we add/remove features one by one. We do not look at the quality of individual features. Instead, in each round of the selection procedure, we add/ remove such one feature so that the new set of features becomes the best.

Greedy Forward Selection
We are looking for the approximately best set of features of given cardinality, say, k features out of all M ones. In this scheme, we start with an empty set, and then we keep adding one feature in each iteration. Assume that we have l features already selected. Then, for every feature f from the set of $M - l$ nonselected ones, evaluated is the set with feature f added, that is, the set of $l + 1$ features. After this, the set with the best evaluation is taken as the set of $l + 1$ selected features. Finally, the feature f included in

a newly created set is deleted from the set of nonselected ones. This method is formally described in Algorithm 1.8.

Algorithm 1.8
Greedy forward selection of the best subset of features
Data: SofF (Set of Features)
           NofFtoS (Number of Features to Select)
           Learning Set of Patterns
           either Index for Feature Set Evaluation
           or Classification Method and Classifier Evaluation Method
Algorithm: **initiate** SofSF (Set of Selected Features) as the empty set
           initiate SofRF (Set of Remaining Features) as SofF
           while cardinality of SofSF is less than NofFtoS **do**
           begin
               for every feature f **in** SofRF **do**
                   evaluate the set SofSF $\cup\{f\}$ of features
                       using Algorithm 1.4 or 1.5 or 1.6
                   choose the feature f_{max} for which SofSF $\cup\{f\}$ gets the best
                       evaluation
                   add f_{max} to SofSF
                   remove f_{max} from SofRF
           end
Results: the subset of features of cardinality Number of Features to Select
           (NofFtoS)

Greedy forward selection is an iterative process running until the set of selected features reaches the required cardinality. In practice, this condition may be replaced with another one based on an intuitive expectation that the quality of the classification model will be increasing along with growing cardinality of the set of selected features. Based on this assumption, the stopping criterion would involve a very intuitive condition based on the quality of the classification model evaluated on the learning set, that is, the quality measured on the training set or on the testing set or on a combination of both measures. Once the required quality has been reached, the iterative process can be stopped.

Greedy Backward Selection
Algorithm 1.9 formulates greedy backward selection. Instead of starting from an empty set and then incrementally adding features, this method starts with a full set of features and iteratively reduces its size. At the beginning, we give the full set of M features. Then, in each turn, we evaluate all subsets obtained by removing one feature from the current set. After that, the current set of features is replaced with the subset with the best evaluation. This process is repeated until we skim the set of features to a desired size or until another stopping condition is satisfied. After each iteration we inspect the quality of features. When the quality starts to decrease substantially, the process should be halted without waiting for other stopping criteria to be met. Greedy backward selection can be used in a case when the initial full set of features is not very large. When we want to select a small subset of features out of a large set of all features, forward selection is less run time consuming than backward selection.

It will be shown later on that, in general, reducing a big set of features increases quality evaluation on the test set. This is a consequence of the phenomenon of over-fitting. Therefore, the halting condition based on quality evaluation would be carefully defined, when applied to big sets of features.

Algorithm 1.9
Greedy backward selection of the best subset of features
Data: SofF (Set of Features)
  NofFtoS (Number of Features to Select)
  Learning Set of Patterns
  either Feature Evaluation Method
  or Classification Method with Classifier Evaluation Method
Algorithm: **initiate** SofSF (Set of Selected Features) as SofF (Set of Features)
  while cardinality of SofSF is less than NofFtoS **do**
  begin
    for every feature f **in** SofSF **do**
      evaluate the set SofSF $-\{f\}$ of features
        using Algorithm 1.4 or 1.5 or 1.6
      choose the feature f_{max} for which SofSF $-\{f\}$ gets the best evaluation
      remove f_{max} from SofSF
    end
Results: the subset of features of cardinality Number of Features to Select (NofFtoS)

Greedy Forward Selection with Limited Expansion
Greedy forward selection with limited expansion extends simple forward selection. In each iteration of this algorithm, l the best sets of k features are processed, where l is the expansion width. Each such set is incremented with each available feature and then the l best sets are selected. Specifically, for each set of cardinality k, out of the l best ones, $M - k$ sets are created by inserting one not selected feature. In total, $l*(M - k)$ new sets of cardinality $k + 1$ are obtained. Then, after deleting duplicated sets, new k best sets are selected. The details are presented in Algorithm 1.10.

Algorithm 1.10
Greedy forward selection with limited expansion
Data: SofF (Set of Features)
  NofFtoS (Number of Features to Select)
  l—Width of Limited Expansion
  Stop Condition
  Learning Set of Patterns
  either Feature Evaluation Method
  or Classification Method with Classifier Evaluation Method
Algorithm: **for** $i = 1$ to l **do**
    Initialize SofSF$_i$ (i-th Set of Selected Features) **as** empty set
  while Stop Condition is not satisfied **do**
  begin

> **initialize** PLofSofF (Pending List of Sets of Features) **as** empty list
> **for** $i = 1$ to l **do**
> **begin**
>   **initialize** SofRF (Set of Remaining Features) **as** SofF − SofSF$_i$
>   **for** each feature f **in** SofRF **do**
>   **add** SofSF$_i \cup \{f\}$ to the PLofSofF
>
> **end**
> **delete** duplicates from the PLofSofF
> **evaluate** sets of features from the PLofSofF
>   **using** Algorithm 1.4 or 1.5 or 1.6
> **for** $i = 1$ to l **do**
>   **replace** SofSF$_i$ (i-th Set of Selected Features) **by**
> i-th top set from PLofSofF
> **end**

Results: the l best subset of selected features

Notice that for the parameter $l = 1$, the greedy forward search with limited expansion turns into a simple greedy forward search. On the other hand, when at each turn the parameter l is equal to $M - k$—that is, it is equal to the number of features not used yet in the current set of selected features—this algorithm turns into the naïve selection.

Notes on Computational Complexity
As mentioned earlier, computational complexity of the presented search algorithms significantly differs, which determines their usefulness in practical applications. Let us take a closer look at this problem. Evaluation of created subsets of features is the dominant operation in methods formulated in Algorithms 1.7–1.10. Therefore, run time of such an algorithm is proportional to the number of executions of this operation. Evaluation of a set of features is performed by Algorithm 1.4 (index-based) or Algorithm 1.5 (classifier, no cross-validation) or Algorithm 1.6 (classifier with cross-validation). Execution of Algorithm 1.6 is, roughly speaking, about r times longer than execution of Algorithm 1.5, where r is the number of cross-validation folds.

Having this in mind, let us estimate the number of executions of different subset search algorithms that in fact is equivalent to asymptotical complexity:

- Greedy selection by rank of single features requires each feature to be evaluated once. Hence complexity is of rank M, that is, $O(M)$.
- Naïve (brute force) selection requires evaluation of each subset of the set of features, which makes complexity to be exponential, that is, $O(2^M)$.
- In greedy forward selection and greedy backward selection in each turn consecutively $M, M-1, M-2, M-3, \ldots$, sets of features are evaluated, which raises (pessimistic) square complexity: $O(M^2)$.
- Greedy forward selection with limited expansion requires evaluation of k times more sets than simple greedy forward selection for constant k and $k << M$, so then the rank of complexity is $O(kM^2)$, where k is width of limited expansion.

Therefore, from the perspective of complexity, any of the previously mentioned methods is a reasonable choice except the naïve method, which may be applied only for small sets of features.

Illustrative Experiment

In this experiment, Algorithm 1.10 was employed to test methods for features selection. We tested the run time for different evaluation methods and then the quality of a classifier built on selected features. The experiment was carried out on the data set of handwritten digits (LeCun *et al.*, 1998).

Besides the previously mentioned asymptotical complexity estimation, in practice important are run time of a dominant operation and proportion between the run time of the dominant and all operations. In our case, the evaluation of a set of features is the dominant operation. We used two types of evaluators: classifiers and clustering indices. In Table 1.6, the run times of Algorithm 1.10 are given. This algorithm was executed for different classification models and evaluation methods for the purpose of the characteristics outlined in Figures 1.10 and 1.11, that is, in each case (evaluation method) selected were sets of features of consecutive cardinality 1, 2, 3, 4, …, 100. In the case of the k-NN method, we took 1 neighbor and there was no cross-validation involved. In the case of the SVM classifier, the following parameters were set: Gaussian kernel function, $\gamma = 0.0625$ and $C = 1$, with 10-fold cross-validation, cf. Section 2.3.

In practice, the final set of features is of cardinality 20–40, much less than the set of 100 features, considered according to the prepared characteristics given in Figures 1.10 and 1.11. Therefore, the computation time of forward selection methods should be divided by a proper factor, say, 5 for the SVM selector and 2.5 for other selectors. Anyway, it is worth drawing attention to the fact that the range of time is huge even if these factors are taken into account. These ranges begin with one/a few hours (ANOVA F-test), 15 or so hours (k-NN and indices), and up to 15 or so days (SVM). Hence, we can conclude that performance time is an important factor that should be taken into consideration. Of course, the choice of an evaluation method determines not only its run time but also the quality of the final classifier constructed with selected features.

In Figure 1.9 we present values of indices for the best sets of features for three values of expansion limit and for cardinality of features sets ranging from 1 to 100. Results concern a dataset of handwritten digits. Displayed are values of ANOVA F-test, PBM index, GDI-41 index and accuracy values of SVM classifier built at

TABLE 1.6   Performance time of Algorithm 1.10 on the MNIST dataset (LeCun *et al.*, 1998) for three clustering indices and two classifiers

	ANOVA	MCR	PBM	k-NN	SVM
$k = 1$	1.5	15	17	7	135
$k = 3$	4.5	47	50	20	400
$k = 5$	7.5	70	80	33	650

Given is performance time in hours needed to compute data for characteristics outlined in Figures 1.10 and 1.11.

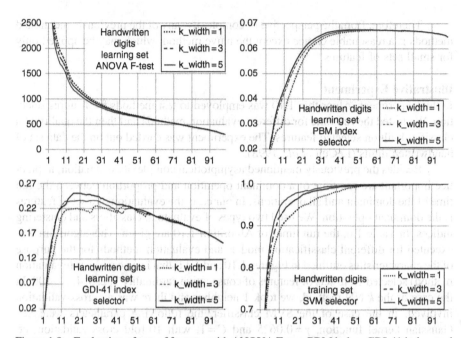

Figure 1.9   Evaluation of sets of features with ANOVA F-test, PBM index, GDI-41 index, and SVM classifier. Sets of features were selected with greedy forward with expansion limited to 1, 3, and 5. Displayed are scores of indices at the whole learning set of patterns and SVM classifier accuracy at the training set of patterns (vertical axis) as a function of cardinality of features sets ranging from 1 to 100 (horizontal axis). Results concern a dataset of handwritten digits.

the best sets of features selected with three values of expansion limit (1, 3, and 5) (cf. Algorithm 1.10).

Let us recall that the higher the value of indices/accuracy, the better the quality of the set of features is. In cases of PBM and GDI-41 Indices and SVM classifiers, quality is growing along with increasing expansion limit and cardinality of best sets of features and, amazingly, ANOVA F-test behaves in the opposite way: quality falls at the same time. In all four cases, around interval [20,30], there is an inflection region (ANOVA F-test, PBM index, and SVM) and maximum region (GDI-41 index).

In Figures 1.10 and 1.11, the quality of classifiers constructed on the basis of features sets selected with various methods is displayed. For each feature set an SVM classifier with parameters set to Gaussian kernel function, $\gamma = 0.0625$ and $C = 1$, with 10-fold cross-validation, was constructed. Results at the training set and at the test set are given for selected indices and for the SVM classifier used for feature sets evaluation.

Note that SVM parameters were the same for each tested set of features, that is, no parameters tuning was done. Parameters tuning is performed empirically. This in fact means that additional repetitions of model construction procedures would be

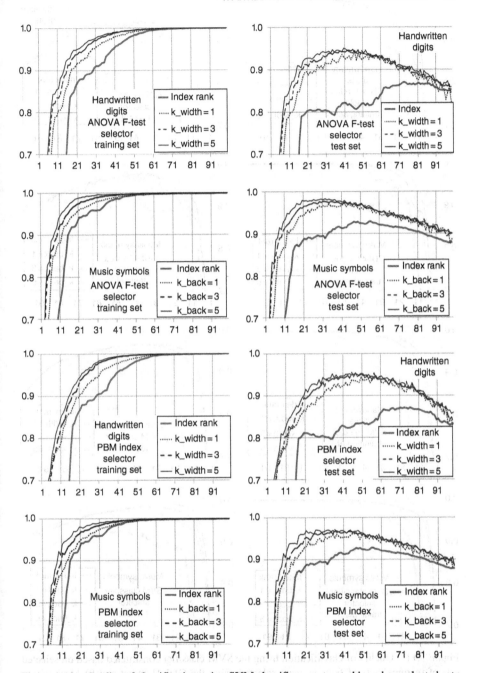

Figure 1.10   Quality of classification using SVM classifier constructed based on selected sets of features. Sets of features were selected using the ANOVA F-test and PBM index. Classification accuracy (vertical axis) is measured at the training set and at the test set for three values of expansion limit and for the best features in the individual index rank, cardinality of features sets ranging from 1 to 100 (horizontal axis). The first two rows of graphs concern F-ANOVA; the last two rows—PBM. Results concern handwritten digits and musical symbols datasets.

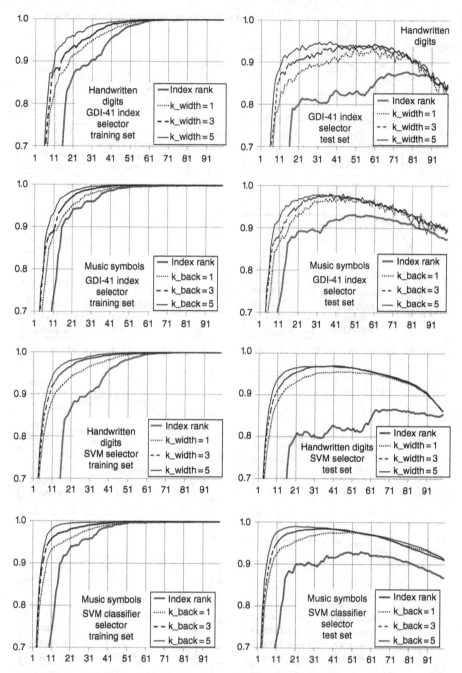

Figure 1.11 Quality of classification using the SVM classifier constructed based on selected sets of features. Sets of features were selected using the GDI-41 index and SVM classifier. Classification accuracy (vertical axis) is measured at the training set and at the test set for three values of expansion limit and for the best features in the individual index rank, cardinality of features sets ranging from 1 to 100 (horizontal axis). The first two rows of graphs concern F-ANOVA; the last two rows, PBM. Results concern handwritten digits and musical symbols datasets.

needed to select optimal parameters. This additional effort did not seem to be reasonable when the focus of this section was not on SVM itself, but on feature selection. Of course, tuned parameters may increase the quality of the final classifier, and it is recommended to perform parameters optimization for final model construction.

It is clearly visible that the SVM classifier based on features selected with the SVM classifier outperforms classifiers constructed based on indices. First, the maximal accuracy for the SVM selector overheads the other selectors by a few percent points. The second highlight is that the maximal accuracy is gained for smaller sets of features than in other cases. Third, characteristics for the SVM selector are more regular and smoother than for the other selectors, especially at the test set. However, time needed to execute a feature search with SVM as an evaluation method is huge. Observing characteristics displayed in Figures 1.10 and 1.11, we compare accuracy at the training set ranging from 0.9 to 0.98. What we can see is that cardinalities of optimal feature sets computed with the PBM and the ANOVA F-test are about 1.5 times greater than the cardinality of a set extracted with the SVM selector.

Also, the PBM and the ANOVA F-test-based models achieve maximal accuracy at the test set for feature sets that are twice as large as for the SVM selector. Having this in mind and referring to Table 1.6, we see that the run time of a procedure utilizing McClain and PMB is 10 times greater than a procedure based on the ANOVA F-test selector. Moreover, the run time of a procedure utilizing SVM is 50 (!) times greater than that of a procedure based on the ANOVA F-test selector. Therefore, we can conclude that the choice of features selector is a matter of a trade-off between model quality and its construction cost.

As an outcome of the experiments concerning greedy feature search, we have discovered suitable sets of features describing the datasets of handwritten digits and symbols of printed music notation. For the handwritten digits, we selected a set made up of 24 features, whereas for the musical symbols, they are described by 20 features. All these features are listed in Appendix 1.B. The order in which the features appear is according to the ranking realized by the SVM classifier used for evaluation of features.

1.5   CONCLUSIONS

It is needless to say that selecting proper features is an elementary and necessary assignment in any pattern recognition task. It determines the quality of models produced at later stages. In visual pattern recognition, feature selection is heavily determined by the kind of processed data. Here, we focus on characters and presented methodology that fits well in this category of problem. However, the discussed approaches may not be the best if we would wish to process substantially different data coming from a different domain, for instance, high-resolution colored scans obtained by high-end medical equipment. As it is for many other technical aspects of pattern recognition, the choice of methods is usually data driven.

APPENDIX 1.A

The list of 159 features used in experiments with handwritten digits and symbols of music notation recognition. Constant features were removed from the full list of 171 features

1   Projection V—Raw—Min—Value
2   Projection V—Raw—Min—Position
3   Projection V—Raw—Max—Value
4   Projection V—Raw—Max—Position
5   Projection V—Raw—Mean
6   Projection V—Raw—First moment
7   Projection V—Raw—Peaks count
8   Projection V—Differential—Min—Value
9   Projection V—Differential—Min—Position
10   Projection V—Differential—Max—Value
11   Projection V—Differential—Max—Position
12   Projection V—Differential—Mean
13   Projection V—Differential—First moment
14   Projection V—Differential—Peaks count
15   Projection H—Raw—Min—Value
16   Projection H—Raw—Min—Position
17   Projection H—Raw—Max—Value
18   Projection H—Raw—Max—Position
19   Projection H—Raw—Mean
20   Projection H—Raw—First moment
21   Projection H—Raw—Peaks count
22   Projection H—Differential—Min—Value
23   Projection H—Differential—Min—Position
24   Projection H—Differential—Max—Value
25   Projection H—Differential—Max—Position
26   Projection H—Differential—Mean
27   Projection H—Differential—First moment
28   Projection H—Differential—Peaks count
29   Histogram V—Raw—Min—Position
30   Histogram V—Raw—Max—Value
31   Histogram V—Raw—Max—Position
32   Histogram V—Raw—Mean
33   Histogram V—Raw—First moment
34   Histogram V—Raw—Peaks count
35   Histogram V—Differential—Min—Value
36   Histogram V—Differential—Min—Position
37   Histogram V—Differential—Max—Value
38   Histogram V—Differential—Max—Position

39   Histogram V—Differential—First moment
40   Histogram V—Differential—Peaks count
41   Histogram H—Raw—Min—Position
42   Histogram H—Raw—Max—Value
43   Histogram H—Raw—Max—Position
44   Histogram H—Raw—Mean
45   Histogram H—Raw—First moment
46   Histogram H—Raw—Peaks count
47   Histogram H—Differential—Min—Value
48   Histogram H—Differential—Min—Position
49   Histogram H—Differential—Max—Value
50   Histogram H—Differential—Max—Position
51   Histogram H—Differential—First moment
52   Histogram H—Differential—Peaks count
53   Cumulative Histogram V—Raw—Min—Value
54   Cumulative Histogram V—Raw—Max—Value
55   Cumulative Histogram V—Raw—Max—Position
56   Cumulative Histogram V—Raw—Mean
57   Cumulative Histogram V—Raw—First moment
58   Cumulative Histogram V—Raw—Peaks count
59   Cumulative Histogram H—Raw—Min—Value
60   Cumulative Histogram H—Raw—Max—Value
61   Cumulative Histogram H—Raw—Max—Position
62   Cumulative Histogram H—Raw—Mean
63   Cumulative Histogram H—Raw—First moment
64   Cumulative Histogram H—Raw—Peaks count
65   Transitions V—Raw—Min—Value
66   Transitions V—Raw—Min—Position
67   Transitions V—Raw—Max—Value
68   Transitions V—Raw—Max—Position
69   Transitions V—Raw—Mean
70   Transitions V—Raw—First moment

71 Transitions V—Differential—Min—
  Value
72 Transitions V—Differential—Min—
  Position
73 Transitions V—Differential—Max—
  Value
74 Transitions V—Differential—Max—
  Position
75 Transitions V—Differential—First
  moment
76 Transitions H—Raw—Min—Value
77 Transitions H—Raw—Min—Position
78 Transitions H—Raw—Max—Value
79 Transitions H—Raw—Max—Position
80 Transitions H—Raw—Mean
81 Transitions H—Raw—First moment
82 Transitions H—Differential—Min—
  Value
83 Transitions H—Differential—Min—
  Position
84 Transitions H—Differential—Max—
  Value
85 Transitions H—Differential—Max—
  Position
86 Transitions H—Differential—First
  moment
87 Offsets L—Raw—Min—Value
88 Offsets L—Raw—Min—Position
89 Offsets L—Raw—Max—Value
90 Offsets L—Raw—Max—Position
91 Offsets L—Raw—Mean
92 Offsets L—Raw—First moment
93 Offsets L—Raw—Peaks count
94 Offsets L—Differential—Min—Value
95 Offsets L—Differential—Min—Position
96 Offsets L—Differential—Max—Value
97 Offsets L—Differential—Max—Position
98 Offsets L—Differential—Mean
99 Offsets L—Differential—First moment
100 Offsets L—Differential—Peaks count
101 Offsets R—Raw—Min—Value
102 Offsets R—Raw—Min—Position
103 Offsets R—Raw—Max—Value
104 Offsets R—Raw—Max—Position
105 Offsets R—Raw—Mean
106 Offsets R—Raw—First moment
107 Offsets R—Raw—Peaks count
108 Offsets R—Differential—Min—Value
109 Offsets R—Differential—Min—Position
110 Offsets R—Differential—Max—Value

111 Offsets R—Differential—Max—Position
112 Offsets R—Differential—Mean
113 Offsets R—Differential—First moment
114 Offsets R—Differential—Peaks count
115 Offsets T—Raw—Min—Value
116 Offsets T—Raw—Min—Position
117 Offsets T—Raw—Max—Value
118 Offsets T—Raw—Max—Position
119 Offsets T—Raw—Mean
120 Offsets T—Raw—First moment
121 Offsets T—Raw—Peaks count
122 Offsets T—Differential—Min—Value
123 Offsets T—Differential—Min—Position
124 Offsets T—Differential—Max—Value
125 Offsets T—Differential—Max—Position
126 Offsets T—Differential—Mean
127 Offsets T—Differential—First moment
128 Offsets T—Differential—Peaks count
129 Offsets B—Raw—Min—Value
130 Offsets B—Raw—Min—Position
131 Offsets B—Raw—Max—Value
132 Offsets B—Raw—Max—Position
133 Offsets B—Raw—Mean
134 Offsets B—Raw—First moment
135 Offsets B—Raw—Peaks count
136 Offsets B—Differential—Min—Value
137 Offsets B—Differential—Min—Position
138 Offsets B—Differential—Max—Value
139 Offsets B—Differential—Max—Position
140 Offsets B—Differential—Mean
141 Offsets B—Differential—First moment
142 Offsets B—Differential—Peaks count
143 Directions—0
144 Directions—135
145 Directions—90
146 Directions—45
147 Directions—WE—Y
148 Directions—NS—X
149 Raw moments—First—m10
150 Raw moments—First—m01
151 Central moments—Second—m20
152 Central moments—Second—m11
153 Central moments—Second—m02
154 Height/width
155 Blackness level
156 Eccentricity
157 Euler number 4
158 Euler number 8
159 Euler number 6

APPENDIX 1.B

Lists of features selected for in experiments with handwritten digits recognition (left list) and symbols of music notation (right list)

1	Raw moments—First—m01		1	Height/width
2	Central moments—Second—m02		2	Offsets T—Raw—Min—Position
3	Offsets L—Differential—Max—Position		3	Central Moments—Second—m11
4	Euler number 4		4	Projection V—Raw—Max—Value
5	Offsets R—Raw—Min—Position		5	Offsets R—Raw—First moment
6	Offsets L—Differential—Min—Position		6	Offsets R—Raw—Mean
7	Directions—45		7	Euler number 4
8	Offsets L—Differential—Max—Value		8	Central moments—Second—m02
9	Offsets R—Differential—First moment		9	Directions—135
10	Offsets T—Differential—Max—Value		10	Transitions H—Raw—Max—Value
11	Raw Moments—first—m10		11	Central moments—Second—m20
12	Height/width		12	Offsets R—Raw—Min—Position
13	Cumulative Histogram V—Raw—First moment		13	Directions—45
14	Offsets L—Differential—Min—Value		14	Offsets L—Raw—Max—Value
15	Offsets L—Differential—First moment		15	Projection V—Differential—Max—Value
16	Central Moments—Second—m20		16	Projection H—Differential—Peaks count
17	Offsets T—Raw—Min—Position		17	Raw moments—First—m01
18	Projection V—Raw—Max—Value		18	Offsets L—Raw—Mean
19	Offsets R—Differential—Max—Value		19	Cumulative histogram V—Raw—first Moment
20	Cumulative Histogram V—Raw—Max—Position		20	Cumulative histogram H—Raw—Peaks Count
21	Projection H—Differential—Min —Value			
22	Directions—90			
23	Offsets B—Differential—Min—Position			
24	Offsets L—Raw—Max—Value			

REFERENCES

S. Bandyopadhyay, M. Pakhira, and U. Maulik, Validity index for crisp and fuzzy clusters, *Pattern Recognition* 37, 2004, 487–501.

H. Bay, A. Ess, T. Tuytelaars, and L. Van Gool, Speeded-up robust features (SURF), *Computer Vision and Image Understanding* 110(3), 2008, 346–359.

M. Charrad, NbClust: An R package for determining the relevant number of clusters in a data set, *Journal of Statistical Software* 61(6), 2014, 1–36.

B. Desgraupes, Clustering indices, *Report*, University Paris Ouest, Lab Modal'X, 2013.

B. Desgraupes, Package clusterCrit for R, *R Documentation*, 2016.

J. C. Dunn, A fuzzy relative of the ISODATA process and its use in detecting compact well-separated clusters, *Journal of Cybernetics* 3(3), 1973, 32–57.

N. O. F. Elssied, O. Ibrahim, and A. H. Osman, A novel feature selection based on one-way ANOVA F-test for e-mail spam classification, *Research Journal of Applied Sciences* 7(3), 2014, 625–638.

I. Guyon, J. Weston, S. Barnhill, and V. Vapnik, Gene selection for cancer classification using support vector machines, *Machine Learning* 46 (1–3), 2002, 389–422.

K. Haris, S. N. Efstratiadis, N. Maglaveras, and A. K. Katsaggelos, Hybrid image segmentation using watersheds and fast region merging, *IEEE Transactions on Image Processing* 7(12), 1998, 1684–1699.

W. Homenda, A. Jastrzebska, and W. Pedrycz, *The web page of the classification with rejection project*, 2017, http://classificationwithrejection.ibspan.waw.pl (accessed October 5, 2017).

W. Homenda, A. Jastrzebska, W. Pedrycz, and R. Piliszek, Classification with a limited space of features: Improving quality by rejecting misclassifications. In: *Proceedings of the 4th World Congress on Information and Communication Technologies (WICT 2014)*, Malacca, Malaysia, December 8–11, 2014.

W. Homenda, M. Luckner, and W. Pedrycz, *Classification with rejection: concepts and formal evaluations, Knowledge, Information and Creativity Support Systems: Recent Trends, Advances and Solutions*, Advances in Intelligent Systems and Computing 364, Switzerland, Springer International Publishing, 2016, 413–425.

M. K. Hu, Visual pattern recognition by moment invariants, *IRE Transactions on Information Theory* IT-8, 1962, 179–187

C. L. Huang and C. J. Wang, A GA-based feature selection and parameters optimization for support vector machines, *Expert Systems with Applications* 31(2), 2006, 231–240.

R. Kohavi and G. H. John, Wrappers for feature subset selection, *Artificial Intelligence* 97(1–2), 1997, 273–324.

S. Krig, *Computer Vision Metrics. Survey, Taxonomy, and Analysis*, New York, Springer, 2014.

M. Kudo and J. Sklansky, Comparison of algorithms that select features for pattern classifiers, *Pattern Recognition* 33(1), 2000, 25–41.

Y. LeCun, C. Cortes, and C. J. C. Burges, *The MNIST database of handwritten digits*, 1998, http://yann.lecun.com/exdb/mnist/ (accessed October 5, 2017).

D. G. Lowe, Distinctive image features from scale-invariant keypoints, *International Journal of Computer Vision* 60(2), 2004, 91–110.

B. S. Manjunath and W. Y. Ma, Texture features for browsing and retrieval of image data, *IEEE Transactions on Pattern Analysis and Machine Intelligence* 18(8), 1996, 837–842.

J. O. McClain and V. R. Rao, CLUSTISZ: A program to test for the quality of clustering of a set of objects, *Journal of Marketing Research* 12(4), 1975, 456–460.

P. Mitra, C. A. Murthy, and S. K. Pal, Unsupervised feature selection using feature similarity, *IEEE Transactions on Pattern Analysis and Machine Intelligence* 24(3), 2002, 301–312.

H. C. Peng, F. H. Long, and C. Ding, Feature selection based on mutual information: Criteria of max-dependency, max-relevance, and min-redundancy, *IEEE Transactions on Pattern Analysis and Machine Intelligence* 27(8), 2005, 1226–1238.

Y. Saeys, I. Inza, and P. Larranaga, A review of feature selection techniques in bioinformatics, *Bioinformatics* 23(19), 2007, 2507–2517.

M. Sonka, V. Hlavac, and R. Boyle, *Image Processing, Analysis and Machine Vision*, Pacific Grove, CA, PWS Publishing, 1998.

J. H. Sossa-Azuela, R. Santiago-Montero, M. Pérez-Cisneros, and E. Rubio-Espino, Computing the Euler number of a binary image based on a vertex codification, *Journal of Applied Research and Technology* 11(3), 2013, 360–370.

R. W. Swiniarski and A. Skowron, Rough set methods in feature selection and recognition, *Pattern Recognition Letters* 24(6), 2003, 833–849.

X. Tan and B. Triggs, Enhanced local texture feature sets for face recognition under difficult lighting conditions, *IEEE Transactions on Image Processing* 9(6), 2010, 1635–1650.

O. D. Trier, A. K. Jain, and T. Taxt, Feature extraction methods for character recognition – A survey, *Pattern Recognition* 29(4), 1996, 641–662.

M. Turk and A. Pentland, Eigenfaces for recognition, *Journal of Cognitive Neuroscience* 3(1), 1991, 71–86.

CHAPTER 2

PATTERN RECOGNITION: CLASSIFIERS

In the previous chapter, we have discussed the introductory steps in the overall process of pattern recognition. In this chapter, we proceed with a detailed discussion and present selected algorithms that operate in the feature space and facilitate classification mechanisms. We discuss architectures and design approaches to fundamental classifiers such as selected probabilistic classifiers, classifiers based on feature space geometry, and ensemble classifiers.

This chapter is structured as follows. First, we present elementary notation needed to proceed with further elaborations. Next, we present selected classifiers: k-nearest neighbors (k-NN), support vector machines (SVMs), decision trees, ensemble classifiers, and a particular example of ensemble classifiers—random forests and naïve Bayes classifier.

The motivation for this selection is to introduce the reader to the wealth of various approaches to pattern recognition varying in terms of their discriminatory capabilities and computing overhead. The content of this chapter aims at discussing standard methods, so that in the later discussion we can build upon the presented algorithms.

2.1 CONCEPTS

Let us recall that an essence of a *standard* pattern recognition problem is to split a set of patterns $O = \{o_1, o_2, o_3, \ldots\}$ into C subsets $O_1, O_2, \ldots, O_C$, which include patterns belonging to the same class such that these subsets are pairwise disjoint:

$$O = \bigcup_{i=1}^{C} O_i \text{ and } (\forall i, j \in \{1, 2, \ldots, C\}, i \neq j) \ O_i \cap O_j = \emptyset \tag{2.1}$$

This task is defined by a mapping called *classifier* $\Psi : O \to \Theta$ where $\Theta = \{O_1, O_2, \ldots, O_C\}$ is the set of classes. For the sake of simplicity, we assume that

Pattern Recognition: A Quality of Data Perspective, First Edition. Władysław Homenda and Witold Pedrycz.

the mapping Ψ takes on values from the set of class indices $\Theta = \{1, 2, \ldots, C\}$, that is, class labels, instead of classes themselves. Class labels (indices) may, of course, be different than the numbers $1, 2, \ldots, C$. For instance, for a two-class problem classes, we may be labeled as -1 and 1. In this chapter, the default class labeling is $1, 2, \ldots, C$, and in case when the labeling is different, we explicitly state this.

Pattern recognition is usually performed on observed features characterizing patterns rather than on the patterns directly. Therefore, we distinguish between a mapping from the space of patterns O into the space features X, $\varphi: O \rightarrow X$. This mapping is called a *feature extractor*. Then, we consider a mapping from the space of features into the space of classes, $\psi: X \rightarrow \Theta$. Such a mapping is called a *classifier*. It is important to notice that the term *classifier* is used in different contexts: classification of patterns and classification of features, or more precisely, classification of points coming from the feature space. The meaning of this term can be easily concluded from the context. Therefore, we will not be distinguishing explicitly between different meanings of this term. The composition of the aforementioned two mappings constitutes the classifier $\Psi = \psi \circ \varphi$. In other words, the mapping

$$O \xrightarrow{\Psi} \Theta \tag{2.2}$$

is decomposed to

$$O \xrightarrow{\varphi} X \xrightarrow{\psi} \Theta \tag{2.3}$$

In general, a classifier Ψ is not known, that is, we do not know the class a given pattern belongs to. However, in pattern recognition problems, it is assumed that the classifier Ψ is known for some subset of a set of all patterns, in particular on a subset called a *learning set*. The learning set is a subset of a set of all patterns $L \subset O$, for which classes are known, that is, for any pattern from the learning set $o \in L$, the value of $\Psi(o)$ is given. Constructing a classifier Ψ with the aid of the learning set is an ultimate objective of pattern recognition. In summary, we explore pattern recognition problems searching for a classifier, that is, a mapping

$$\Psi: O \rightarrow \Theta \tag{2.4}$$

assuming that this mapping is known for $L \subset O$. The classifier is decomposed to a feature extractor

$$\varphi: O \rightarrow X \tag{2.5}$$

and a (features) classifier (or in other words a classification algorithm)

$$\psi: X \rightarrow \Theta \tag{2.6}$$

Given a learning set of patterns $O \supset L = \{l_1, l_2, \ldots, l_N\}$ and its split into classes,

$$L = \bigcup_{i=1}^{C} L_i \text{ and } (\forall i \in \{1, 2, \ldots, C\}) \, L_i \subset O_i \tag{2.7}$$

implies that learning classes are also pairwise disjoint:

$$(\forall i, j \in \{1, 2, \ldots, C\}, i \neq j) \, L_i \cap L_j = \emptyset \tag{2.8}$$

and obviously the learning set should contain samples from all classes in O.

Furthermore, each learning class is usually split into two pairwise disjoint nonempty subsets called training and test subsets:

$$L_i = Tr_i \cup Ts_i \text{ where } Tr_i \cap Ts_i = \emptyset \text{ for } i = 1, 2, ..., C \qquad (2.9)$$

Quite often, one splits the learning set into three pairwise disjoint subsets called training, test, and validation sets.

Machine learning and pattern recognition are well established and still rapidly growing areas of research with numerous applications. There are a number of representative textbooks one may consult with this regard. In this chapter, in particular we refer to the textbooks by Bishop (2006), Duda *et al.* (2001), Frank *et al.* (2001), Hastie *et al.* (2009), Mitchell (1997), and Webb and Copsey (2001). In addition, we would like to draw attention to a couple of representative machine learning textbooks published in Polish (in case of Polish-speaking readers) authored by Koronacki and Ćwik (2005) and Stapor (2011).

2.2   NEAREST NEIGHBORS CLASSIFICATION METHOD

Let us assume that a space of features is a subset of the Cartesian product of real numbers, that is, $X = X_1 \times X_2 \times \cdots \times X_M \subset R^M$. M denotes the number of features (the dimensionality of the feature space). The values of the features are usually numeric (integers or real numbers) coming from a given interval. Therefore, in practice, we may assume that $X = I_1 \times I_2 \times \cdots \times I_M \subset R^M$, where $I_j = [l_j, r_j]$ is a closed interval with left and right endpoints l_j and r_j, $j = 1, 2, ..., M$.

Subsequently, let us assume that patterns are mapped to the space of features $X_1, X_2, ..., X_M$ and that two patterns denoted p and r are characterized by the following values of features: $\varphi(p) = \mathbf{x} = (x_1, x_2, ..., x_M)^T \in R^M$ and $\varphi(r) = \mathbf{y} = (y_1, y_2, ..., y_M)^T \in R^M$.

The nearest neighbor (NN) classifier is based on a notion of *similarity* of patterns. In this study, the similarity is expressed with an inverted distance function (an inverted metric) in the space R^M, meaning that the smaller the distance between two patterns is, the more similar these patterns are. There are numerous examples of distance functions (or metrics) in R^M. The most popular metrics include Euclidean, Manhattan, and Chebyshev. They are expressed as follows:

$$d_E(\mathbf{x}, \mathbf{y}) = \sqrt{(x_1 - y_1)^2 + (x_2 - y_2)^2 + \cdots + (x_M - y_M)^2} \quad - \text{ Euclidean}$$

$$d_H(\mathbf{x}, \mathbf{y}) = |x_1 - y_1| + |x_2 - y_2| + \cdots + |x_M - y_M| \quad - \text{ Manhattan} \qquad (2.10)$$

$$d_C(\mathbf{x}, \mathbf{y}) = \max\{|x_1 - y_1|, |x_2 - y_2|, ..., |x_M - y_M|\} \quad - \text{ Chebyshev}$$

for any pair of points $\mathbf{x}, \mathbf{y} \in R^M$.

Say, we have a labeled training set of patterns. We obtain a new, unknown pattern that we want to classify. It is reasonable to look for a pattern coming from the training set that is the closest (in terms of the assumed distance function) to the pattern to be classified and assign the class label of this pattern to the pattern of discussion. The similarity of two patterns is typically expressed, as we already mentioned, as an inverted distance between them in the space of features: the closest patterns are the most similar. Therefore, the following formula defines the NN rule:

$$\Psi(p) = \Psi(r_{min}) \quad \text{where} \quad r_{min} = \arg\min_{r \in Tr}\{d(p, r)\} = \arg\min_{r \in Tr}\{d(\mathbf{x}, \mathbf{y})\} \quad (2.11)$$

where p is a pattern to be classified, r is a pattern from the training set, $\mathbf{x}$ and $\mathbf{y}$ are vectors of features characterizing patterns p and r, respectively, and the r_{min} is the pattern from the training set closest to the unknown pattern p.

The NN method can be generalized to the k-nearest neighbors method, k-NN for short. Having a pattern p with the features vector denoted by $\mathbf{x}$, which is subjected to classification, we take k patterns from the training set that are the closest to it. In order to perform k-NN classification, we consider a sphere in the space R^M centered in $\mathbf{x}$, which includes k patterns from the training set Tr. The pattern p is assigned to the class that is the most frequently observed in this sphere.

In a case when two or more classes have a maximal number of patterns, we may consider involving a secondary choice decision rule. The simplest secondary rule may be just to draw a class label randomly from those most frequently observed classes. Besides random selection, we can consider other factors, which would be more appropriate. We may apply the NN choice for patterns from classes with maximal representation in the sphere. In other words, we consider only patterns from classes with the maximal representation in the sphere, and we check which one has the closest pattern to the pattern p. We may also take this class with maximal representation for which the sum of distances of patterns is the smallest.

The k-NN method is illustrated in Figure 2.1. Let us consider an unknown pattern marked with a black star. It is located at the center of the plotted circles, which make a sphere in which we will be counting patterns. The NN of the star pattern belongs to the class of plus signs, so then for $k = 1$ the star pattern is classified to the class of plus signs. For $k = 3$, two square patterns and one plus sign pattern are included in the disk; hence the star pattern is assigned to the squares class. For $k = 5$ two squares, two triangles, and one plus fall into the disk; therefore we must apply some secondary choice factor. It is obvious that the closest pattern out of these four is a square, so assuming this as the secondary choice factor, we can classify the star to squares. In contrast, an alternative secondary choice rule can be the sum of distances. We see in Figure 2.1 that the sum of distances from the star to these two squares is smaller than the sum of distances from the star to triangles. Therefore, this rule also suggests the star should be classified to the square class. For $k = 7$ the closest to the star are three triangles, two squares, and two plus signs. Hence, the star pattern is accounted into the class of triangles.

On the other hand, the other two unknown patterns in Figure 2.1 marked with hash and at signs will be certainly classified to the classes of plus signs and squares, respectively. We see that in these two cases class assignment does not depend on the value of the k parameter for a wide range of values. The discussed classification cases let us draw a conclusion that the effectiveness of the k-NN classifier depends on the value of the k parameter (the number of neighbors). In addition, we note that classification characteristics (*style*) of the k-NN is different, more stable when we classify a pattern located inside a region occupied mainly by patterns from the same class, and

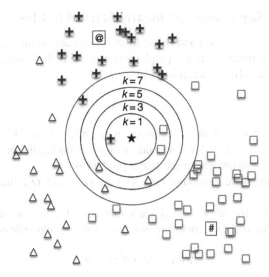

Figure 2.1   Illustration of the k-NN classifier. The pattern marked with a black star is being classified depending on the value of k: for $k = 1$ to the class of plus signs, for $k = 3$ to the class of squares, for $k = 5$ to the class of squares, and for $k = 7$ to the class of triangles. We may conclude that the classification outcome for the pattern marked with a star greatly depends on parameter k. In contrast, it is clear that classification of two patterns marked with a hash (#) and an at (@) sign is independent of the parameter k for quite wide ranges of k.

not that stable when we consider areas close to borders of regions occupied by various classes or areas of low density.

The k-NN method does not require any learning: the entire training set is the model. Hence, we call it a *lazy* classifier. Despite its simplicity, a number of more sophisticated methods rely on the idea of k-NN, for instance, Denoeux (1995), Hu *et al.* (2008), Tan *et al.* (2005), and Zouhal and Denoeux (1998).

For more on k-NN classifiers, one may consult textbooks by Hastie *et al.* (2009, section 13.3), Mitchell (1997, section 8.2), and Duda *et al.* (2001, chapter 4). We would also like to point at an article very relevant to the k-NN method authored by Friedman (1997).

2.3   SUPPORT VECTOR MACHINES CLASSIFICATION ALGORITHM

Support Vector Machines, SVMs, are an algorithm invented by C. Cortes and V. Vapnik (1995). SVMs in its basic form are a non-probabilistic binary linear classifier used in supervised machine learning in order to split a set of patterns $O = \{o_1, o_2, \ldots, o_N\}$ into two (disjoint) classes labeled -1 and 1, that is, $\Theta = \{O_{-1}, O_1\}$. Let us assume that features characterize patterns $\phi: O \to X; X \subset R^M$, and for simplicity of considerations, we are looking for a mapping $\psi: X \to \{-1, 1\}$.

2.3.1 Linear Separation of Linearly Separable Classes

Let us assume that classes O_{-1} and O_1 are linearly separable in the Euclidean space R^M of features, that is, there exists a hyperplane separating both classes. Assume that the following formula defines such a hyperplane H':

$$\mathbf{w}'^T \mathbf{x} + b' \equiv \mathbf{w}' \cdot \mathbf{x} + b' = 0 \tag{2.12}$$

where $\mathbf{x}$ denotes points of the hyperplane in Euclidean space R^M, $\mathbf{w}'$ is a normal vector to the hyperplane, $b' \in R$ is a scalar, and $\mathbf{w}'^T \mathbf{x}$ denotes the product of matrices that, for vectors, is equivalent to the scalar product $\mathbf{w}' \cdot \mathbf{x}$.

If this hyperplane separates both classes O_{-1} and O_1, then the following inequalities hold:

$\mathbf{w}' \cdot \mathbf{x}_i + b' < 0$ for all $\mathbf{x}_i = \varphi(o_i)$ such that $o_i \in O_{-1}$ and $\mathbf{w}' \cdot \mathbf{x}_i + b' > 0$ for all $\mathbf{x}_i = \varphi(o_i)$ such that $o_i \in O_1$. Therefore, we have a simple classification rule:

$$\psi(x_i) = \operatorname{sgn}(\mathbf{w}' \cdot \mathbf{x}_i + b') \tag{2.13}$$

As we can see in a simple illustration in Figure 2.2, there might be not a single one, but a set of separating hyperplanes. Let us consider distances between a separating hyperplane and patterns from both classes and then, for a given class, take a pattern for which the distance to the separating hyperplane is minimal in this class. Patterns realizing minimal distances, called support vectors (i.e., vectors beginning in the origin of the coordinates system and ending in the pattern) or support patterns, can be used to construct two border hyperplanes H'_{-1} and H'_1, one for each class. It is obvious that the area between border hyperplanes, called a separating margin, does not contain any pattern. Maximizing the distance between border hyperplanes, that is, enlarging the width of the separating margin, determines border hyperplanes uniquely. So then,

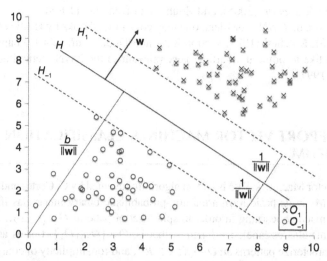

Figure 2.2 Illustration of the SVM algorithm: a linearly separable case.

placing a separating hyperplane in the middle of the margin area also determines it in a unique way. Assuming that the separating hyperplane is defined by

$$\mathbf{w'} \cdot \mathbf{x} + b' = 0 \qquad (2.14)$$

we get the following inequalities

$$\mathbf{w'} \cdot \mathbf{x} + b' \leq -c \ \text{ and } \ \mathbf{w'} \cdot \mathbf{x}_i + b' \geq c \qquad (2.15)$$

for patterns $\mathbf{x}_i$ coming from class O_{-1} and O_1, respectively. Finally, after rescaling we come up with the following equalities for the separating hyperplane H

$$\mathbf{w} \cdot \mathbf{x} + b = 0 \qquad (2.16)$$

and border hyperplanes H_{-1} and H_1

$$\mathbf{w} \cdot \mathbf{x} + b = -1 \ \text{ and } \ \mathbf{w} \cdot \mathbf{x} + b = 1 \qquad (2.17)$$

and inequalities for patterns from both classes

$$\mathbf{w} \cdot \mathbf{x} + b \leq -1 \ \text{ and } \ \mathbf{w} \cdot \mathbf{x}_i + b \geq 1 \qquad (2.18)$$

where, of course, $\mathbf{w} = \mathbf{w'}/c$ and $b = b'/c$.

Both inequalities can be combined in

$$y_i(\mathbf{w} \cdot \mathbf{x}_i + b) \geq 1 \qquad (2.19)$$

for patterns from both classes, where $y_i = \psi(\mathbf{x}_i) = \psi(\varphi(o_i))$ is the class label for o_i.

The distance between the separating hyperplane H and border hyperplanes H_{-1} and H_1 is equal to $1/\|\mathbf{w}\|$ so then the distance between the two border hyperplanes is equal to $2/\|\mathbf{w}\|$. Therefore, an optimal placement of the separating hyperplane, that is, a placement maximizing the margin, requires maximization of the aforementioned functional or, equivalently, minimization of the following functional:

$$\frac{\|\mathbf{w}\|^2}{2} \qquad (2.20)$$

assuming the following constraints:

$$y_i(\mathbf{w} \cdot \mathbf{x}_i + b) \geq 1 \qquad (2.21)$$

The task defined by the aforementioned two formulas is a well-known optimization problem with a square objective function ($\|\mathbf{w}\|^2 = \mathbf{w} \cdot \mathbf{w} = w_1^2 + w_2^2 + w_3^2 + \cdots + w_M^2$) with linear constraints. Such optimization problem can be solved with the Lagrange function

$$L(\mathbf{w}, b, \boldsymbol{\alpha}) = \frac{1}{2}\|\mathbf{w}\| - \sum_{i=1}^{N} \alpha_i[y_i(\mathbf{w} \cdot \mathbf{x}_i + b) - 1] \qquad (2.22)$$

where $\boldsymbol{\alpha} = (\alpha_1, \alpha_2, \ldots, \alpha_N)$ is a vector of nonnegative Lagrange multipliers. The solution is a saddle point, where the Lagrange function reaches its maximum with regard to multipliers and minimum with regard to $\mathbf{w}$ and b. Since partial derivatives must vanish in such a point, we will use this condition to find it.

Meanwhile, let us notice that the expression $\alpha_i[y_i(\mathbf{w} \cdot \mathbf{x}_i + b) - 1]$ is nonnegative, so then $L(\mathbf{w}, b, \alpha)$ will reach its maximum supposing the following condition:

$$\alpha_i[y_i(\mathbf{w} \cdot \mathbf{x}_i + b) - 1] = 0 \quad \text{for} \quad i = 1, 2, \ldots, N \tag{2.23}$$

which implies that $\alpha_i = 0$ if $\mathbf{x}_i$ is not a support pattern. For $\mathbf{x}_i$ being a support pattern $\alpha_i \geq 0$ due to $y_i(\mathbf{w} \cdot \mathbf{x}_i + b) - 1 = 0$.

A vanishing gradient with regard to the vector $\mathbf{w}$

$$\frac{\partial}{\partial \mathbf{w}} L(\mathbf{w}, b, \alpha) = \mathbf{w} - \sum_{i=1}^{N} \alpha_i y_i \mathbf{x}_i = 0$$

directly implies

$$\mathbf{w} = \sum_{i=1}^{N} \alpha_i y_i \mathbf{x}_i \tag{2.24}$$

and with regard to the parameter b implies

$$\frac{\partial}{\partial b} L(\mathbf{w}, b, \alpha) = \sum_{i=1}^{N} \alpha_i y_i = 0 \tag{2.25}$$

Rewriting (2.22) we come to

$$L(\mathbf{w}, b, \alpha) = \frac{1}{2} \mathbf{w}^T \cdot \mathbf{w} - \sum_{i=1}^{N} \alpha_i y_i \mathbf{w} \cdot \mathbf{x}_i - b \sum_{i=1}^{N} \alpha_i y_i + \sum_{i=1}^{N} \alpha_i$$

and substituting (2.24) we get

$$L(\mathbf{w}, b, \alpha) \equiv L(\alpha) = \frac{1}{2} \sum_{i=1}^{N} \sum_{j=1}^{N} \alpha_i \alpha_j y_i y_j \mathbf{x}_i^T \cdot \mathbf{x}_j - \sum_{i=1}^{N} \sum_{j=1}^{N} \alpha_i \alpha_j y_i y_j \mathbf{x}_i^T \cdot \mathbf{x}_j - b \sum_{i=1}^{N} \alpha_i y_i + \sum_{i=1}^{N} \alpha_i$$

and finally we get the Lagrange function for minimization

$$L(\alpha) = \sum_{i=1}^{N} \alpha_i - \frac{1}{2} \sum_{i=1}^{N} \sum_{j=1}^{N} \alpha_i \alpha_j y_i y_j \mathbf{x}_i^T \cdot \mathbf{x}_j \tag{2.26}$$

with constraints

$$\alpha_i \geq 0, \quad i = 1, 2, \ldots, N, \quad \sum_{i=1}^{N} \alpha_i y_i = 0 \tag{2.27}$$

Let us recall (2.23) and the conclusion that α_i must vanish for all $i = 1, 2, \ldots, N$, for which $\mathbf{x}_i$ does not belong to border hyperplanes. Therefore, all summations in (2.26) and (2.27) are restricted to indices of points $\mathbf{x}_i$ located on border hyperplanes, that is, restricted to indices of support vectors.

Summarizing, the solution of the optimization problem provides optimal Lagrange multipliers' values $\alpha^0 = (\alpha_1^0, \alpha_2^0, \ldots, \alpha_N^0)$ and the following formula for the separating hyperplane:

$$\sum_{i \in SV} \alpha^0{}_i y_i \mathbf{x} \cdot \mathbf{x}_i + b^0 = 0 \tag{2.28}$$

where SV is the set of indices of support vectors. Coefficient b^0 must satisfy (2.23), which allows to compute its value with the use of one support vector. However, in practice its value is the average based on all support vectors:

$$b^0 = \frac{1}{|SV|} \left[\sum_{\mathbf{x}_i \in SV} y_i - \sum_{\mathbf{x}_i \in SV} \left(\sum_{\mathbf{x}_j \in SV} \alpha_i \mathbf{x}_j \right) \cdot \mathbf{x}_i \right] \tag{2.29}$$

Recalling (2.24) and having in mind that $\alpha_i = 0$ for such i that $\mathbf{x}_i$ is not a support vector, we come to the formula

$$\mathbf{w} = \sum_{\mathbf{x}_i \in SV} \alpha^0_i y_i \mathbf{x}_i \tag{2.30}$$

and finally the decision formula (2.13) is expressed as

$$\psi(\mathbf{x}) = \mathrm{sgn}(\mathbf{w} \cdot \mathbf{x}) = \mathrm{sgn} \left(\sum_{\mathbf{x}_i \in SV} \alpha^0_i y_i \mathbf{x} \cdot \mathbf{x}_i + b^0 \right) \tag{2.31}$$

2.3.2 Linear Separation of Classes Linearly Not Separable

Let us recall that up to this point, we have assumed that patterns belonging to the two classes are linearly separable. Yet, real problems rarely satisfy this assumption and classes are not linearly separable, which leads to misclassification errors. For instance, one may visualize a case when patterns belonging to different classes are located very close to each other (in some sense of closeness), and then it is not possible to separate both classes with a hyperplane. In such a case inequality (2.19) is turned to

$$y_i(\mathbf{w} \cdot \mathbf{x}_i + b) \geq 1 - \xi_i \tag{2.32}$$

where $\xi_i \geq 0$, $i = 1, 2, \ldots, N$

For $0 < \xi_i < 1$ the pattern $\mathbf{x}_i$ is inside the margin, though still on the *correct* side of the separating hyperplane, that is, it is between the separating hyperplane and the margin hyperplane corresponding to a pattern's class. For $1 < \xi_i$, the pattern $\mathbf{x}_i$ is on the incorrect side of the separating hyperplane (cf. Figure 2.3). Minimization of problems (2.20) and (2.21) is turned to minimization of the following objective function:

$$\frac{\|\mathbf{w}\|^2}{2} + C \sum_{i=1}^{N} \xi_i \tag{2.33}$$

assuming the constraints (2.32).

A mechanism that allows only a partial separation is known as a soft margin. The C parameter (consult (2.33)), also called the *regularization parameter*, controls the shape of the margin by adjusting penalty for misclassification. The C parameter is a positive real number, its high values might cause overfitting, that is, the constructed classifiers may be very accurate for a training set, but not general enough to correctly classify patterns outside the training set.

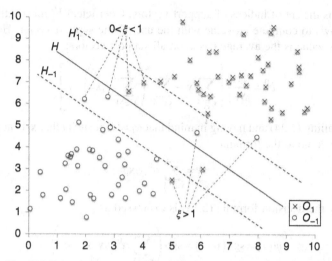

Figure 2.3   The SVM algorithm: a case when classes O_1 and O_{-1} are not linearly separable.

The *regularization parameter C* indicates a cost of incorrect classification. The greater the value of this parameter is, the greater the loss of incorrect classification. Hence, the greater C is, the narrower the margin becomes. On the other hand, the smaller the value of C, the greater capability of model generalization we obtain.

Summarizing, the construction of a separating hyperplane for a problem, which is linearly not separable, relies on the minimization of the objective function (2.33) with constraints (2.32). Such an optimization problem can be solved as before, that is, with the Lagrange function:

$$L(\alpha) = \sum_{i=1}^{N} \alpha_i - \frac{1}{2}\sum_{i=1}^{N}\sum_{j=1}^{N} \alpha_i \alpha_j y_i y_j \mathbf{x}_i^{T} \cdot \mathbf{x}_j \qquad (2.34)$$

with modified constraints

$$0 \le \alpha_i \le C, \quad i = 1, 2, \ldots, N, \quad \sum_{i=1}^{N} \alpha_i y_i = 0 \qquad (2.35)$$

2.3.3   Nonlinear Separation of Classes Linearly Not Separable

In the previously presented formulas, for instance, in (2.31), we see a scalar product as a fundamental operation applied to the pattern to be classified. As explained before, an ideal classification (with a zero classification error) could be performed for the two linearly separable classes. However, we can deliberately consider transformation of patterns from an M-dimensional space to a space of higher dimensionality so that we can apply a scalar product to patterns in this extended space. More specifically, a feature vector $\mathbf{x} \in I^M$ can be transformed to a vector $\mathbf{g}(\mathbf{x}) \in I^K$ using some vector function $\mathbf{g}: I^M \rightarrow I^K$, where $I = [a, b] \subset R$ is a closed interval of real numbers and

$M < K$; usually $M << K$. Then, a scalar product $\mathbf{x} \cdot \mathbf{x}_i$ may be replaced with $\mathbf{g}(\mathbf{x}) \cdot \mathbf{g}(\mathbf{x}_i)$, and, in such case, a separating hyperplane in the space I^K corresponds to a more complex hypersurface in the original features space I^M. Such a hypersurface can easily separate classes of patterns that are linearly not separable.

Let us consider a feature transformation to polynomials of a degree not greater than two. For a vector of features made up of two features $\mathbf{x} = (x_1, x_2)$, we have six basic monomials $1, x_1, x_2, x_1x_2, x_1^2, x_2^2$, so then, in this case, $M = 2$ and $K = 6$. For more features and polynomials of a higher degree, the number of monomials grows very fast, and therefore it may bring an unacceptable complexity. For instance, for M features we have $\begin{pmatrix} M+d-1 \\ d \end{pmatrix}$ monomials of degree d and $\begin{pmatrix} M+d \\ d \end{pmatrix} = \begin{pmatrix} M+d \\ M \end{pmatrix}$ monomials of degree at most d, which is a polynomial in M of degree d. Anyway, the idea of feature space transformation to a space of a higher dimensionality and then the use of a scalar product in such space is straightforward. Considering the case of two features and the transformation to polynomials of a degree at most two, we have $\begin{pmatrix} 2+2 \\ 2 \end{pmatrix} = 6$ monomials of a degree at most two and the transformation:

$$\mathbf{g}(\mathbf{x}) = \begin{bmatrix} g_1 \\ g_2 \\ g_3 \\ g_4 \\ g_5 \\ g_6 \end{bmatrix} (x_1, x_2) = \begin{bmatrix} g_1(x_1, x_2) \\ g_2(x_1, x_2) \\ g_3(x_1, x_2) \\ g_4(x_1, x_2) \\ g_5(x_1, x_2) \\ g_6(x_1, x_2) \end{bmatrix} = \begin{bmatrix} 1 \\ x_1 \\ x_2 \\ x_1 x_2 \\ x_1^2 \\ x_2^2 \end{bmatrix} \tag{2.36}$$

Finally, the decision formula (2.13) is turned to

$$\psi(\mathbf{x}) = \text{sgn}(\mathbf{g}(\mathbf{w}) \cdot \mathbf{g}(\mathbf{x})) = \text{sgn}\left(\sum_{i \in SV} \alpha^0{}_i y_i \mathbf{g}(\mathbf{x}) \cdot \mathbf{g}(\mathbf{x}_i) + b^0 \right) \tag{2.37}$$

For a polynomial transformation, dimensionality of the target space is finite, but we may also consider target spaces of an infinite dimensionality. More importantly, instead of computing such a transformation and then computing a scalar product, we can employ a special function in a space of features I^M instead of a scalar product in an image space I^K, where I is a closed interval of real numbers, that is, we can replace a scalar product $\mathbf{g}(\mathbf{x}) \cdot \mathbf{g}(\mathbf{x}_i)$ by a scalar function $K(\mathbf{x}, \mathbf{x}_i)$, called a *kernel function*:

$$\mathbf{g}(\mathbf{x}) \cdot \mathbf{g}(\mathbf{x}_i) = K(\mathbf{x}, \mathbf{x}_i) \tag{2.38}$$

Therefore, decision formula takes the form

$$\psi(\mathbf{x}) = \text{sgn}\left(\sum_{i \in SV} \alpha^0{}_i y_i K(\mathbf{x}, \mathbf{x}_i) + b^0 \right) \tag{2.39}$$

It is important to notice that we even do not need to know the transformation $g: I^M \rightarrow I^K$. It is enough to apply a kernel function in the decision formula (2.39). However, a question is raised how to find a kernel function K for a given transformation g. This question can be rephrased, and we can ask whether a given scalar function $K: I^M \times I^M \rightarrow R$ corresponds to a scalar product $g(\mathbf{x}) \cdot g(\mathbf{x}_i)$ for some transformation $g: I^M \rightarrow I^K$ (correspondence is in the sense of (2.38)). The answer to such question is based on the Mercer theorem developed in the area of functional analysis. A symmetric function $T: I^M \times I^M \rightarrow R$ is a kernel function if it satisfies some general conditions. A sufficient condition for a symmetric function is that it is continuous and nonnegative, though these conditions could be weakened.

There are several kernel functions that can be successfully used in real-world applications:

- Linear kernel with one parameter c, which cannot be used for nonlinear separation:

$$K(\mathbf{x}, \mathbf{y}) = \langle \mathbf{x}, \mathbf{y} \rangle + c$$

- Polynomial kernel function of degree d, which is recommended for normalized data:

$$K(\mathbf{x}, \mathbf{y}) = (\langle \mathbf{x}, \mathbf{y} \rangle + c)^d$$

- Gaussian kernel function, also called radial basis function (RBF), which is the fundamental kernel function due to its very good results reported in a multitude of empirical classification studies:

$$K(\mathbf{x}, \mathbf{y}) = \exp\left(-\gamma \|\mathbf{x} - \mathbf{y}\|^2\right) \tag{2.40}$$

We would also like to point toward a paper in which we find a comparison of SVMs with Gaussian kernels to RBF classifiers (an older technique) (Scholkopf *et al.*, 1997).

- Hyperbolic tangent with γ, which is the steepness of the curve:

$$K(\mathbf{x}, \mathbf{y}) = \tanh(\gamma \cdot \langle \mathbf{x}, \mathbf{y} \rangle + c)$$

- Laplace function:

$$K(\mathbf{x}, \mathbf{y}) = \exp(-\gamma \cdot |\mathbf{x} - \mathbf{y}|)$$

- Sinc (cardinal sine function) kernel function:

$$K(\mathbf{x}, \mathbf{y}) = \text{sinc}(|\mathbf{x} - \mathbf{y}|) = \frac{\sin(|\mathbf{x} - \mathbf{y}|)}{|\mathbf{x} - \mathbf{y}|}$$

- Sinc2 kernel function:

$$K(\mathbf{x}, \mathbf{y}) = \text{sinc}\left(\|\mathbf{x} - \mathbf{y}\|^2\right) = \frac{\sin\left(\|\mathbf{x} - \mathbf{y}\|^2\right)}{\|\mathbf{x} - \mathbf{y}\|^2}$$

- Quadratic kernel function:

$$K(\mathbf{x}, \mathbf{y}) = 1 - \frac{\|\mathbf{x} - \mathbf{y}\|^2}{\|\mathbf{x} - \mathbf{y}\|^2 + c}$$

- Minimum kernel function:

$$K(\mathbf{x}, \mathbf{y}) = \sum_i \min(x_i, y_i) \tag{2.41}$$

where $\langle \mathbf{x}, \mathbf{y} \rangle$ is used to denote a scalar product for the sake of better readability, $\|\mathbf{x} - \mathbf{y}\| = \sqrt{\sum_i (x_i - y_i)^2}$ is the Euclidean norm, and $|\mathbf{x} - \mathbf{y}| = \sum_i |x_i - y_i|$ is the Manhattan norm.

Additional material on SVM and kernel methods can be found also in Bishop (2006, chapter 6), Webb and Copsey (2001, chapter 5), Hastie *et al.* (2009, section 12.3), and Frank *et al.* (2001, section 12.3).

SVM is well known for its superior performance. Hence, it has been successfully applied to many areas of science, including satellite images classification (Huang *et al.*, 2002), drug design (Burbidge *et al.*, 2001), spam detection (Drucker *et al.*, 1999), pattern recognition (Burges, 1998), microarray data classification (Brown *et al.*, 2000; Furey *et al.*, 2000), biometrics (Osuna *et al.*, 1997; Dehak *et al.*, 2011), and many more.

2.4 DECISION TREES IN CLASSIFICATION PROBLEMS

A tree is a generic data structure known also as an acyclic-connected graph. This definition does not distinguish any node of such a graph. It is also important that there are no cycles in a tree and that every two nodes in a tree are connected, that is, there exists exactly one path between any two nodes.

For the sake of discussion in this study, an alternative definition of a tree will be more convenient. A tree is an undirected graph $T = (V, E)$, where V is a finite set of nodes and $E \subset \{\{u, v\} : u, v \in V\}$ are undirected edges in this tree. Moreover, the following definition and relationships hold:

- There is a distinguished node $v \in V$, called the root of the tree.
- All other nodes (apart from the root) are split into k or less pairwise disjoint subsets.
- Each such subset is a tree with its own root, and we can call it a subtree.
- For each subtree there is an (undirected) edge joining the root v and the root of this subtree. The node v is called the parent of the subtree root. Roots of subtrees are called children of the node v.

A subset with exactly one node creates a (degenerated) tree without subtrees. Such tree is called a leaf. A tree is called a k-tree assuming that at least one of its nodes has k children and, of course, each of its nodes has no more than k children.

Obviously, the aforementioned definition implies that a tree is an undirected acyclic-connected graph. For any node, there is a unique path from the root to this node. Similarly, we may say that each node, of course except the root of the tree, has exactly one parent. The number of edges in a path is called the length of the path. The height of a tree is the length of the longest path from the root to a leaf.

2.4.1 Decision Tree at a Glance

A decision tree is a tree, which could be used as a classifier. In a decision tree, every node has a subset of training patterns assigned to it. The process of pairing tree nodes with patterns occurs at the stage of tree construction. Tree construction is, in fact, classifier training. During the tree construction, we may use knowledge about patterns assigned to a given node to form a label to identify this node. The root has the entire training set assigned. If a node v is not a leaf, then its set of training patterns is split into subsets, which are assigned to children of this node (at the same time, a parent keeps all patterns that are assigned to its children). The set is split into subsets with the use of a chosen feature. The splitting process is completed when some termination condition has been satisfied. One of such conditions is when all patterns that are assigned to a node belong to the same class. In this case, when all patterns in the node belong to a single class, this node becomes a leaf and it is labeled with a class label of these patterns.

Having constructed such a tree, we can use it for classification of new patterns. A given new pattern traverses from the root to a leaf according to the values of the features, that is, for each node the pattern will be moved to a child according to the value of a feature used to split this node.

Before we initiate a decision tree construction, we must answer the following questions:

- How do we choose a feature in a given node in order to perform a split?
- How do we find the number of subsets for the given feature in a node, that is, the number of children of the given node?
- When do we stop growing the tree, that is, when to stop the process of splitting the set of training patterns?
- Which class label should be assigned for a given leaf, especially which class label should be assigned when a leaf contains patterns coming from more than a single class?

Let us now briefly discuss the aforementioned questions as they are discussed in detail later on.

The last two questions were already tackled earlier. Let us recall that the splitting process is stopped when for a given node all patterns that were assigned to it belong to the same class. Unfortunately, this solution often leads to fine-grained splitting, which entails overfitting. This problem has to be tackled in detail, and more sophisticated solutions have to be constructed to arrive at some stopping criteria.

The answer to the last question can be formulated as follows: "which class label should be assigned for a given leaf" is trivial, assuming that in this leaf we have patterns from one class only. In a case when more than a single class is represented in a leaf, the most frequent class should be assigned. This rule is applied to any node of the tree, not only to the leaves.

The answer to the first question is not obvious. Intuitively, it would be reasonable to choose such a feature, which delivers a good separation among the classes. The choice of a feature is related to the second question, namely, how to find the number of subsets for the given feature (in other words: how many children should be assigned to the parent?). Choosing a feature, which allows splitting into subsets in a way that each subset contains samples from one class only, would be the perfect solution. In such a case, a necessary (but in some cases insufficient) condition, required to split a feature so that the result provides complete classification, is to have the number of subsets equal to the number of classes, say, C. In such case, we get a subtree of height one, with its root with a set of training patterns assigned and each of its C children with patterns assigned belonging to one class. Note that for certain features it may happen that patterns belonging to different classes will have the same value of the feature. Since usually we build a classification model with more than one feature, typically a split is performed into the number of subsets fewer than the number of classes. If a split is done for fewer subsets than the number of classes represented in the set of patterns, then some subset(s) will include patterns belonging to more than one class. Subsequently, these subsets will be recursively subjected to further splitting. Repeating this process finally leads to the formation of subsets including patterns from one class only. In such case, the height of the whole decision tree is equal to the depth of the recursion.

Unfortunately, such perfect solution that each subset assigned to a leaf includes patterns from only one class is quite impossible in practice. Therefore, we will be looking for a *good* solution. A solution is not immediately apparent, that is, a model designer cannot choose a feature for the split just by looking at the data or by computing some simple statistics. Also, the number of subsets into which one should make a split is not directly visible. Hence, in this overview, we assume binary splits and then, on the basis of this assumption, the best feature will be chosen.

An illustrative example of a tree is presented in Figure 2.4. To build this tree, we used the *wine* dataset available in the UCI Machine Learning Repository, Wine Dataset. The dataset $O = \{o_1, o_2, \ldots, o_{178}\} = O_1 \cup O_2 \cup O_3$ includes 178 patterns from three classes characterized by 13 numerical features. Initial data inspection revealed that one feature was correlated with two others, and therefore it was removed. Finally, we had 12 features enumerated by integers from 1 to 13 with the 7th feature removed due to its correlations. The set was split to a training set and a test set in proportion $7:3$. The training set included 43, 52, and 35 patterns in classes O_1, O_2, and O_3, respectively, 43-52-35 for short. The test set included 16-19-13 patterns belonging to the three classes. The tree was built using the training set as follows:

- The training set 43-52-35 was assigned to the root.
- 13th feature and the value 730 were chosen to split the set of patterns to two subsets.
- Both subsets, which include 0-46-31 and 43-6-4 patterns, respectively, were assigned to the children of the root.

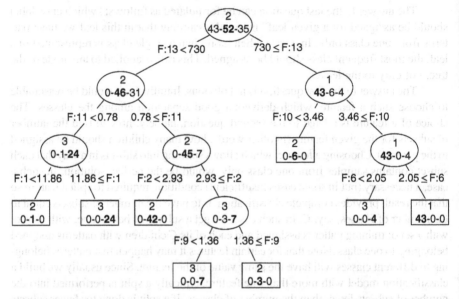

Figure 2.4   The decision tree built for the training set of the wine dataset. Nodes are illustrated with ellipses (non-leaf nodes) and rectangles (leaf nodes). Notation x-y-z (e.g., 43-**52**-35 in the root) informs about the number of patterns from classes O_1, O_2, and O_3, respectively, that were assigned to a given node. A single number inside tree nodes, positioned above the x-y-z notation, informs about the label of the majority class in a node. The majority class is also indicated with a number of patterns in boldface, for example, in the root we have 43-**52**-35 that says that the second class is the majority class in this node. Inequalities on the edges indicate the split condition of the father's set of patterns and the feature that was used to perform this split.

- 11th feature and the value 0.78 were taken to split the set of patterns 0-46-31 into two subsets, which were assigned to the children of this node (node 0-46-31).

- 10th feature and the value 3.46 were taken to split the set of patterns 43-6-4 to two subsets, which were assigned to the children of the current node.

The splitting process was halted when a set of patterns in a node included patterns from one class only. Therefore, each leaf represents one class.

The tree is shown in Figure 2.4. Internal nodes (ellipses) and leaves (rectangles) contain two rows of numbers. The second row contains three integers written in the convention x-y-z, where x says how many elements belonging to class O_1 were assigned to a given node, y says how many samples from class O_2 were assigned to a given node, and z indicates about the number of patterns from class O_3 that were assigned to this node. In addition, in each node, the maximal integer (in other words, a majority class) is shown in boldface. The first row in each node contains a single integer that tells about the majority class—we use notations 1, 2, and 3, which correspond to classes O_1, O_2, and O_3, respectively. For instance, in the root of the tree in the second row, we have three numbers 43-**52**-35, which inform us that there are 43 patterns from class number 1 (O_1), 52 patterns from class number 2 (O_2), and 35 patterns from class number 3 (O_3). Number 2 in the first row of the tree root node indicates the majority class (O_2). Then, the set of training patterns is split according to a chosen feature. Feature number 13 was used in the tree root (it is denoted F : 13 in the figure), and the value 730 was used to perform the split. Patterns with the 13th feature value smaller than 730 were moved to the left child, and the remaining patterns were moved to the right child. We labeled the edges with information about the feature number and split value. This procedure was continued for each left and right subtree until a set of patterns assigned to a node contained elements of one class only. A node containing patterns belonging to the same class became a leaf of the tree. For all nodes, internal nodes, and leaves of the tree, we identified the majority class in this node. This ended the tree construction procedure. The outcome is that the trained decision tree can be applied to predict class membership for previously unseen patterns. In the case of this experiment, unseen patterns are patterns from the test set. Note that we did not employ any additional stopping criteria to be used while the tree was being built. The tree was constructed in the complete, full form. The tree construction was performed with the use of the training data, and since it is a full tree (and each leaf contains samples from one class only), classification accuracy on the training set is 100%.

Once the tree has been constructed, we can take any pattern and check which class the tree will classify it to. In order to do so, we can pass it starting from the root down the tree until we reach a leaf. Then, we inspect which class during the tree construction procedure based on the training set was a majority class in this leaf, and we assume that this pattern belongs to this majority class. A decision tree is a classifier that after the construction procedure can be used for patterns outside of the training set. Naturally, we have to have the training set to build it beforehand. The decision tree in Figure 2.4 is a perfect classifier for the training set, meaning that all 130 patterns from the training set of the wine dataset receive correct class labels when we use this tree for classification. However, if we use a tree to classify patterns located outside of the training dataset, the classification result will not necessarily be error free.

The classification result for the test set of the wine dataset is displayed in Figure 2.5. Note that the structure of this tree was discovered during the tree construction procedure, and hence it is the same as presented in Figure 2.4. In particular, class label assignments for nodes and split features and split values form an important outcome of the tree construction. We see that test patterns are not classified perfectly, that is, three out of 48 patterns from the test set were misclassified. Also, we see that not all branches of the tree were used when we were passing down test set patterns: four leaves and one internal node were not involved in the test set classification process. Moreover, there is one internal node and one leaf, where the majority of samples from the test set do not match expected class assignments, discovered before for the training set. Hence, we can conclude that the structure of the tree could be simplified by deleting some nodes.

A simplified tree is shown in Figure 2.6. Here there are two trees: on the left side we see results for the training set, and on the right side we see results for the test set. The simplified tree structure was obtained by pruning: we cut a few branches of the full tree and in this way turned a few internal nodes into the leaves. The rule that we applied to prune the full tree was that an internal node was turned to a leaf if at this node the set of training patterns included less than 10% incorrectly classified patterns.

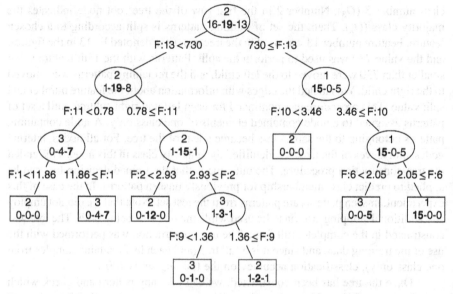

Figure 2.5 Classification procedure for the test set using the decision tree constructed in Figure 2.4. Numbers inside nodes (ellipses and rectangles) depict numbers of patterns from each class that fall into a given node. The majority class number (an integer in the first row in each node) concerns the set of training data. The majority class for the test set is indicated with a number written in boldface in the second row.

2.4.2 Splitting Feature

We start the tree building procedure in order to select a feature to split the root at which we consider the entire set of features. However, at lower levels of the tree, we may consider two options. In the first option to do a split in each node, we consider the entire set of features. In such case, the same feature may be used more than once at a path from the root to a leaf. The second option is that, after a feature was used at a given node, it is removed from the set of features-candidates for a split in children of this node. In such case, features are not repeated on any path moving from the root to a leaf. If we decide to go for the second option, the maximal length of a path in a tree cannot exceed the number of all features, and, as a consequence, the number of features naturally limits the height of the tree.

Which option of the aforementioned two should be chosen? It depends on the problem and the type of data (features). If a split of a set of patterns is based on a feature with a few distinct values and each value creates its own subset of patterns, then such feature should not be used again. However, when the number of children is (much) smaller than the number of distinct feature values, reusing such feature may be very effective. On the other hand, a *good* method of feature selection should eliminate the already employed features that are not effective in further tree construction.

As it was mentioned, a question about the number of subsets into which we split the set of patterns is another important issue determining the choice of the features.

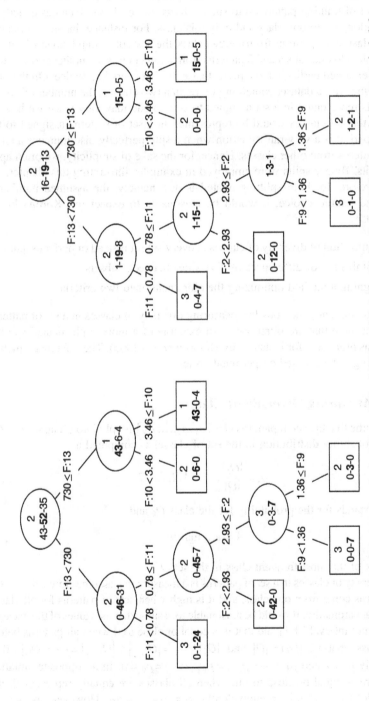

Figure 2.6 A simplified decision tree. In the full tree outlined in Figure 2.4, internal nodes for which the set of training patterns included less than 10% incorrectly classified patterns were turned to leaves. Classification results for the simplified (pruned) tree are displayed for the training set (left graph) and the test set (right graph).

There is no universally correct answer to this question. The best solution would be to split the set of training patterns into such subsets that each one includes exactly all patterns belonging to one class and only this class. For instance, having 15 patterns from one class and 5 patterns from another class, the best split should move 15 patterns from one class to one subset and 5 patterns from another class to another subset. However, as mentioned earlier, such a perfect case rarely occurs in practice. On the other hand, dealing with a dataset containing more than two classes, the number of distinct class labels represented in a set assigned to some node may vary since we have no guarantee that all classes would be represented in a set of patterns assigned to this node. Second, it is a very rare situation when a split perfectly discriminates a single class of patterns from other classes. Hence, for the sake of simplicity, we often apply binary splits. Binary splits were employed in examples illustrating this chapter.

Therefore, for both splitting-related issues, namely, the assumed number of subsets and feature choice, it would be reasonable to expect (cf. Koronacki and Ćwik, 2005):

- Minimization of diversity of classes at every subset instead of perfect separation
- Maximization of differences of diversities between subsets
- Designing a method optimizing the aforementioned two criteria

There are many methods for measuring diversity of classes in a set of patterns. We present three that are often used. An example of a more sophisticated splitting method was proposed, for instance, by Hothorn *et al.* (2006). The referenced method uses splitting criteria based on permutation tests.

2.4.3 Measuring Diversity of Classes

Let us assume that the set of patterns $O = \{o_1, o_2, \ldots, o_N\}$ is split into C classes $O_1, O_2, \ldots, O_C$. Probability distribution in the set of classes is computed as

$$p_k = \frac{|O_k|}{|O|}, \quad k = 1, 2, \ldots, C \tag{2.42}$$

that is, p_k stands for the probability for the class O_k and

$$\hat{k} = \arg \max_k p_k \tag{2.43}$$

is an index of the most frequent class in the set O.

Diversity of classes in a set of patterns is a measure that assumes low values when most patterns come from one class, and it is high when sets of patterns for all classes have similar cardinality. It would be reasonable to assume that the values of this measure are in the unit interval $[0, 1]$ and that it is equal or close to 0 when all patterns belong to one class, that is, $|O_{\hat{k}}| = |O|$ and $|O_1| = \cdots = |O_{\hat{k}-1}| = |O_{\hat{k}+1}| = \cdots = |O_C| = 0$ or equivalently $p_{\hat{k}} = 1$ and $p_1 = \cdots = p_{\hat{k}-1} = p_{\hat{k}+1} = \cdots = p_C = 0$. In an opposite situation, the measure is equal or close to one when all classes are equally represented, that is, $|O_1| = |O_2| = \cdots = |O_C|$ or equivalently $p_1 = p_2 = \cdots = p_C$. However, as we can see, it is more convenient not to restrict these values to the unit interval.

It is easy to design many measures satisfying the aforementioned intuitive insights. We will present three of the most popular measures.

Index of Incorrect Classification
This index is a simple diversity measure that expresses the ratio of the number of patterns not belonging to the majority class to the number of all patterns:

$$R(O) = p_1 + \cdots + p_{\hat{k}-1} + p_{\hat{k}+1} + \cdots + p_C = 1 - p_{\hat{k}} \tag{2.44}$$

This index is equal to 0 when all patterns come from a single class, and it is equal to $1 - 1/C$ when patterns are equally distributed across all classes. Assuming that $C = 2$ and $p_1 = p_2 = 0.5$ and assuming $\hat{k} = 1$, we get $R(O) = 1 - p_1 = 0.5$.

It assumes a minimal value when only one class, say, O_1, is represented in the set O, $p_1 = 1$, $p_2 = 0$, and it is maximal when both classes are evenly represented in the set O, $p_1 = 1/2 = p_2$ (cf. Figure 2.7)

$$R(O) = 1 - 1 = 0$$
$$R(O) = 1 - 1/2 = 1/2 \tag{2.45}$$

Let us investigate a case when $C = 8$. We consider a situation when all patterns belong to only one class, say, O_1, $p_1 = 1$, $p_2 = \ldots p_8 = 0$ and another situation when patterns are evenly distributed between classes, $p_1 = \cdots = p_8 = 1/8$:

$$R(O) = 1 - 1 = 0$$
$$R(O) = 1 - 1/8 = 7/8 \tag{2.46}$$

For example, for the training set of the wine dataset, the index of incorrect classification is

$$R(O) = p_1 + p_3 = 1 - p_2 = \frac{43}{130} + \frac{35}{130} = 1 - \frac{52}{130} = 0.6 \tag{2.47}$$

Entropy
Let us return to the question about the optimal choice of a feature and its split value. There is a method whose application seems to be very straightforward. Assuming that a feature is numerical, as, for instance, all features in the case of the wine dataset, we simply consider all possible splits of the feature into two intervals. Subsequently, we need a measure for information dispersion, and a prominent example of such a measure is entropy, which is minimal when we are most certain about class membership (there is no uncertainty). Hence, we compute entropy for each split and take the split with the minimal entropy. Then, we find the split with minimal entropy for every feature, and finally we choose this feature and its split that has the minimal entropy.

In our discussion, entropy of information in the set of patterns $O = \{o_1, o_2, \ldots, o_N\}$ is equivalent to entropy of the random variable defined by (2.42), that is, of the probability distribution on the set of classes. Entropy of information in this set is defined by the following formula:

$$E(O) = -\sum_{i=1}^{C} \frac{|O_i|}{|O|} \log_2 \frac{|O_i|}{|O|} = -\sum_{i=1}^{C} p_i \log_2 p_i \tag{2.48}$$

And since $\log_2 0$ is undefined, we assume that $\log_2 0 = \lim_{p \to 0^+} \log_2 p = 0$.

Note that the greater the entropy is, the greater the diversity of the data becomes. This measure is equal to 0 when all patterns come from one class, and it assumes higher values when patterns tend to be equally distributed among classes. Let us illustrate these properties of entropy by using a simple example. Let us assume that $C = 2$. We get $p_1 = 1 - p_2$ and $0 \le p_1 \le 1$. Hence, entropy becomes

$$E(O) = -p_1 \log_2 p_1 - p_2 \log_2 p_2 = -p_1 \log_2 p_1 - (1 - p_1) \log_2 (1 - p_1) \tag{2.49}$$

and assumes values in the unit interval $[0, 1]$ (cf. Figure 2.7).

It attains its minimum when only one class, say, O_1, is represented in the set O, $p_1 = 1$, $p_2 = 0$, and it achieves its maximum when both classes have the equal number of patterns in the set O, $p_1 = 1/2 = p_2$:

$$\begin{aligned} E(O) &= -1 \cdot \log_2 1 - 0 \cdot \log_2 0 = 0 \\ E(O) &= -1/2 \cdot \log_2 1/2 - 1/2 \cdot \log_2 1/2 = 1 \end{aligned} \tag{2.50}$$

Let us now inspect a case when $C = 8$. We are interested in two border examples: when all patterns in this set come from one class only, say, O_1, $p_1 = 1$, $p_2 = \ldots p_8 = 0$ and when patterns are evenly distributed between classes, $p_1 = \cdots = p_8 = 1/8$:

$$\begin{aligned} E(O) &= -1 \cdot \log_2 1 - 0 \cdot \log_2 0 - \cdots - 0 \cdot \log_2 0 = 0 \\ E(O) &= -1/8 \cdot \log_2 1/8 - \cdots - 1/8 \cdot \log_2 1/8 = 3 \end{aligned} \tag{2.51}$$

Let us look at entropy of the training set of the wine dataset. It is equal to

$$E(O) = -\frac{43}{130} \log_2 \frac{43}{130} - \frac{52}{130} \log_2 \frac{52}{130} - \frac{35}{130} \log_2 \frac{35}{130} = 1.57 \tag{2.52}$$

The Gini Index

The Gini index measures inequality among values of probability distribution on the set of classes:

$$G(O) = \sum_{i \ne j} p_i p_j = \sum_{k=1}^{C} p_k (1 - p_k) \tag{2.53}$$

Like for the previous measures, the greater the value of the Gini index is, the greater the diversity of data is. The Gini index equal to 0 expresses a perfect inequality of cardinality of patterns in classes. In other words, it describes a case when all patterns in a dataset come from one class only. For a perfectly equal distribution of patterns between classes, the value of the Gini index is equal to $1 - 1/C = (C-1)/C$, for C classes. Let us justify this observation with a simple example. Let us assume that we have $p_1 = 1 - p_2$ and $0 \le p_1 \le 1$. Hence, the Gini index is computed as

$$G(O) = p_1 p_2 + p_2 p_1 = 2 p_1 p_2 = 2 p_1 (1 - p_1) \tag{2.54}$$

and it ranges in the unit interval $[0, 1]$ (cf. Figure 2.7).

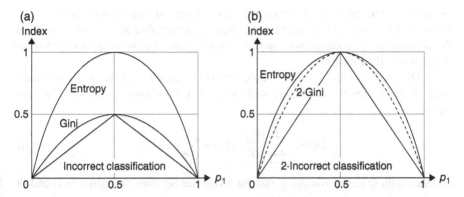

Figure 2.7   Graphs of entropy, the Gini index, and index of incorrect classification for a case of two classes. (a) True values of indexes and (b) values of indexes scaled to the unit interval.

The Gini index assumes its minimal value when only one class, say, O_1, is represented in the set O, $p_1 = 1$, $p_2 = 0$, and it assumes its maximum when both classes are evenly represented in the set O, $p_1 = 1/2 = p_2$:

$$G(O) = 1 \cdot 0 + 0 \cdot 1 = 0$$
$$G(O) = 1/2 \cdot 1/2 + 1/2 \cdot 1/2 = 1/2 \tag{2.55}$$

Let us investigate an example of $C = 8$ classes. We are interested in the two extreme scenarios: (i) when all patterns come from one class only, say, from O_1, $p_1 = 1$, $p_2 = \ldots p_8 = 0$ and (ii) when patterns are evenly distributed cross all classes, $p_1 = \cdots = p_8 = 1/8$:

$$G(O) = 1 \cdot 0 + 0 \cdot 1 + \cdots + 0 \cdot 1 = 0$$
$$G(O) = 1/8 \cdot 7/8 + \cdots + 1/8 \cdot 7/8 = 7/8 \tag{2.56}$$

Lastly, let us compute the Gini index for the training set of the wine dataset:

$$G(O) = \frac{43}{130} \cdot \left(1 - \frac{43}{130}\right) + \frac{52}{130} \cdot \left(1 - \frac{52}{130}\right) + \frac{35}{130} \cdot \left(1 - \frac{35}{130}\right) = 0.66 \tag{2.57}$$

2.4.4   Choosing a Splitting Feature

Let us consider a set of patterns $O = \{o_1, o_2, \ldots, o_N\}$ and its split into pairwise disjoint subsets $\Theta = \{Q_1, Q_2, \ldots, Q_r\}$, that is, $\cup_{i=1}^{r} Q_i = O$ and $Q_i \cap Q_j = \emptyset$ for $i, j = 1, 2, \ldots, r$ $i \neq j$. Note that this split of the training set of patterns is different from the split to classes $O_1, O_2, \ldots, O_C$. In Section 2.4.2 we discussed how to measure diversity of classes in sets of patterns, so we are equipped with tools necessary to compute for the set O and for each member Q_i of the family $\Theta = \{Q_1, Q_2, \ldots, Q_r\}$. First, we want to measure diversity of classes in the family Θ of subsets. Then, having such measure, we can find the difference between diversity of classes in the set O and its r subsets. Of course, it is desirable to decrease diversity, that is, we wish to increase the difference. In consequence, such a feature would be preferred for which the difference is the highest.

Diversity of classes in the whole family of subsets is defined as a weighted sum of diversities in all subsets with weights proportional to the size of each subset

$$D(\Theta) = \sum_{i=1}^{r} \frac{|Q_i|}{|O|} D(Q_i) = \sum_{i=1}^{r} q_i D(Q_i) \qquad (2.58)$$

and then gain in class diversity is just the difference between diversities in O and Θ:

$$\Delta D(O, \Theta) = D(O) - D(\Theta) \qquad (2.59)$$

where D is a measure of class diversity, for instance, index of incorrect classification, entropy, or the Gini index.

A set of values of any feature is finite since the set of training patterns is finite. Therefore, we can consider all possible splits of the set of values of a feature F and then splits of patterns corresponding to splits of values of the feature. However, such method, though theoretically possible, is useless due its exponential complexity. Considering only binary splits, that is, splits to a subset and its complement, we come to the following number of such dual pairs:

$$\frac{2^{|F|} - 2}{2}$$

where $F = \{f_1, f_2, \ldots, f_{|F|}\}$ is the set of values of a feature F, and, of course, $|F| \le |O| = N$, since in the extreme case, there are as many values of the feature as the number of patterns in the set O. This is the cardinality of the family of all nontrivial subsets of F (the empty set and the whole set F are not considered). Therefore, there are more splits for families, which have more than two members.

Instead, assuming that all feature values are linearly ordered (e.g., they are numbers), say, $f_1 < f_2 < \cdots < f_{|F|}$, we consider splits into subintervals covering the whole interval of feature values $[f_1, f_{|F|}]$ and such that each subinterval includes at least one pattern. For instance, a split to k subintervals is

$$\left[f^{(0)}, f^{(1)} \right], \left(f^{(1)}, f^{(2)} \right], \ldots, \left(f^{(k-1)}, f^{(k)} \right]$$

where $f_1 = f^{(0)}, f^{(k)} = f_{|F|}$ and such that $\left[f^{(0)}, f^{(1)} \right] \cap F \ne \emptyset$ and $\left(f^{(i-1)}, f^{(i)} \right] \cap F \ne \emptyset$ for $i = 2, \ldots, k$.

The splits to subintervals could anticipate a distribution of feature values for a set O of patterns split to classes $O_1, O_2, \ldots, O_C$, because, for a given class of patterns, the range of values distribution is more likely to be narrow. Even the simplest type of splits, binary splits, would separate patterns from two subsets of classes. Of course, in practice, the mentioned anticipation of values distribution is too optimistic. Nevertheless, such unrealistic anticipation illustrates well the existing problem.

2.4.5 Limiting Tree Structure

There are three approaches that we can apply in order to limit an expansion (growth) of the constructed decision tree:

- Self-evident termination of expansion. This approach is utilized the following cases: (i) the set of patterns is empty, (ii) all patterns belong to the same class, and (iii) there are no more features to be used.
- Early termination. This means checking a predefined condition and if this condition is satisfied, we stop growing the tree. Avoiding overfitting or abridging the structure of the tree are examples of such conditions.
- Tree pruning (limits the structure of a tree rather than tree expansion) could be employed instead of early termination criteria; we start with a full tree, and then we cut weaker branches (we need a criterion to evaluate branches or nodes).

The first case is more *technical* than theoretically justified. When we run the procedure of tree construction, we use a training set to develop the tree structure. There are three aforementioned cases when there is no justification for further branch expansion. The first one (i) is when we do not have any patterns to subject to splitting (the set of patterns is empty). The second case (ii) is when we arrive at a node to which assigned are patterns that belong to one class only. We are unable to perform a split since there are no distinct classes for which we can maximize diversity with a split. In the third case, (iii) there is no feature to be used. This may happen when we decided not to repeat features along paths from the root to leaves.

The second and the third methods, namely, the early stopping criteria and tree pruning, are well-known and frequently applied strategies improving decision tree quality. Let us stress that if we do not apply any procedures for tree simplification, we will end up with a full-grown tree. Such model is perfectly adjusted to classify patterns from the training set—in the illustrative example with the wine dataset, we have shown that accuracy on the training set for a full tree is 100%. However, usually, such a tree becomes overfitted, that is, classification efficiency of patterns not belonging to the training set is low. In order to enhance the ability of generalization, we need to modify (simplify) the tree structure and make it less fine grained. In order to do so, we can either grow a full tree and then prune it or stop the tree construction procedure earlier. In the classification tree illustrated in Figure 2.4, expansion was limited using the case of patterns from one class only. Then, pruning applied to this tree limited its structure; see Figure 2.6.

Decision trees are among most elementary machine learning algorithms. One can find more elaborate discussions in many sources; good examples are the following books: Frank *et al.* (2001, chapter 6), Duda *et al.* (2001, sections 8.2–8.4), Webb and Copsey (2001, chapter 7), Hastie *et al.* (2009, section 9.2), and Mitchell (1997, chapter 3). It is also worth to recall general surveys on decision tree methodology (Safavian and Landgrebe, 1991; Murthy, 1998).

The research on decision trees has produced a number of valuable algorithms, including few well-known strategies for tree construction such as such as CHAID (short for chi-square automatic interaction detectors) (Sonquist and Morgan, 1964),

CART (short for classification and regression trees) (Breiman *et al.*, 1983), and ID3/ 4.5/5.0 (Quinlan, 1986) (Utgoff, 1989; Wu *et al.*, 2003). The intensity of research plateaued, but then decision trees regained attractiveness with the era of hybrid and ensemble classifiers.

2.5   ENSEMBLE CLASSIFIERS

Strengths and weaknesses of various classifiers and classification schemes brought up an idea that perhaps combining several classification models by building their ensemble could help improve classification accuracy. This has resulted in a development of methods of so-called *ensemble classification* or *ensemble classifiers*. An ensemble classifier is an arrangement of (i) a set of single classifiers, say, *weak classifiers*, and (ii) a method of aggregating results of single classifiers into a final classification decision. An intuitive way of aggregating classification decisions is through simple voting, that is, we have constructed a few classifiers, used them to predict a class label for a pattern, and then assign this class label that was most frequently suggested by the classifiers. In a case of more than one majority class, the final decision could be made with a secondary criterion, for instance, random choice or some kind of weighting. Of course, we expect that an ensemble classifier should be *better* than each weak classifier. Let us consider a two-class classification problem. Having weak classifiers slightly better than random class assignment (i.e., the ratio of correctly classified patterns is somewhat greater than 0.5), we expect that an ensemble made of such weak classifiers will do better, that is, the ratio of correctly classified patterns would be much greater than 0.5. Other advantages of an ensemble classifier are that it may be more stable and it may have a smaller variance than weak classifiers. It would be desirable if an ensemble classifier had a smaller bias than single classifiers, though it may not be the case. Note that we avoid focusing on any type of particular weak classifiers. Indeed, the idea is that—within reason—any weak classifier can be employed and then aggregated into an ensemble classifier. However, the choice of decision trees as weak classifiers is very popular. What is important is that ensemble classification cannot be limited only to the technical issue of forming several classifiers, but it also requires adaptation of the classification scheme. It calls for a reconsideration of the usage of the training/validation/test set.

There are different methods for aggregation of weak classifiers. Later, we discuss bagging and boosting as commonly encountered examples of weak classifiers aggregation methods. We also discuss the AdaBoost algorithm as a particular example of the boosting method.

Appearance of ensemble classification did not slow down the pace of research on standard, single-model classifiers. Comparing efficiency of single and composite classifiers is a practical and well-established issue. The reader may refer to a survey on this topic (Lim *et al.*, 2000).

2.5.1   Bagging

Breiman (1996) proposed a very simple method for joining single classifiers. The method is called *bagging* (acronym of *bootstrap aggregating*). It assumes building

B training sets of patterns $O^{(1)}, O^{(2)}, ..., O^{(B)}$, say, bagging training sets, based on a given training set $O = \{o_1, o_2, ..., o_N\}$. Then, for each bagging training set, a weak classifier is built. Each bagging training set $O^{(i)}$ is collected by running random sampling with replacement from the training set of patterns O. Sampling is performed based on a uniform probability distribution in the training set $O = \{o_1, o_2, ..., o_N\}$, that is, any pattern can be picked up with an equal probability $1/N$. Note that each training set $O^{(1)}, O^{(2)}, ..., O^{(B)}$ may have some patterns repeated and some missing. Statistically, each of them has about 1/3 patterns missing (more specifically $1/e \cong 0.368$). Details are given in the following algorithm:

Algorithm 2.1
Building weak classifiers with bootstrap aggregating
Data: Training Set of Patterns $O = \{o_1, o_2, ..., o_N\}$
           Classification Method (weak classifier)
           B—number of weak classifiers
Algorithm: **assume** weights $w_i = 1/N$ for $i = 1, 2, ..., N$
           for $k = 1$ **to** B **do**
           begin
               build bagging training set $O^{(k)}$ by random sampling with replacement
                from the training set $O = \{o_1, o_2, ..., o_N\}$, use weights w_i for
                sampling
               build weak classifier $\psi^{(k)}$ for the training set $O^{(k)}$
           end
Result: the set of weak classifiers $\psi^{(1)}, \psi^{(2)}, ..., \psi^{(B)}$

For a given pattern **x**, a bagging ensemble classifier assigns the class obtained with simple voting of the weak classifiers $\psi^{(1)}, \psi^{(2)}, ..., \psi^{(B)}$:

$$\psi(\mathbf{x}) = \arg \max_{1 \leq k \leq C} \sum_{i=1}^{B} \delta_{k, \psi^{(i)}(\mathbf{x})} \tag{2.60}$$

where $\delta_{i,j}$ is the Kronecker delta function. In other words, a pattern **x** is assigned class label k such that number of weak classifiers voting for it is maximal.

2.5.2 Boosting

Another, more general method called *boosting* was proposed by Shapire (1990) and then improved by Freund and Schapire (1997). Boosting was invented independently of Breiman's bagging, but it can be seen as a generalization of bagging. The construction of training sets $O^{(1)}, O^{(2)}, ..., O^{(B)}$ for bagging and boosting employs random sampling from a given training set $O = \{o_1, o_2, ..., o_N\}$ with a given probability distribution defined in O. Random sampling for all bagging training sets is based on a fixed uniform probability distribution. However, boosting adapts probability distribution for consecutive training sets. Probability of choosing patterns incorrectly classified by the current weak classifier gets increased (as such patterns are *difficult* to learn and hence we want to emphasize their role when designing the classifier). As a consequence of increasing weights for such *difficult* patterns, a chance of selecting *easy* patterns decreases, that is, the patterns that were classified correctly by the current weak classifier are less likely to occur in the new training set.

There are different methods to define probability distribution. We discuss here the most popular one called *AdaBoost* (acronym of *Adaptive Boosting*) (Freund and Schapire, 1997). AdaBoost begins with a uniform probability distribution to select the first training set $O^{(1)}$ and then for the next training set $O^{(i+1)}$ increased are weights of patterns incorrectly classified by the current weak classifier ψ^i, $i = 1,2..., B-1$, and, of course, probability of choosing correctly classified patterns decreases. So then, incorrectly classified patterns get a better chance to be picked up for the next boosting training set than those correctly classified.

Algorithm 2.2

Building weak classifiers for boosting

Data: Training Set of Patterns $O = \{o_1, o_2, ..., o_N\}$

           x_i—vector of features of pattern o_i

           y_i—class label of pattern o_i

           Classification Method

           B—number of weak classifiers

Algorithm: **assume** weights $w_i^{(1)} = 1/N$ for $i = 1, 2, ..., N$

           for $k = 1$ **to** B **do**

           begin

              build the bagging training set $O^{(k)}$ by weighted random sampling with replacement of the training set $O = \{o_1, o_2, ..., o_N\}$

              build the weak classifier $\psi^{(k)}$ for the training set $O^{(k)}$

$$\text{compute weighted error rate } wer^{(k)} = \sum_i^N w_i^{(k)} \bar{\delta}_i, \quad \bar{\delta}_i = \left(1 - \delta_{y_i, \psi^{(k)}(x_i)}\right)$$

$$\text{compute correction factor } \chi^{(k)} = \frac{1 - wer^{(k)}}{wer^{(k)}}$$

$$\text{adjust weights } w_i^{(k+1)} = \begin{cases} w_i^{(k)} \left(\chi^{(k)}\right)^{\delta_i} \Big/ \sum_{j=1}^N w_j^{(k)} \left(\chi^{(k)}\right)^{\bar{\delta}_i} & \chi^{(k)} \geq 1 \\ 1 & \text{otherwise} \end{cases}$$

              for $i = 1, 2, ..., N$

           end

Result: the set of weak classifiers $\psi^{(1)}, \psi^{(2)}, ..., \psi^{(B)}$

Finally, unknown patterns are classified by all weak classifiers, and then the majority class is assigned to the pattern by a weighted voting scheme of the weak classifiers $\psi^{(1)}, \psi^{(2)}, ..., \psi^{(B)}$. For a given pattern **x**, boosting ensemble classifier assigns the following class:

$$\psi(\mathbf{x}) = \arg \max_{1 \leq k \leq C} \sum_{i=1}^B \left(\delta_{k, \psi^{(i)}(\mathbf{x})} \cdot \ln \chi^{(k)}\right) \tag{2.61}$$

where $\delta_{i,j}$ is the Kronecker delta function.

2.5.3 Random Forest

Random forest is a type of an ensemble classifier developed by Breiman (2001). Random forests take decision trees as their weak classifiers. Classification quality of random forests relies on the quality and independence of weak classifiers, that is, on the quality of decision trees. In general, for a given classification problem, we build a set of weak classifiers. Unlike in the cases of bagging and boosting, where any type of classifier can be used, random forests must employ decision trees as weak classifiers. Aggregation of weak classifier results for random forests is made based on simple voting, like in the case of bagging.

Algorithm 2.3
Building weak classifiers for random forests
Data: Training Set of Patterns $O = \{o_1, o_2, ..., o_N\}$
              Set of features $F = \{f_1, f_2, ..., f_M\}$
              B—the number of weak classifiers (decision trees)
              p—the number of features used to build weak classifiers
Algorithm: **assume** weights $w_i = 1/N$ for $i = 1, 2, ..., N$
              for $k = 1$ **to** B **do**
              begin
                  build bagging training set $O^{(k)}$ by random sampling with
                      replacement of the training set $O = \{o_1, o_2, ..., o_N\}$, use weights w_i
                      for sampling
                  construct a decision tree for $\psi^{(k)}$ for the training set $O^{(k)}$ **as follows:**
                  begin
                      for each node of the constructed tree **do**
                      if the node is not a leaf **do**
                      begin
                        select p features from the set F of all features by random
                          sampling without replacement, assume $p << M$
                        find the best feature from the selected p ones
                        find the best split of the set of training samples
                        use each subset of the split to construct respective node of
                        the tree
                    end
                  end
                end
Result: the set of decision trees (weak classifiers) $\psi^{(1)}, \psi^{(2)}, ..., \psi^{(B)}$

For a given pattern **x**, apply simple voting of the weak classifiers $\psi^{(1)}, \psi^{(2)}, ..., \psi^{(B)}$ to assign the class

$$\psi(\mathbf{x}) = \arg \max_{1 \le k \le C} \sum_{i=1}^{B} \delta_{k, \psi^{(i)}(\mathbf{x})} \qquad (2.62)$$

where $\delta_{i,j}$ is the Kronecker delta function.

The process of tree expansion does not use locally all features; instead a small fraction of all features is involved in the tree construction procedure. This distinguishes trees grown in a forest from *standard* decision trees, where the splitting feature is selected from the entire set of features. The subset of features used for consideration when we construct a tree in a forest is obtained by random sampling of p features without replacement from the entire set of features. The same parameter p is used for construction of all decision trees in the forest. In order to select the proper value of p, literature suggests taking a square root of the number of all features as the value of the parameter p ($p \approx \sqrt{M}$). When p is too large, we may increase predictive strength but at the risk of higher correlation (Breiman, 2001). Alternatively, parameter tuning could be done for a particular classification problem. This formula suggests that the set of features used to build trees should be reduced significantly. What this entails is that trees in a random forest are indeed *weak*. A thorough study on bias in feature selection for random forest has been discussed in Strobl *et al.* (2007). The cited article formulates a conclusion that for data of varying types, it is very beneficial to rethink random forest construction procedure.

In addition, random forests are very suitable and convenient for application in classification problems with a big number of features, say, even thousands. In other words, random forests incorporate their own randomized method for feature selection. In addition, feature selection as in random forest algorithm is applied as a stand-alone mechanism in other algorithms (Genuer *et al.*, 2010).

Random forests (and, in general, bagging as well) allow employing an interesting method for quality evaluation. As we noticed in Section 2.5.1, for each bagging training set $O^{(k)}$, about 0.368 out of all training patterns is not included in it, $k = 1, 2, ..., B$. Therefore, there is a similar proportion of bagging training sets not including a given training pattern. As a consequence, a given training pattern $\mathbf{x}$ can be classified with each tree that was constructed based on the bagging training set not including that pattern $\mathbf{x}$. Finally, the pattern $\mathbf{x}$ is assigned the majority class pointed out by these trees. Having results of such classification for all training patterns, we can compute the ratio of correctly classified patterns to all training ones. This value is a good estimation of the quality of the constructed random forest. Since each training pattern was subjected to classification only by a subgroup of such trees, which were constructed without employing this pattern, the estimation is unbiased. Of course, such evaluation concerns also a bagging ensemble classifier employing any weak binary classifier.

Random forests and other ensemble classifiers based on decision trees have been successfully applied to many areas such as predictions in ecology (Prasad *et al.*, 2006; Elith *et al.*, 2008), microarray analysis (Diaz-Uriarte and de Andres, 2006), modeling in cheminformatics (Svetnik *et al.*, 2003), remote sensing classification (Ham *et al.*, 2005; Pal, 2005), computer-aided medical analysis (Wu *et al.*, 2003), and many more.

For more elaboration on ensemble classification, one may consult Hastie *et al.* (2009, chapters 10, 15, and 16), Duda *et al.* (2001, section 9.5), Bishop (2006, chapter 14), and Frank *et al.* (2001, chapter 8).

2.6   BAYES CLASSIFIERS

In this section, we discuss classification based on a known probability distribution in a set of classes and a probability distribution in feature space for a given class.

Probability distribution in the set of classes is called *a priori* probability. Probability distribution in feature space for a given class is called *class conditional* probability. Then, for a given pattern characterized by features values for each class, we compute probability that this pattern belongs to a given class. Those probabilities are called *a posteriori*. Finally, the pattern is assigned to the class with the highest *a posteriori* probability. This specific classifier minimizes misclassification probability and is called the *Bayes classifier*. Generalization of this method offers classifications with conditions different than minimization of misclassification probabilities.

2.6.1   Employing Bayes Theorem

Formalizing this intuitive idea, let us assume that for a given classification problem, we have information about fractions of patterns belonging to each class and characteristics of features for each class. Such information could be gained on the basis of collected observations, for example, based on a training set. Let us assume that we know a probability $P(O_i)$ that a given pattern falls into the class O_i, for $O_i \in \Theta$, that is, we have the class probability distribution

$$p: \Theta \to [0, 1], \quad \sum_{O_k \in \Theta} p(O_k) = 1 \tag{2.63}$$

where lowercase p denotes probability density and uppercase P stands for probability of an event. Class probability distribution is defined in the set of classes, that is, in a finite set. Values of the distribution are either known or may be determined with a fraction of patterns from a given class in the training set. In this case, values of probability distribution $p(O_i)$ are equal to the probability that a given pattern falls into this class $P(O_i)$. Note that the training set may not reflect a real-world problem well. For instance, a training set may include a similar number of data describing healthy patients and ones suffering from some disease, while, in the real world, the fraction of ill patients may be very small. Therefore, in similar cases, it is a challenge to have a probability distribution that reflects a real-world problem.

Next, let us assume that a pattern from the class O_k is characterized by a vector $\mathbf{X}$ from the feature space R^M, that is, $\mathbf{X} \in R^M$. Let us also assume that we either know class conditional probability $P(\mathbf{X}|O_k)$ or can estimate it in some way; usually the former case is quite rare. Estimation of class conditional probabilities is not as direct as for *a priori* probabilities. To continue the discussion on the Bayes classification, let us assume that the density function $p(\mathbf{x}|O_k)$ of class conditional probability is given as a continuous function. Later on we will consider class conditional probabilities in detail. The density function $p(\mathbf{x}|O_k)$ defines probability distribution in the space of the features for the class O_k. Based on the aforementioned assumptions, *a posteriori* probabilities, using the Bayes formula, can be written in the form

$$p(O_k|\mathbf{x}) = \frac{p(\mathbf{x}|O_k) \cdot P(O_k)}{p(\mathbf{x})} \tag{2.64}$$

where $\mathbf{x} \in R^M$, $p(\mathbf{x})$ is the unconditional normalization factor, that is, the (unconditional) density function given by the formula

$$p(\mathbf{x}) = \sum_{k=1}^{C} p(\mathbf{x}|O_k) \cdot P(O_k) \qquad (2.65)$$

The normalization factor ensures that a posteriori probabilities $p(\mathbf{x}|O_k)$ add up to one:

$$\sum_{k=1}^{C} p(O_k|\mathbf{x}) = 1 \qquad (2.66)$$

We also assume that, based on this distribution, probabilities $P(\mathbf{X}|O_k)$ of features values $\mathbf{X} \in R^M$ for classes O_k, $k = 1, 2, ..., C$ are known.

Now, assuming that for the given pattern with features values $\mathbf{X}$, conditional probabilities $P(\mathbf{X}|O_k)$ and a priori probabilities $P(O_k)$ are known, conditional probability $P(O_k|\mathbf{X})$, that is, probability that this pattern is assigned the class O_k assuming the feature vector $\mathbf{X}$, is defined by the formula corresponding to (2.64):

$$P(O_k|\mathbf{X}) = \frac{P(\mathbf{X}|O_k) \cdot P(O_k)}{P(\mathbf{X})} \qquad (2.67)$$

where $P(\mathbf{X})$ is again the unconditional normalization factor computed as follows:

$$P(\mathbf{X}) = \sum_{k=1}^{C} P(\mathbf{X}|O_k) \cdot P(O_k) \qquad (2.68)$$

Finally, we can provide the formula for the Bayes classifier as

$$\Psi(\mathbf{X}) = \arg \max_k P(O_k|\mathbf{X}) \qquad (2.69)$$

In other words, the Bayes classifier assigns an unknown pattern characterized by features vector $\mathbf{X}$ to the class O_k if

$$P(O_k|\mathbf{X}) > P(O_j|\mathbf{X}) \quad \text{for all} \quad j = 1, 2, ..., C, \ j \neq k \qquad (2.70)$$

which, having in mind that normalization factor is constant and independent of classes, implies

$$P(\mathbf{X}|O_k) \cdot P(O_k) > P(\mathbf{X}|O_j) \cdot P(O_j) \quad \text{for all} \quad j = 1, 2, ..., C, \ j \neq k \qquad (2.71)$$

This is a rule for assigning each point of the feature space to one of the C classes. This means that the Bayes classifier divides the feature space into decision regions D^1, $D^2, ..., D^C$ defined by (2.70) and corresponding to classes $O_1, O_2, ..., O_C$, that is, each pattern characterized by the feature vector $\mathbf{X} \in D^k$ is assigned to one of the classes O_k, $k = 1, 2, ..., C$. It is important to notice that each of these regions does not need to be connected and may be divided into disjoint parts related to the same class. The boundaries between decision regions are called decision boundaries or decision surfaces.

2.6.2 Minimizing Misclassification Probability

As mentioned earlier, the Bayes classifier guarantees minimal misclassification probability. Now, we present an intuitive justification of this claim considering the case

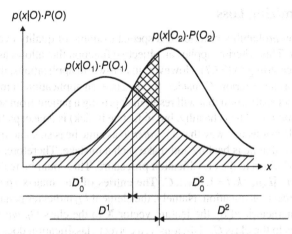

Figure 2.8   Illustration of the Bayes classifier. Minimal misclassification error is achieved for the crossing point of joined probability densities. Minimal misclassification error is shown as a lined area; reducible error is shown as a double lined area.

of two classes and one feature with continuous class conditional density function. Figure 2.8 schematically illustrates joint *a priori* and class conditional probabilities, that is, the product of them, for such case. Of course, such joint probability is proportional to an *a posteriori* probability with normalization factor being the proportion factor. The vertical lines define decision boundaries in the feature space, which is the axis x. We have two different divisions for decision regions: D^1, D^2 and D_0^1, D_0^2. Note that for a given decision boundary, misclassification probability can be expressed as the area under respective joint probability curves, that is, under the curve corresponding to the class O_1 in the decision region of the class O_2 and under the curve corresponding to the class O_2 in the decision region of the class O_1 (cf. lined areas in Figure 2.8).

For decision regions D^1, D^2, misclassification probability is shown in Figure 2.8 with joint areas lined and double lined. Alike, for decision regions D_0^1, D_0^2, misclassification probability is shown with a (single) lined area. We can formally express it as follows:

$$P(\text{missclassif.}) = \int_{D^1} P(x|O_2) \cdot P(O_2) dx + \int_{D^2} P(x|O_1) \cdot P(O_1) dx$$

and   (2.72)

$$P_0(\text{missclassif.}) = \int_{D_0^1} P(x|O_2) \cdot P(O_2) dx + \int_{D_0^2} P(x|O_1) \cdot P(O_1) dx$$

It is easy to notice that we get the minimal misclassification error for the division for decision regions defined by the crossing point of probability distributions for both classes. Any other choice of decision regions brings reducible error shown as the double lined region in Figure 2.8.

2.6.3 Minimizing Loss

Misclassification probability criterion is a special example of quality evaluation of the Bayes classifier. This criterion applies an objective function that allows assigning a pattern to a class according to (2.67). However, in real-world applications, this may not be the most appropriate criterion. Consider, for instance, an application in medicine. Here, the cost of a misclassification that will result in depriving a patient from some treatment (because *the system* said he is healthy, but in fact he is sick) is unacceptable. Hence, we wish to weigh decisions in a way that it is *safer* to assume he is sick and make a mistake here than assume that he is healthy and make such a mistake. Therefore, it is desired to include some priorities to the classification procedure. This could be realized by defining a *loss matrix* $[L_{kl}]$, $k, l = 1, 2, \ldots, C$. The entries of this matrix specify penalties related to incorrect classification. Namely, the element L_{kl} indicates penalty for assigning a pattern characterized by the feature vector $\mathbf{X}$ to the class O_l, while this pattern actually belongs to the class O_k. Obviously, a correct classification does not result in a loss, that is, $L_{kk} = 0$, $k = 1, 2, \ldots, C$. Then, we can compute an average loss of incorrect classification for such pattern and for each class O_k:

$$R'_k(\mathbf{X}) = \sum_{l=1}^{C} L_{kl} P(O_l|\mathbf{X}) = \sum_{l=1}^{C} L_{kl} \frac{P(\mathbf{X}|O_l) \cdot P(O_l)}{P(\mathbf{X})} \qquad (2.73)$$

Since the normalization factor $P(\mathbf{X})$ is constant and independent on classes, we can simplify the aforementioned formula and come to

$$R_k(\mathbf{X}) = \sum_{l=1}^{C} L_{kl} P(\mathbf{X}|O_l) \cdot P(O_l) \qquad (2.74)$$

Finally, the pattern is assigned to class O_k if

$$R_k(\mathbf{X}) < R_l(\mathbf{X}) \quad \text{for all} \quad j, k = 1, 2, \ldots, C, \quad j \neq k \qquad (2.75)$$

or equivalently the classifier is defined by the formula

$$\Psi(\mathbf{X}) = \arg \min_k R_k \qquad (2.76)$$

When elements of a loss matrix are equal to $L_{kl} = 1 - \delta_{kl}$, $k, l = 1, 2, \ldots, C$ (where δ_{kl} is the Kronecker delta function), that is, 0 s are at the main diagonal and 1 s elsewhere, then the minimal loss is equivalent to the minimal misclassification probability. However, for different applications, elements of loss matrix may significantly differ. For instance, for the aforementioned medical applications, we require the loss value for incorrect classifications of an ill patient as a healthy one to be much higher than the loss for assuming that a healthy patient is ill. Such values could be either specified by experts (here from the medical domain) or determined on the basis of trusted data.

2.6.4 Rejecting Uncertain Patterns

We may expect that, in Bayes classification, most incorrectly classified patterns will be located in such areas of feature space where joint *a priori* and class conditional

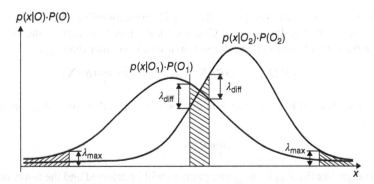

Figure 2.9   Illustration of rejecting regions of the Bayes classifier. A posteriori probabilities for both classes are smaller than δ_{max} on the left and the right, while the absolute difference between *a posteriori* probabilities is smaller than δ_{diff} in the middle.

probabilities (*a posteriori* probabilities) are close to each other, that is, around decision surfaces. Another potential area of misclassifications is where these probabilities are small. Let us refer to Figure 2.9. In this illustrative example, potential misclassification errors could be expected for feature values either falling around the crossing point of both probability curves or being far away from this point. Such patterns are burdened with high uncertainty with regard to classification, and due to this fact, they should be excluded from the classification procedure. This observation would be important for problems for which it is better to avoid classification rather than to risk an error. Excluded patterns may be subsequently subjected to further processing, either automatic or manual. We may formally describe conditions that need to be satisfied in order to exclude a pattern from the classification process by setting two thresholds. The first threshold λ_{diff} limits the difference between the maximal and the second maximal *a posteriori* probabilities, which reveals an uncertainty in choosing a class with the maximal a posteriori probability. The second threshold λ_{max} restricts all a posteriori probabilities, which can be seen as a hesitation in assigning any class label. More exactly, classification will not be executed when the difference between the maximal and the second maximal a posteriori probability is smaller than λ_{diff}:

$$\max_{l} P(O_l|\mathbf{X}) - \max_{l,\, l \neq k} P(O_l|\mathbf{X}) < \lambda_{diff} \quad k = \arg \max_{i} P(O_i|\mathbf{X}) \qquad (2.77)$$

or when the maximal a posteriori probability (so then all of them are) is smaller than λ_{max}:

$$\max_{l} P(O_l|\mathbf{X}) < \lambda_{max} \qquad (2.78)$$

The greater the threshold values are, the more patterns will be rejected and the fewer patterns will be subjected to classification. Figure 2.9 illustrates rejecting regions for the case of one feature and two classes.

Instead of *a posteriori* probabilities with corresponding thresholds λ_{diff} and λ_{max}, we may employ loss values. Classification cannot be done if the difference between the second minimal and the minimal loss is smaller than λ_{diff}:

$$\min_{l, l \neq k} R_l(\mathbf{X}) - \min_l R_l(\mathbf{X}) < \lambda_{\text{diff}} \quad k = \arg \min_l R_l(\mathbf{X}) \tag{2.79}$$

or when the minimal loss value and, in consequence, all other loss values are greater than λ_{max}:

$$\min_l R_l(\mathbf{X}) > \lambda_{\text{max}} \tag{2.80}$$

The greater the λ_{diff} is, the more patterns will be rejected and the fewer patterns will be classified. The smaller the λ_{max} is, the more patterns will be rejected and the fewer patterns will be classified.

2.6.5 Class Conditional Probability Distribution

We assumed that *a priori* and class conditional probability distributions are given in some way. Usually, these distributions are not known. As mentioned earlier, a priori probability distribution could be learned on the ground of observations collected in the past. If this is not possible, we can estimate this distribution based on the training set of patterns, assuming that the training set is representative for the entire space of possible patterns

$$p(O_k) = \frac{|O_k|}{|O|} \quad k = 1, 2, ..., C \tag{2.81}$$

Finding class conditional probability distributions $p(\mathbf{x}|O_k)$ $k = 1, 2, ..., C$ is not so straightforward. Typically, we cannot induce these distributions, as we are only provided with a sample of patterns. Therefore, we apply a training set of patterns to estimate them. In general, we distinguish two cases. In the first case, if we know a family of probability distribution, we need to find parameters of distribution functions. In contrast, if a type of distribution function is unknown, we can apply nonparametric methods. It is important to notice that we will be looking for smooth enough density functions. For practical reasons, such functions should have second derivatives, which is useful for finding optimal points of functions. For the sake of simplicity and since the discussion on these distributions is the same for each class, we skip class index denoting a set of considered patterns (class) simply $p(x)$ and denoting the number of patterns in this set (class) N.

Parametric Estimation
Parametric estimation of a density function requires that we know the type (family) of this function, meaning that we know a specific function definition and we know what parameters are governing it. The values of parameters could be determined with the help of a given dataset, say, using a training set of patterns. The values of parameters are typically subjected to some optimization process. For instance, optimized is the mean square error of estimated values compared with a given training set in order to find the best fit of the estimated function to the training set. The most widely utilized

density function and perhaps the simplest one is the Gaussian function. We restrict the discussion to this function, as our objective is just to explain the basic principles of parametric estimation of a density function.

Now, let us recall basic facts with regard to normal distribution. The normal density function of one variable is written as

$$p(x) = \frac{1}{\sqrt{2\pi\sigma^2}} \exp\left(-\frac{(x-\mu)^2}{2\sigma^2}\right) \tag{2.82}$$

where μ and σ are parameters of the function. μ is called *expectation* or *mean value* and σ^2 is called *variance* (it is the second central moment). σ is called *standard deviation*. The scaling factor (viz., $1/\sqrt{2\pi\sigma^2}$) guarantees that the function integrates to 1. Of course, the mean value and variance of the normal distribution (2.82) can be calculated using general formulas

$$\mu = E[x] = \int_{-\infty}^{\infty} xp(x)dx$$

$$\tag{2.83}$$

$$\sigma^2 = E\left[(x-\mu)^2\right] = \int_{-\infty}^{\infty} (x-\mu)^2 p(x)dx$$

where $E[x]$ denotes the expected value of the random variable x. Notice that the variance σ^2 is the expected value of the random variable $(x-\mu)^2$.

The M-dimensional normal distribution function takes the form

$$p(\mathbf{x}) = \frac{1}{\sqrt{(2\pi)^M |\Sigma|}} \exp\left(-\frac{1}{2}(\mathbf{x}-\mu)^T \Sigma^{-1}(\mathbf{x}-\mu)\right) \tag{2.84}$$

where $\mathbf{x}$ and μ are points (vectors) in R^M, μ is an M-dimensional mean vector, and Σ is an $M \times M$ *covariance matrix*, while $|\Sigma|$ is the determinant of the covariance matrix and Σ^{-1} is the inverse of the covariance matrix. By analogy to (2.83), the (M-dimensional) mean vector and the ($M \times M$) covariance matrix of the density function (2.84) could be computed with the use of the following formulas

$$\mu = E[\mathbf{x}]$$

$$\Sigma = E\left[(\mathbf{x}-\mu)(\mathbf{x}-\mu)^T\right] \tag{2.85}$$

Note that the covariance matrix in (2.85) is symmetric, and since its elements are real numbers, then it is positive defined (we do not attempt to talk about degenerated normal distributions with positive semi-defined covariance matrices). The function appearing in the exponent in (2.84)

$$D_M(\mathbf{x}, \mu) = \sqrt{(\mathbf{x}-\mu)^T \Sigma^{-1}(\mathbf{x}-\mu)} \tag{2.86}$$

is called the Mahalanobis distance. Since the covariance matrix is positive defined, surfaces defined by constant distance of points $\mathbf{x}$ from the center μ are hyperellipsoids.

Therefore, this is a straightforward corollary that surfaces of constant value of the density function (2.84) are hyperellipsoids as well. Eigenvalues and eigenvectors of the covariance matrix describe the hyperellipsoid. Let us recall that eigenvalues and eigenvectors satisfy the equation $\Sigma \mathbf{u}_i = \lambda_i \mathbf{u}_i$. Eigenvectors correspond to principal axes; they define semiaxes, that is, line segments from the origin μ to the point on the surface. Eigenvalues define lengths of the corresponding semiaxes.

It is important to notice that since the covariance matrix Σ is symmetric, it has $M(M+1)/2$ independent elements. Of course, the mean vector μ is a vector of M independent elements. So then a complete estimation of the Gaussian density function requires $M(M+3)/2$ parameters whose values are to be determined.

If Σ is a diagonal matrix with diagonal elements σ_l^2, $l = 1, 2, \ldots, M$, then (2.84) becomes the product of one-dimensional Gaussian densities

$$p(\mathbf{x}) = \prod_{l=1}^{M} \frac{1}{\sqrt{2\pi\sigma_l^2}} \exp\left(-\frac{(x_l - \mu_l)^2}{2\sigma_l^2}\right) \tag{2.87}$$

where $\mathbf{x} = [x_1, x_2, \ldots, x_M]^T$. So then, one-dimensional probability distributions are independent. In this case, estimation of a density function requires determination of $2M$ parameters. Of course, the simplest case $\Sigma = \sigma^2 \mathbf{I}$, where $\mathbf{I}$ is the identity matrix, turns (2.84) to

$$p(\mathbf{x}) = \frac{1}{(2\pi\sigma^2)^{M/2}} \prod_{l=1}^{M} \exp\left(-\frac{(x_l - \mu_l)}{2\sigma^2}\right) \tag{2.88}$$

and it requires $M + 1$ parameters to be determined.

Nonparametric Estimation: One-Dimensional Case

Nonparametric class conditional probability distribution is employed if no assumption about the type of density function can be applied. Let us discuss the simplest case—a histogram of one-dimensional feature with numeric values in the whole set of real numbers R. Of course, when we have $M > 1$ features, we cope with each feature separately. A direct method to estimate class conditional probability is just to employ subsets of the training set of patterns to construct an approximation of probability distribution. For instance, in order to form a histogram, which is a kind of approximation of density distribution, we need to divide the domain of this feature (a set of real numbers) into intervals. There are different methods to split the set of feature values, for instance, we may cluster feature values or split them according to a division of the set of real numbers to intervals. The set of real numbers can be split into intervals of the same length (this is called equilength split), or it can be split into intervals including the same or similar numbers of patterns (this is called equidense split). The equilength split is a simple one, though with comparably higher sensitivity to data deviations. The equidense method is similar, but it requires some technical details to be fixed. The equidense method is more reliable compared with the equilength split. Since our aim is to present an idea, we discuss the equilength method and its variation that has the characteristics of feature clustering. It is obvious that since training sets are finite, the range of feature values is finite, and so then only a finite number of intervals include one or more patterns.

In order to split a feature space R, we need to fix an origin point X^0 of the sequence of intervals and intervals' length h. All intervals have the same length and are called histogram bins,

$$\dots, \left(X^0 - 2h, X^0 - h\right], \left(X^0 - h, X^0\right], \left(X^0, X^0 + h\right], \left(X^0 + h, X^0 + 2h\right], \dots$$

and then we assume a distribution function to be fixed in each interval

$$p(x) = \frac{1}{hN} \sum_{i=1}^{N} \delta\left(X_i, \left(X^0 + kh, X^0 + (k+1)h\right]\right) \quad \text{for} \quad x \in \left(X^0 + kh, X^0 + (k+1)h\right]$$

$$(2.89)$$

where X_i is the feature value of the i-th pattern and the mapping δ is defined as follows

$$\delta(X, (a, b]) = \begin{cases} 1 & X \in (a, b] \\ 0 & X \notin (a, b] \end{cases} \tag{2.90}$$

that is, distribution function is fixed in a given interval (bin) and it is proportional to the number of patterns falling into this bin. Numbers of patterns in intervals are normalized by dividing by the cardinality of the training set of patterns and interval length. Note that this is a well-defined distribution function on the whole domain R

$$\int_R p(x)dx = \frac{1}{hN} \sum_{k=-\infty}^{+\infty} \sum_{i=1}^{N} \delta\left(X_i, \left(X^0 + kh, X^0 + (k+1)h\right]\right) = 1 \tag{2.91}$$

Simplicity of the direct density estimation with a histogram is its advantage. Once the distribution function has been constructed, we can discard source data and only preserve the origin point X^0, the length of intervals h, and the finite list of intervals, where the distribution function does not vanish. However, such a function is not smooth, that is, it is a piecewise constant function and it is discontinuous in a finite number of ending points of intervals.

An example of a histogram-based estimation of density estimation is shown in Figure 2.10. The example is based on the wine dataset and its first feature (cf. Section 2.4.1). The equilength split of the feature domain is employed to estimate density function according to (2.89).

As mentioned earlier, for an equilength split of feature space R, we need to fix an origin point X^0 of a sequence of intervals and intervals' length h. Therefore, the split depends on both parameters. A choice of the origin X^0 has a low impact on characteristics of density function, while the influence of the interval length h is important. Short intervals produce a density function reflecting details of a problem, but it is of low reliability for data fluctuations. In contrast, longer intervals reflect general tendencies and give a smoother and more reliable density (cf. Figure 2.10). It is worth stressing that (2.89) must be applied cautiously: decreasing the length of intervals h too much may result in values of the density function greater than 1.

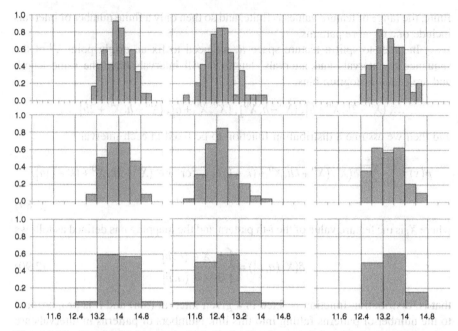

Figure 2.10 Density function estimation for the first feature of the wine dataset. We present estimation of density function for classes 1, 2, and 3 in the first, second, and third columns, respectively. Results concern interval length h equal to 0.2, 0.4, and 0.8 in the first, second, and third rows, respectively.

There is an interesting modification of a histogram-based estimation of a distribution function. Instead of finding a value $p(x)$ of density function in a fixed interval, the interval can be shifted to place its center at the point x

$$p(x) = \frac{1}{hN} \sum_{i=1}^{N} \delta(X_i, (x-h/2, x+h/2]) \qquad (2.92)$$

This function is like the aforementioned one based on a histogram (cf. (2.89)) that is, it is piecewise constant, it has nonzero values in a finite number of finite intervals, and it is discontinuous in a finite number of ending points of the intervals. Of course, it is easy to store full information defining such a function; it is just a (finite) list of intervals, where the density function does not vanish, with a function value in each such interval. However, in this case, intervals of function steadiness are not as regular as in the case of a histogram, that is, they differ in length. Also, there are as many intervals as the cardinality of the training set.

Another interesting modification of a histogram-based density estimation concerns NNs. We take the smallest interval centered in the point x, that is, $(x-h/2, x+h/2]$, such that it includes k patterns (more precisely, at least k patterns), say, k-NN interval, and then apply the following formula to compute density estimation

$$p(x) = p(x|O) = \frac{k}{k'h} \qquad (2.93)$$

where k' is the number of all training patterns included in the interval and k is the number of patterns from a set (class) O. Note that k is a parameter of the k-NN method and k' is the number of patterns included in an interval. k' is usually equal to k, but it might be greater than k in a case when more than one pattern lies at the end of the k-NN interval.

The function $\delta(X_i, (x-h/2, x+h/2])$ (cf. (2.92)) does not vanish if and only if $x-h/2 < X_i \le x+h/2$, that is, $-h/2 < X_i - x$ and $X_i - x \le h/2$. So then, we can rewrite (2.92) in the form

$$p(x) = \frac{1}{hN} \sum_{i=1}^{N} \kappa\left(\frac{x-X_i}{h}\right) \tag{2.94}$$

where X_i is feature value of the i-th pattern, and, in order to keep consistency with the previous assumption about intervals, after swapping strict and not strict inequalities, we get

$$\kappa(u) = \begin{cases} 1 & u \in (-1/2, 1/2] \\ 0 & u \notin (-1/2, 1/2] \end{cases}$$

Formula (2.94) is a special case of the following density estimation

$$p(x) = \frac{1}{hN} \sum_{i=1}^{N} K\left(\frac{x-X_i}{h}\right) \tag{2.95}$$

where K is called a kernel function.

Let us point out that the density function (as defined in (2.95)) is a linear combination of kernel functions (for different values of X_i). Since kernel κ is a rectangle function, then density function (defined in (2.94)) is discontinuous. We can regard a smooth kernel function K in order to get a smooth density function (2.95). A common example of such function is the Gaussian function

$$K(u) = \frac{1}{\sqrt{2\pi}} \exp\left(-\frac{u^2}{2}\right)$$

or its parameterized version

$$K(x) = \frac{1}{\sqrt{2\pi\sigma^2}} \exp\left(-\frac{(x-\mu)^2}{2\sigma^2}\right) \tag{2.96}$$

where parameters μ and σ play a similar role to X^0 and h, respectively, in a histogram-based density estimation. This kernel gives the following density function

$$p(x) = \frac{1}{\sqrt{2\pi}hN} \sum_{i=1}^{N} \exp\left(-\frac{(x-X_i)^2}{2h^2}\right) \tag{2.97}$$

An example of density estimation is given in Figure 2.10. As in the histogram-based density estimation, the example is based on the wine dataset and its first feature (cf. Section 2.4.1). Estimations employed density (as defined in (2.97)) for three classes and for intervals of three different lengths. Density functions are smooth

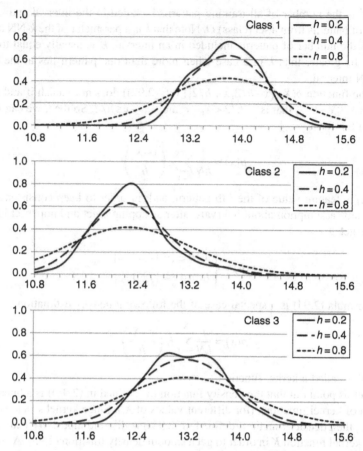

Figure 2.11 Estimation of density function with the Gaussian kernel function for the first feature of the wine dataset. We present the estimation of the density function for classes 1, 2, and 3 and for interval length (h) equal to 0.2, 0.4, and 0.8, respectively.

since the kernel function is smooth and the density is a linear combination of kernel functions. Again, short intervals produce density function reflecting details of the problem, but it is of low reliability to data variations. In contrast, longer intervals impose generality and give a smoother density more reliable to data variability (cf. Figure 2.11).

Nonparametric Estimation: Multidimensional Case
So far we discussed density estimation based on a histogram of one dimension feature. Roughly speaking, generalization of a one-dimensional case to a multidimensional one, say, M-dimensional, is straightforward: we shall split feature space R^M into M-dimensional bins (hypercubes or hyperrectangles), which will play the role of bins D. Next, we count patterns falling in each bin and then, by analogy to (2.89), normalize these numbers by dividing them by the volume of a bin and the total number of patterns.

Generalizing the aforementioned discussion, let us assume that we have a density function $p(\mathbf{x})$ defined in feature space R^M. Then, the probability that a pattern falls into a decision region $D \subset R^M$ is

$$P(D) = \int_D p(\mathbf{x})d\mathbf{x} \qquad (2.98)$$

If the region D is closed and the density function $p(\mathbf{x})$ is smooth enough (sufficiently if continuously differentiable) in D, then due to the mean value theorem there exists a point $\mathbf{x} \in D$ such that $P(D) = p(\mathbf{x}) \int_D d\mathbf{x} = p(\mathbf{x}) \cdot V(D)$. If the region D is small, which is manifested with its small volume $V(D) = \int_D d\mathbf{x}$, then we can assume density $p(\mathbf{x})$ to be approximately constant in D. Therefore, we obtain the following density approximation

$$p(\mathbf{x}) = \frac{1}{V(D) \cdot N} \sum_{i=1}^{N} \delta(\mathbf{X}_i, D) \quad \text{for some } \mathbf{x} \in D \qquad (2.99)$$

where, by analogy to (2.90), we come to

$$\delta(\mathbf{X}, D) = \begin{cases} 1 & \mathbf{X} \in D \\ 0 & \mathbf{X} \notin D \end{cases} \qquad (2.100)$$

It is reasonable to estimate the region D with a simple geometric figure. We can take an ellipsoid or even a sphere; we can also choose a hyperrectangle or just a hypercube. Of course, we visualize that patterns are located inside a region bounded by such a figure (or figures). Let us also make a side note that both a sphere and a hypercube are spheres in the respective metrics: Euclidean and Chebyshev (cf. (2.10)). We can also consider spheres in other metrics, for instance, in the Manhattan metric.

However, in order to get a sensible density function, we need a much greater number of patterns to construct a model. Otherwise, the density function will vanish in most bins. For instance, the (one-dimensional, $M = 1$) unit interval split into four subintervals gives four histogram's bins. The unit hypercube in five-dimensional setting, $M = 5$, split into four subintervals in each dimension, gives $4^5 = 1024$ histogram bins. In the first case, we need at least four patterns to guarantee nonzero density function in each bin. In the second case, we need at least $4^5 = 1024$ patterns to achieve the same goal. As we see, the need for training patterns grows exponentially. This exponential growth, called the *curse of dimensionality*, makes this approach to density estimation impractical.

Kernel Functions: Multidimensional Case

The one-dimensional Gaussian function regarded as an example of a continuous kernel function is given in (2.96). Now, let us recall (2.84). Here, we are generalizing the Gaussian density function to M-dimensional case

$$p(\mathbf{x}) = \frac{1}{\sqrt{(2\pi)^M |\Sigma|}} \exp\left(-\frac{1}{2}(\mathbf{x}-\mu)^T \Sigma^{-1}(\mathbf{x}-\mu)\right) \tag{2.101}$$

where $\mathbf{x}$ and μ are points (vectors) in R^M, μ is an M-dimensional mean vector, Σ is an $M \times M$ *covariance matrix*, it corresponds to variance in one-dimensional density, $|\Sigma|$ is the determinant of covariance matrix, and Σ^{-1} is the inverse of covariance matrix.

Since parameters $P(\mathbf{x}|O_l)$ and $P(O_l)$ of the Bayes classifier (see (2.64)) are usually not known, they should be determined on the basis of a training set. Here we discuss the case of the Gaussian density function, specified in (2.101), as class conditional probability. Hence, we need to estimate the center of the Gaussian function μ and covariance matrix Σ based on training patterns from class O_l. As we have already shown in (2.81), *a priori* probability is just a proportion of the number of patterns from a given class to the number of all patterns in the training set. Estimation of Gaussian function parameters, based on a method such as least squares or maximal reliability, is given by the following formulas

$$\mu = \frac{1}{N}\sum_{i=1}^{N}\mathbf{X}_i$$

and $\tag{2.102}$

$$\Sigma = \frac{1}{N-1}\sum_{i=1}^{N}(\mathbf{X}_i-\mu)(\mathbf{X}_i-\mu)^T$$

where $\mathbf{X}_i$ is the i-th training pattern from the class O_l and N is the number of training patterns in this class. It is worth recalling that the covariance matrix Σ is symmetric. Therefore, it has $M(M+1)/2$ independent elements. Of course, the center of the Gaussian function μ is a vector of M independent elements. Finally, the complete estimation of the Gaussian density function requires $M(M+3)/2$ parameters to be determined. If Σ is a diagonal matrix with diagonal elements σ_l^2, $l=1,2,...,M$, then (2.101) becomes a product of one-dimensional Gaussian densities

$$p(\mathbf{x}) = \prod_{l=1}^{M}\frac{1}{\sqrt{2\pi\sigma_l^2}}\exp\left(-\frac{(x_l-\mu_l)^2}{2\sigma_l^2}\right) \tag{2.103}$$

where $x=[x_1,x_2,...,x_M]^T$. As such, one-dimensional probability distributions are independent. In this case, estimation of a density function requires determination of $2M$ parameters. Of course, the simplest case $\Sigma = \sigma^2\mathbf{I}$, where $\mathbf{I}$ is the identity matrix, turns (2.103) to

$$p(\mathbf{x}) = \frac{1}{(2\pi\sigma^2)^{C/2}}\prod_{l=1}^{C}\exp\left(-\frac{(x_l-\mu_l)}{2\sigma^2}\right) \tag{2.104}$$

and it requires $M+1$ parameters to be determined. Therefore, a given training set of patterns can be used for each dimension separately curing problems imposed by the curse of dimensionality.

Naïve Bayes Classifier

In order to overcome the course of dimensionality, it is often assumed that coordinates of an M-dimensional density function are statistically independent. This assumption allows a decomposition of an M-dimensional probability distribution $p: R^M \rightarrow [0, 1]$ to a product of one-dimensional distributions $p_l: R \rightarrow [0, 1]$, $l = 1, 2, \ldots, M$

$$p(\mathbf{x}) = \prod_{l=1}^{M} p_l(x_l) \quad \mathbf{x} = (x_1, x_1, \ldots, x_M)^T \tag{2.105}$$

and then we can approximate density function in each dimension separately and, in this way, avoid the course of dimensionality. However, statistical independence of one-dimensional probability distributions is hardly ever satisfied, so then such decomposition would be rather difficult. Anyway, in practice, the requirement for statistical independence is ignored, and decomposition outlined in (2.105) is utilized with success. The Bayes classifier employing this decomposition without the requirement of independence is called the naïve Bayes classifier.

Having in mind that (2.101) is used as class conditional probability $p(\mathbf{x}|O_i)$, we can employ the natural logarithm to (2.68) and then drop constant term $C/2 \cdot \ln(2\pi)$ in order to get a simpler form of (2.71) for class indexes $l = 1, 2, \ldots, M$

$$d_l(\mathbf{x}) = -\frac{1}{2}(\mathbf{x} - \mu_l)^T \Sigma_l^{-1}(\mathbf{x} - \mu_l) - \frac{1}{2}\ln|\Sigma_l| + \ln P(O_l) \tag{2.106}$$

Note that the natural logarithm applied to inequalities as defined in (2.71) does not change decision regions. Therefore, from (2.106) we conclude that separation surfaces are quadratic hypersurfaces (ellipsoids, paraboloids, hyperboloids).

Bayesian classification is a thriving subject. For more detailed discussion, one may consult the textbooks by Hastie *et al.* (2009, chapter 8), Mitchell (1997, chapter 6), Duda *et al.* (2001, chapter 3), and Bishop (2006, chapter 2).

2.7   CONCLUSIONS

Substantial efforts in the pattern recognition area have resulted in the formulation of a wide range of classification algorithms. The variety of real-world applications called for new solutions and this quest is still ongoing. There are an impressive number of other well-established approaches, such as neural networks, sophisticated nonparametric probabilistic approaches, and so on, that were not covered in this chapter. In addition, classification itself brings to life a plethora of other topics directly related to pattern analysis and classification. Selected problems appearing as a consequence and accompanying a *standard* pattern recognition task are covered in the subsequent chapters of this book.

REFERENCES

C. M. Bishop, *Pattern Recognition and Machine Learning*, New York, Springer, 2006.
L. Breiman, Bagging predictors, *Machine Learning* 24(2), 1996, 123–140.
L. Breiman, Random forests, *Machine Learning* 45(1), 2001, 5–32.

L. Breiman, J. H. Friedman, R. A. Olshen, and C. J. Stone: *CART: Classification and Regression Trees*, Belmont, CA, Wadsworth, 1983.

M. P. S. Brown, W. N. Grundy, D. Lin, N. Cristianini, C. W. Sugnet, T. S. Furey, M. Ares, and D. Haussler, Knowledge-based analysis of microarray gene expression data by using support vector machines, *Proceedings of The National Academy of Sciences of The United States of America* 97(1), 2000, 262–267.

R. Burbidge, M. Trotter, B. Buxton, and S. Holden, Drug design by machine learning: Support vector machines for pharmaceutical data analysis, *Computers & Chemistry* 26(1), 2001, 5–14.

C. J. C. Burges, A tutorial on support vector machines for pattern recognition, *Data Mining and Knowledge Discovery* 2(2), 1998, 121–167.

C. Cortes and V. Vapnik, Support-vector networks, *Machine Learning* 20(3), 1995, 273–297.

N. Dehak, P. J. Kenny, R. Dehak, P. Dumouchel, and P. Ouellet, Front-end factor analysis for speaker verification, *IEEE Transactions on Audio Speech and Language Processing* 19(4), 2011, 788–798.

T. Denoeux, A K-nearest neighbor classification rule-based on Dempster-Shafer theory, *IEEE Transactions on Systems Man and Cybernetics* 25(5), 1995, 804–813.

R. Diaz-Uriarte and S. A. de Andres, Gene selection and classification of microarray data using random forest, *BMC Bioinformatics* 7, 2006, article number 3.

H. Drucker, D. H. Wu, and V. Vapnik, Support vector machines for spam categorization, *IEEE Transactions on Neural Networks* 10(5), 1999, 1048–1054.

R. O. Duda, P. E. Hart, and D. G. Stork, *Pattern Classification*, 2nd ed., New York, John Wiley & Sons, Inc., 2001.

J. Elith, J. R. Leathwick, and T. Hastie, A working guide to boosted regression trees, *Journal of Animal Ecology* 77(4), 2008, 802–813.

E. Frank, I. H. Witten, and M. A. Hall, *Data Mining: Practical Machine Learning Tools and Techniques*, 3rd ed., London, Morgan Kaufmann Series in Data Management Systems, 2001.

Y. Freund and R. Schapire, A decision-theoretic generalization of on-line learning and an application to boosting, *Journal of Computer and System Sciences* 55(1), 1997, 119–139.

J. Friedman, On bias, variance, 0/1-loss, and the curse-of-dimensionality, *Data Mining and Knowledge Discovery* 1(1), 1997, 55–77.

T. S. Furey, N. Cristianini, N. Duffy, D. W. Bednarski, M. Schummer, and D. Haussler, Support vector machine classification and validation of cancer tissue samples using microarray expression data, *Bioinformatics* 16(10), 2000, 906–914.

R. Genuer, J. M. Poggi, and C. Tuleau-Malot, Variable selection using random forests, *Pattern Recognition Letters* 31(14), 2010, 2225–2236.

J. Ham, Y. C. Chen, M. M. Crawford, and J. Ghosh, Investigation of the random forest framework for classification of hyperspectral data, *IEEE Transactions on Geoscience and Remote Sensing* 43(3), 2005, 492–501.

T. Hastie, R. Tibshirani, and J. Friedman, *The Elements of Statistical Learning*, New York, Springer, 2009.

T. Hothorn, K. Hornik, and A. Zeileis, Unbiased recursive partitioning: A conditional inference framework. *Journal of Computational and Graphical Statistics*, 15(3), 2006, 651–674.

Q. Hu, D. Yu, and Z. Me, Neighborhood classifiers, *Expert Systems with Applications* 34(2), 2008, 866–876.

C. Huang, L. S. Davis, and J. R. G. Townshend, An assessment of support vector machines for land cover classification, *International Journal of Remote Sensing* 23(4), 2002, 725–749.

J. Koronacki and J. Ćwik, *Statistical Machine Learning Systems (in Polish)*, Warszawa, WNT, 2005.

T. S. Lim, W. Y. Loh, and Y. S. Shih, A comparison of prediction accuracy, complexity, and training time of thirty-three old and new classification algorithms, *Machine Learning* 40(3), 2000, 203–228.

T. Mitchell, *Machine Learning*, New York, McGraw Hill, 1997.

S. K. Murthy, Automatic construction of decision trees from data: A multi-disciplinary survey, *Data Mining and Knowledge Discovery* 2(4), 1998, 345–389.

E. Osuna, R. Freund, and F. Girosi, Training support vector machines: an application to face detection. In: *Proceedings of 1997 IEEE Computer Society Conference on Computer Vision and Pattern Recognition*, San Juan, Puerto Rico, 1997, 130–136.

M. Pal, Random forest classifier for remote sensing classification, *International Journal of Remote Sensing* 26(1), 2005, 217–222.

A. M. Prasad, L. R. Iverson, and A. Liaw, Newer classification and regression tree techniques: Bagging and random forests for ecological prediction, *Ecosystems* 9(2), 2006, 181–199.

J.R. Quinlan, Induction of decision trees, *Machine Learning* 1(1), 1986, 81–106.

S. R. Safavian and D. Landgrebe, A survey of decision tree classifier methodology, *IEEE Transactions on Systems Man and Cybernetics* 21(3), 1991, 660–674.

B. Scholkopf, K. K. Sung, C. J. C. Burges, F. Girosi, P. Niyogi, T. Poggio, and V. Vapnik, Comparing support vector machines with Gaussian kernels to radial basis function classifiers, *IEEE Transactions on Signal Processing* 45(11), 1997, 2758–2765.

R. E. Shapire, The strength of weak learnability, *Machine Learning* 5, 1990, 197–227.

J. A. Sonquist and J. N. Morgan, *The Detection of Interaction Effects*, Ann Arbor, MI, Survey Research Center, Institute for Social Research, The University of Michigan, 1964.

K. Stapor, *Methods of Objects Classification in Computer Vision (in Polish)*, Warszawa, WN PWN, 2011.

C. Strobl, A. L. Boulesteix, A. Zeileis, and T. Hothorn, Bias in random forest variable importance measures: Illustrations, sources and a solution, *BMC Bioinformatics* 8, 2007, article number 25.

V. Svetnik, A. Liaw, C. Tong, J. C. Culberson, R. P. Sheridan, and B. P. Feuston, Random forest: A classification and regression tool for compound classification and QSAR modeling, *Journal of Chemical Information and Computer Sciences* 43(6), 2003, 1947–1958.

X. Y. Tan, S. C. Chen, Z. H. Zhou, and F. Y. Zhang, Recognizing partially occluded, expression variant faces from single training image per person with SOM and soft k-NN ensemble, *IEEE Transactions on Neural Networks* 16(4), 2005, 875–886.

UCI Machine Learning Repository, https://archive.ics.uci.edu/ml/index.php (accessed October 11, 2017).

P. E. Utgoff, Incremental induction of decision trees, *Machine Learning* 4(2), 1989, 161–186.

A. R. Webb and K. D. Copsey, *Statistical Pattern Recognition*, 3rd ed., New York, John Wiley & Sons, Inc., 2001.

Wine Dataset, https://archive.ics.uci.edu/ml/datasets/Wine (accessed October 11, 2017).

B. L. Wu, T. Abbott, D. Fishman, W. McMurray, G. Mor, K. Stone, D. Ward, K. Williams, and H. Y. Zhao, Comparison of statistical methods for classification of ovarian cancer using mass spectrometry data, *Bioinformatics* 19(13), 2003, 1636–1643.

L. M. Zouhal and T. Denoeux, An evidence-theoretic k-NN rule with parameter optimization, *IEEE Transactions on Systems Man and Cybernetics, Part C-Applications and Reviews* 28 (2), 1998, 263–271.

CHAPTER 3

CLASSIFICATION WITH REJECTION PROBLEM FORMULATION AND AN OVERVIEW

Pattern recognition is one of the main topics in the field of computer science exhibiting an invaluable synergy of theoretical studies and practical ones. For decades, it has been an area of intensive, purely theoretical research inspired by practical needs. The developments in this domain have often been presented in prestigious scientific journals. There are numerous applications; some representative examples include recognition of printed text, manuscripts, music notation, biometric features, voice, recorded music, medical signals, and images, visual object tracking, control and decision support systems, and so on.

In a standard formulation, pattern recognition is concerned with a separation of a set of patterns into subsets of patterns that belong to the same class, that is, every pattern is classified (assigned) to one set, which is included into or equal to one class. The set of classes is either fixed a priori (then it refers to the problem of supervised learning) or determined at the stage of recognizer construction (an unsupervised problem). In both cases, it is assumed that each classified pattern belongs to one of the desired, given classes (either known a priori or determined). However, in practice, this assumption is often too optimistic. It has been observed in important practical applications that in some recognition tasks we are exposed not only to patterns coming from an expected, fixed set of classes but also undesired ones, that is, the patterns not belonging to proper classes. Let us mention the problem of contaminated datasets. Contaminated datasets contain not only proper patterns but also "garbage" patterns that appear there due to some gross error.

The chapter offers a detailed formulation of the problem of classification with rejection, delivers a number of illustrative examples, and elaborates on existing studies carried out in this area.

Pattern Recognition: A Quality of Data Perspective, First Edition. Władysław Homenda and Witold Pedrycz.
© 2018 John Wiley & Sons, Inc. Published 2018 by John Wiley & Sons, Inc.

3.1   CONCEPTS

Undesired patterns, namely, those not belonging to given classes, are usually not known a priori. In other words, they are not known at the stage of the construction of the recognizer, and therefore we cannot assume that they create their own distinct and consistent class(es) and cannot be used at this stage. Otherwise, if samples of patterns not belonging to desired classes are available at the stage of recognizer construction, such a problem can be turned into a standard pattern recognition problem. Having a set of extra patterns, apart from patterns belonging to desired classes, we can group them into one or more classes and then construct a recognizer for all desired and extra classes of patterns. Therefore, in this scenario, patterns of desired classes together with patterns of an extra class known at the stage of recognizer construction will give rise to a multiclass dataset. This set can be used for recognizer construction, and in this perspective there is no difference between these two types of patterns. However, this approach limits the usefulness of the model. In many practical applications, we have very little knowledge about the source and characteristics of extra patterns. Hence, we rather be prepared to receive literally anything as the input of the classifier and reject any extra pattern, not only a pattern that is in some sense similar to other extra patterns that were used at the stage of model construction. It is worth stressing that we will distinguish between patterns known at the stage of recognizer construction and patterns unknown at this stage. We assume that extra patterns are unknown at the stage of recognizer construction and they may come from various sources. Furthermore they may be not similar to one another in any sense.

3.1.1   Native Versus Foreign Patterns

To distinguish between these two types of patterns, the following terms are introduced and used:

- *Native* patterns are all patterns that are known at the stage of recognizer construction. Native patterns are divided into classes, which must be recognizable by the model (classifier).
- *Foreign* patterns are all other patterns, that is, patterns positioned outside of native classes (i.e. sets) determined at the stage of recognizer construction.

Therefore, in the outlined context, an intuitively straightforward and direct method for dealing with foreign patterns is to furnish a standard pattern recognizer with a tool that will allow us to reject foreign patterns. Let us reiterate that we assume that formation of such model has to be completed without knowledge about foreign patterns. Let us reiterate that only native patterns can be used to train a recognizing/rejecting model. The mechanism formed in this manner should be able to:

- Classify native patterns
- Reject foreign patterns

Let us stress that the notation introduced previously suggests that we design a model for solving a supervised classification task with an additional pattern rejection option. This discussion can be easily translated to an unsupervised processing scenario, in which classes are clusters and the action of classification is replaced with clustering. In order to assume a consistent terminology, we will keep using the terms *classification* and *class*, but it is only a matter of nomenclature as the proposed models and ideas are also valid for unsupervised learning. In a classifying/rejecting model, we can distinguish two distinct actions: recognition and rejection. However, it should be noted that the order of taken actions can vary: first recognition then rejection, first rejection then recognition, or we can split a problem into some subproblems and employ recognition and/or rejection in some sort of a multilayered processing scenario. What is important and what combines those approaches together is that as a conclusion, we expect to separate foreign patterns from native ones.

Due to the aforementioned assumption that foreign patterns are not known at the stage of model (classifier) construction, standard recognition methods cannot be directly applied, and we need nonstandard, more sophisticated methods. One may ask why we consider identification of foreign patterns. The answer is that foreign patterns always negatively influence the quality of classification, because they do not belong to *any* class. This situation is illustrated in Figure 3.1. In Figure 3.1a we see a classifier constructed based on native patterns only: here rectangles, circles, stars, …, triangles play role of native patterns. Then, a new dataset comprising native and foreign patterns (being denoted as question marks) is presented as the input of the constructed classifier. All foreign patterns will be incorrectly classified to classes of native patterns since there is no other choice, as one can see in Figure 3.1b. Naturally, in practice classifiers are not fault-free, and therefore some native patterns can also be assigned to an incorrect class.

In order to avoid incorrect classification of a foreign pattern, a classifier should be furnished with a mechanism capable to reject it. In Figure 3.2, we show an idea of a classifier with rejecting option. We would like to underline that this construct is not just a classifier extended to distinguish an extra class of patterns. As we have mentioned, the characteristics of foreign patterns are not known at a stage of classifier construction. Therefore, rejecting mechanism cannot be set up by invoking a potential extra class.

In spite of the obvious importance of the issue of foreign pattern rejection, we have found a relatively modest number of studies focused on this problem. Papers devoted to practical applications of pattern recognition methods ignore the problem of foreign patterns, which may come from insufficient theoretical background studies on this subject and limited abilities of existing rejection methods.

There are significant exceptions that show that the rejection problem cannot be disregarded (De Stefano *et al.*, 2000; Scholkopf *et al.*, 2001; Pillai *et al.*, 2011).

Theoretical foundations of rejection were formulated by Chow (1970). He proposed the now so-called Chow rejection rule. Chow noticed that in many classification problems, some patterns are very difficult to process and it would be safer to reject them than to classify them. In this sense, Chow's research originated from a different position and assumption than the one put forward here. Patterns that he rejects are native ones. In contrast, we reject foreign patterns. What is important is that Chow presented the idea of rejection. In his studies he distinguished an indecision class. Notably, background assumptions behind Chow's research are consistent with our

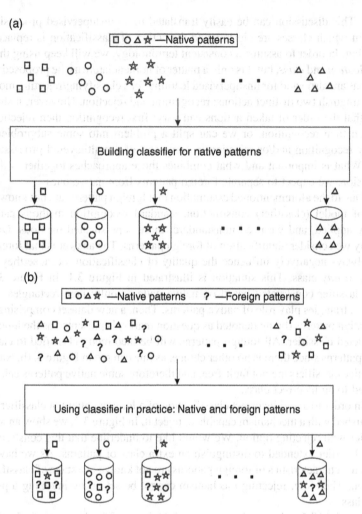

Figure 3.1 An idea of classification without rejection. (a) A classifier is constructed on the basis of a learning set of native patterns. (b) Constructed classifier is used in practice. Native and foreign patterns are presented to the classifier. Foreign patterns do not belong to any class, but anyway they have to be classified to a native class, which decreases the overall quality of data processing. In addition, we see that a few native patterns are classified to incorrect classes.

definition of a recognizing/rejecting mechanism: the model construction is based on native patterns only. In Chow's studies a classification mechanism responds with a decision about class membership (in the form of probability). He proposed to introduce a rejection rule that does not allow assigning class label if a decision is uncertain (probability is low). In turn, if no native class label was assigned to a given pattern, it gets rejected. Rejection threshold is experimentally selected for a given classification problem, and it is formulated as an optimization task as the rule minimizes the error

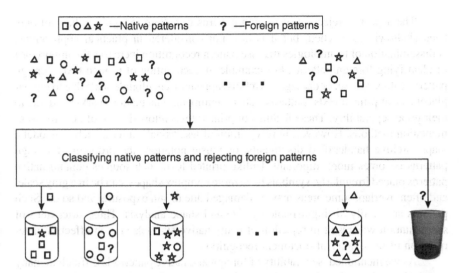

Figure 3.2 An idea of classification with rejection. Now most of foreign patterns are rejected. Please compare this sketch with Figure 3.1b that depicts analogous classification mechanism but without rejection, where foreign patterns are incorrectly classified to native classes.

rate. Chow's idea was to reject such patterns that would be incorrectly classified. A generalization to a multiclass issue was presented by Ha (1997). Roughly speaking, in this study, Chow's rule is determined for all pairs of classes separated by a discrimination plane. There are also distinct solutions for a linear multiclassification task (Fumera and Roli, 2004). However, the majority of theoretical works were limited to a binary case (Mascarilla and Frélicot, 2002; Xie *et al.*, 2006). Chow's rule was redefined to suit practical applications as a rejection rule for support vector machine (SVM) classifiers (Li and Sethi, 2006). The distance function can be also used to define a rejection rule based on dissimilarity between a class pattern and a recognized pattern (Lou *et al.*, 1999). There are a few more attempts based on neural networks. For details refer to Ishibuchi *et al.* (1999) and error-correcting output coding (ECOC) classification systems (Simeone *et al.*, 2012; Dieterich and Bakiri, 1995).

Among other important studies carried out in the area of pattern rejection, one may turn to approaches where classification is performed on the basis of scores. Prominent examples of such methods are fuzzy approaches to classification, in which class membership is expressed in terms of fuzzy set membership functions (degrees of membership). A natural consequence of this approach is that a classification result is not expressed in binary terms (belongs to a class/does not belong to a class); instead it is expressed as a number from [0,1] interval that could be interpreted as a certainty score (membership degree). The general idea behind scores (in a context wider than fuzzy classification methods) is that they determine strength of confidence that a certain pattern belongs to a certain class. If a pattern comes with low scores for all proper classes, this indicates that it may be treated as garbage. Several researchers have investigated the described approach, for example, Elad *et al.* (2001), Koerich (2004), Meel *et al.* (2015), and Homenda *et al.* (2014).

The aspect of rejection of foreign patterns in pattern recognition has not only been shallowly researched, but it is also not considered in practical applications. A dissemination of technologies that use pattern recognition increases the importance of identifying foreign patterns. For example, in recognition of printed texts, foreign patterns (blots, grease, or damaged symbols) appear in a negligible scale due to regular placement of printed texts' patterns (letters, numbers, punctuation marks) and due to their good separability. These features of printed texts allow the use of effective segmentation methods. However, in recognition of less trivial sources, such as geodetic maps, archive handcrafted documents, or music notation, the problem of foreign patterns becomes more important. Unlike printed text, such sources contain native patterns placed irregularly, symbols have more complex shapes, can be in a gray scale, and often overlap, some areas may be damaged due to sun exposure, and so on. Such patterns are hardly distinguishable by size and shape analysis. Thus, strict rules of segmentation will result in rejection of many native symbols and, in effect, in deterioration of the quality of document recognition.

As we mentioned, separability of foreign and native patterns is difficult in many practical problems. In addition, in practice, segmentation of source materials subjected to recognition is often very difficult and ineffective. This is why segmentation criteria for complicated recognition problems (like cutting text from natural scenes) must be more tolerant than for easy cases, as it is, for instance, in printed text processing. If a segmentation process is strict, it may result in rejecting native patterns along with the foreign ones. In consequence, a lenient segmentation will produce many foreign patterns, which are then subjected to the recognition process. In such case, rejecting foreign patterns gets high priority, as input data is highly contaminated and we need to eliminate garbage in order to obtain a sensible outcome. The problem of foreign patterns analysis is highly important, for instance, in domains such as analysis of medical signals and images, recognition of geodetic maps, or recognition of music notations (both printed and handwritten).

It is worth pointing out that foreign patterns should not be confused with the so-called *outliers*. Let us recall that an outlier is a native pattern belonging to one of the native classes, but highly dissimilar from the majority of data in this class. In general, patterns that are distant from other patterns are treated as outliers. The meaning of terms *distant* and *other patterns* is not objective and certainly depends on the case. For the sake of this discussion, let us assume that the term *distant* refers to the standard meaning of this word, that is, to a big distance between two patterns. In the case of a classification problem, distance between the two patterns is understood as a distance between their two features vectors (cf. Section 2.2). The term *other patterns* in the context of outliers may refer to patterns that are concentrated in a relatively compact area, but not necessarily made of patterns coming from the same class. Naturally, other definitions may be applied. For instance, referring to the discussion on clustering in Chapter 8, we can distinguish distant patterns as those that are far away from the center of a cluster. In this case the cluster center is a representative of a larger group of data points, and distance measure informs about the nearness of a certain pattern to the center. This interpretation of distant patterns slightly differs from the standard understanding of the term outlier. An alternative method for discriminating *regular* patterns from outlying ones is to involve some measure of pattern density. In such a context, for a given training dataset forming scatter in the feature space, we may assume that dense areas of patterns are reserved for native data. We can involve some geometrical methods to formally define regions occupied densely by native patterns. In consequence, by an elimination rule, we can assume that a complement of the devised regions of regular patterns in the feature space is where outlying/foreign patterns reside. This concept bears promising research potential, and we investigate closely the idea of geometrical rejection in Chapter 5.

The problem with outliers is that we wish to recognize them, but it is very difficult to do so due to the dissimilarity mentioned previously. Therefore, in some approaches to classification, including Chow's works, we see a suggestion to reject outliers, because classifying correctly such patterns is usually only a matter of luck. The borderline is that outliers are native patterns, and we would wish to have them correctly classified. In contrast, there is no right classification decision when it comes to foreign patterns—the only decision here is to reject.

Another property of foreign patterns is that we do not know their characteristics and we should be prepared to reject any foreign pattern. Hence, we cannot assume that foreign patterns create distinct classes. From the perspective of our approach, we do not see any point in generating a *substitute* foreign class that will be added to the training set and used to reject patterns that will be classified to this substitute class. Such approach is merely an extended classification scheme and does not truly reflect the problem of foreign patterns.

Lastly, let us stress that what causes another problem with foreign pattern rejection is that foreign patterns may have characteristics similar to the native ones. This occurs frequently when recognition concerns a problem in which an object could be decomposed into smaller parts. For instance, in handwritten digits recognition, we may encounter patterns in the form of text in the Latin alphabet, for example, author's sidenotes. A segmentation procedure will extract letters such as handwritten "b," "l" or "O," "B" that look very similar to six, one, zero, and eight, respectively. Rejection of such examples can be very tricky.

3.2 THE CONCEPT OF REJECTING ARCHITECTURES

In this section we would like to discuss rejecting foreign patterns based on adapted standard recognizers. From the perspective of flow of processing, we can distinguish between three different architectures, as discussed in Homenda *et al.* (2015). Interpreting classification of native patterns and rejection of foreign patterns as a series of actions, the process of rejection of foreign patterns can be performed in three ways:

1. Rejection prior to classification of native patterns
2. Rejection after classification of native patterns
3. Rejection simultaneously with classification of native patterns

In the first case, the action of foreign pattern rejection is executed in the first place. Only after rejection can we employ further pattern processing, usually classification. Since rejection occurred first, we can assume that classification is performed

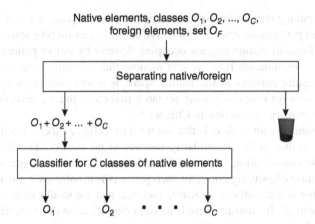

Figure 3.3 The architecture of global rejection. Native and foreign patterns are separated prior to classification of native patterns. Since foreign patterns are eliminated at the first step of recognition with rejection, the set of patterns subjected to classification is assumed to include native patterns only.

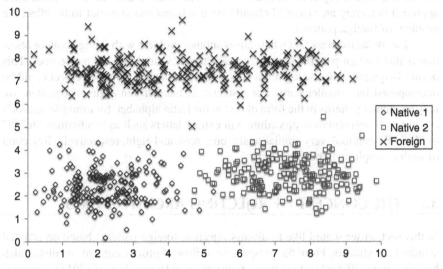

Figure 3.4 An example of a dataset of native patterns contaminated with foreign patterns suitable for global rejection. Both sets of native patterns are easily separable from the set of foreign patterns.

only for native patterns. This scheme is what we call *global rejecting architecture*. It is presented in Figure 3.3.

In Figure 3.4, we illustrate an example of a dataset containing patterns that belong to two classes contaminated with foreign patterns. This dataset is intuitively very suitable to be processed with a global rejecting architecture. In this case, both classes of native patterns are easily separable from foreign patterns. The underlying assumption is that we are able to separate native from foreign patterns in the feature space. Naturally, real-world datasets are not as straightforward as the dataset displayed in Figure 3.4. Usually, native and foreign patterns overlap and they do not form such regular and dense clusters in the feature space. Let us stress that we do not know the characteristics of foreign patterns, but we only know the native ones. In addition, we have only a learning set of native patterns; in real-world experiments we never have information about all native patterns. From this perspective there are two factors that make the task of foreign pattern rejection so difficult.

The second processing scenario is called *local rejecting architecture*. In this case all patterns, native and foreign, are first subjected to classification. As a result, we obtain C subsets of patterns, split into C native classes that we have in a given dataset. At this point, each foreign pattern is (incorrectly) assigned to some class of native patterns. Next, we execute rejection action. Since the initial dataset is already split into classes, rejection has to be performed for each formed subset. Figure 3.5 presents a diagram of pattern recognition with local rejection.

Local rejecting is suitable for problems where foreign patterns and native ones are not easily separable (cf. Figure 3.6). Local rejecting incorporates the idea of divide and conquer. The first action splits the dataset into smaller subsets. Rejection is executed for smaller subsets, in which we may have higher chances for success. The intuition behind this claim is that in smaller subsets native patterns are supposed to be more similar than when this group contains all native patterns. The assumed coherence of native patterns in smaller subsets makes it easier to distinguish foreign patterns.

Global and local rejection architectures can be simply implemented by supplying a standard classifier of native patterns with a tool for rejecting foreign patterns. The task of classification, no matter if executed first as it is in the local rejection architecture or second as it is in the global rejection architecture, is performed for a known set of C classes of native patterns, and naturally, it is trained on a training dataset of

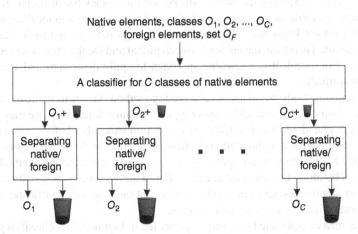

Figure 3.5   The architecture of local rejection. Sets of native and foreign patterns are subjected to classification prior to foreign pattern rejection.

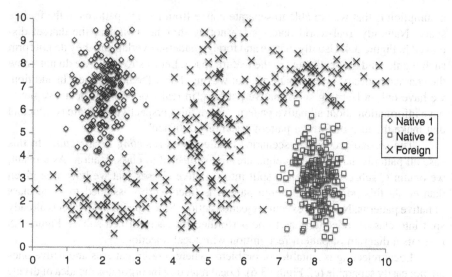

Figure 3.6 An example of a dataset consisting of native and foreign patterns suitable to be processed using the local rejection architecture. Native and foreign patterns are hardly separable. Hence, local rejecting would be more effective than global rejecting.

native patterns. In fact, the same classifier can be used in both architectures, as the difference here is not in the way this classifier is built but in the order in which it is being used. We can employ one of the well-known classification methods, for instance, k-NN, SVMs, random forests (RF), and so on. Alternatively, we can employ an unsupervised data processing method, for example, k-means, DBSCAN, hierarchical clustering, and so on. The use of a particular algorithm depends on the processing objective. For the sake of narration, we focus here on supervised classification, but it is only the matter of narration to switch to an unsupervised case.

Obviously, example datasets could be seen as somewhat artificial. Real problems rarely concern such simple, easy-to-separate data. It is worth recalling here that we usually do not know the characteristics of foreign patterns and their relations to native patterns. Therefore, the choice between global and local architectures is neither obvious nor univocal. It is a sensible approach to apply them both and verify their quality empirically.

Combining both global and local rejecting methods brings us to the third type of rejecting scheme called *embedded rejecting architecture*. Intuitively, we may say that while in the global and local architectures, rejection was either before or after classification, for embedded architecture rejection occurs during the process of classification. Embedded architecture requires an integration of the rejecting method with a classifier. Therefore, the structure of a classifier of native patterns is usually upgraded in order to identify and separate foreign patterns at the stage of classification, that is, inside a classification mechanism's structure.

The idea of embedded rejecting is presented in Figure 3.7. A classifying/rejecting mechanism is constructed with a specific set of binary classifiers; therefore the architecture has the form of a binary tree. In the root of this tree, we have a full set

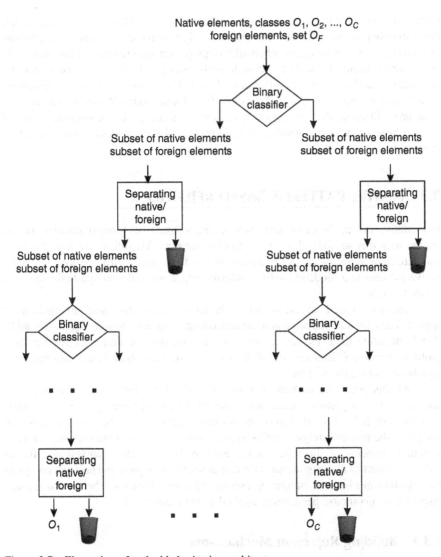

Figure 3.7   Illustration of embedded rejection architecture.

of patterns. When we move down from the root, we split native patterns into subsets, and after each split we separate foreign patterns from native ones. At the lowest levels of the tree, we have leaves in which we have native patterns of a single class. In other words, we have C leaves with native patterns only. However, these are not the only leaves in the tree. In addition, each rejection action is in fact a binary split that separates native patterns from foreign ones. Rejected foreign patterns end up in a node of a tree that is no longer expanded and becomes a leaf as well. The embedded rejection architecture is illustrated in Figure 3.7. In a full variant of the embedded architecture after each split, we perform a rejection, but we may consider different variations of such a structure. For instance, if it happens that at the lowest level, native patterns are not contaminated with foreign ones, we may resign from separation between patterns of both types. However, a case when all foreign patterns are rejected at internal levels of the tree is highly unlikely to occur in real-world problems. Therefore, it may be necessary to add a binary classifier to each leaf, like in local rejecting architecture, which separates native patterns of the class from foreign ones. We may also consider structures different than a binary tree, for instance, a class-contra-class set of binary classifiers may be furnished with regression like a mechanism with a rejecting threshold.

3.3   NATIVE PATTERNS-BASED REJECTION

Essentially, the problem of unknown characteristics of foreign patterns is the reason why it is so difficult to reject foreign patterns. Therefore, we look for non-standard approaches to this task, which we can base on the different kinds of data: native patterns and modified native patterns, which we call semisynthetic data and synthetic data.

Solutions based on native data only have an attractive and straightforward appeal. They are also the most general and versatile, as they do not require any additional substitute data. However, they may not be suitable and effective for each problem. Therefore, later on we will discuss other solutions based on mentioned semisynthetic and synthetic data.

At this point, let us look at some methodologies for constructing rejection mechanisms in a pattern recognition problem. In the succeeding text, we discuss two methods behind the design of the rejection mechanism. The first one relies on assigning the role of foreign patterns to some subset of native patterns and then constructing elements of the rejection mechanism. Dynamically changed assignment (where a subset of native patterns is treated as a set of foreign ones) helps us complete the rejecting mechanism integrating created rejecting elements. The second attempt employs the geometric perspective applied to the feature space.

3.3.1   Building Rejection Mechanisms

Here we present two methods of constructing rejecting mechanism. Both methods rely only on known classes of the training set of patterns and can be used in all rejecting architectures, namely, global, local, and embedded.

Class-Contra-All-Other-Classes
The first method is formulated in the form of Algorithm 3.1. It relies on creating a binary classifier for each class from $\Theta = \{O_1, O_2, \ldots, O_C\}$. We may use different methods for training these classifiers: k-NN, SVM, decision tree, RF, and so on. Each binary classifier ψ_l requires two classes of patterns for training: the mother class O_l and combined all other classes O_l^{contra}. Hence we refer to this rejecting mechanism as *class-contra-all-other-classes*. It is worth stressing that Algorithm 3.1 works well for numerous classes and fails for only two classes.

Algorithm 3.1

Construction of a rejecting mechanism of the type class-contra-all-other-classes

Data: $\Theta = \{O_1, O_2, \ldots, O_C\}$, classes of the training set of patterns
Classification Method,

Algorithm: **for** each class $O_l \in \Theta$ **do**
    begin
        take class O_l and construct *contra* class: $O_l^{contra} = \bigcup\limits_{O_i \in \Theta - \{O_l\}} O_i$
        build binary classifier ψ_l based on a given Classification Method
        and on two sets of patterns: the class O_l and the class O_l^{contra}
    end

Result: the set of C binary classifiers class-contra-other-classes $\psi_1, \psi_2, \ldots, \psi_C$
constituting together the rejection mechanism

A detailed discussion on how to use the class-contra-all-other-classes method for rejection will be presented in the next section.

Naturally, this analysis can be extended to a dataset with no labels, in which clusters of unlabeled patterns correspond to classes. For the sake of clarity, we focus on a supervised case, in which patterns are labeled.

Note that creating a *contra* class can make a class imbalanced with respect to class cardinality. In such case, formation of a binary classifier can be made using an undersampled *contra* class or oversampled *pro* class.

Class-Contra-Class

The second method also relies on creating binary classifiers based on the set of classes $\Theta = \{O_1, O_2, \ldots, O_C\}$, and, as aforementioned, we may use different methods in order to build binary classifiers, say, k-NN, SVM, and so on. Binary classifiers $\psi_{k,l}$ are trained for each pair of classes $\{O_k, O_l\}$, $l < k$ (cf. Algorithm 3.2). Therefore, we get $C \cdot (C-1)/2$ binary classifiers. Since each classifier is trained on two classes, this rejecting mechanism is called *class-contra-class*. As in the case of class-contra-all-other-classes, it is important to stress that Algorithm 3.2 works well for several classes and fails for only two classes.

Algorithm 3.2

Construction of a rejecting mechanism of the type class-contra-class

Data: $\Theta = \{O_1, O_2, \ldots, O_C\}$, classes of the training set of patterns
Classification Method,

Algorithm: **for** each two classes $O_l, O_i \in \Theta$, $l = 1, 2, \ldots, C-1$,
    $i = l+1, l+2, \ldots, C$ **do**
    begin
        build a binary classifier $\psi_{l,i}$ based on a given Classification Method
        and on two sets of patterns: class O_l and class O_i
    end

Result: the set of $C \cdot (C-1)/2$ binary classifiers class-contra-class $\psi_{l,i}$
constituting together the rejection mechanism.

3.3.2   Rejection Mechanisms in Global Rejecting Architecture

Let us recall global rejecting architecture, illustrated in Figure 3.3 and discussed in Section 3.2. In this scheme all incoming patterns are first subjected to native/foreign identification, and then patterns are classified to native classes. In global architecture we can employ both rejecting mechanisms class-contra-all-other-classes and class-contra-class described later on in this section.

Class-Contra-All-Other-Classes Method in Global Rejection Architecture
The intuition behind the idea of rejecting with class-contra-all-other-classes idea is as follows. Let us assume that we have C classes of native patterns. We have constructed C binary classifiers trained in a way class-contra-all-other-classes, one for each native class. Each such classifier assigns one of the two labels. Say that the label is *pro* when a pattern belongs to the single class or *contra* if a pattern belongs to the set of all other patterns. Next, we get a new pattern for which we wish to check if it is native or foreign. We *ask* all C binary classifiers to assign a label. If all binary classifiers assign it a *contra* label, this means that no classifier assumed that this pattern is native, and on this ground we reject it. In other words, if all binary classifiers exclude a pattern, we reject it.

The application of the mechanism of class-contra-all-other-classes rejection is very straightforward for the global rejecting architecture. Let us recall that the class-contra-all-other-classes method assumes that a pattern is native when at least one binary classifier ψ_l says it belongs to a native class. If we incorporate the class-contra-all-other-classes rejection mechanism in the global rejection architecture, we obtain a data processing scheme presented as Algorithm 3.3.

Algorithm 3.3
Global rejection employing the mechanism of the type class-contra-all-other-classes
Data:　　$\psi_1, \psi_2, ..., \psi_C$—a set of binary classifiers constituting a rejecting
　　　　　　mechanism, the set of classifiers is constructed by Algorithm 3.1,
　　　　　　O_X—a set of unknown patterns
Algorithm:　**for** each unknown pattern $\mathbf{X} \in O_X$ **do**
　　　　　　begin
　　　　　　　　for each binary classifier ψ_l,　$l = 1, 2, ..., C$ **do**
　　　　　　　　　　present the pattern $\mathbf{X}$ to the classifier ψ_l and register its output
　　　　　　　　　　the output is either class O_l, or class contra O_l^{contra}
　　　　　　　　if all classifiers ψ_l,　$l = 1, 2, ..., C$ output class *contra* **then**
　　　　　　　　　　assume that the pattern $\mathbf{X}$ is a foreign one
　　　　　　　　else
　　　　　　　　　　assume that the pattern $\mathbf{X}$ is a native one
　　　　　　end
Result:　　unknown patterns of the set O_X are identified either as native or as foreign

The processing scheme occurring in a combination of the class-contra-all-other-classes rejection with global architecture is as follows. On the input to the method, we pass new patterns. At first, they are identified as native or foreign. Next, native patterns are subject to classification.

It is important to notice that the class-contra-all-other-classes rejecting mechanism may also serve as a complete recognizer with rejection. Indeed, not rejected patterns must be assigned a native class by at least one binary classifier ψ_l, so then we can classify the pattern to this class. It may happen that such a pattern is assigned more than one native class label, so then there should be a method for resolving such a conflict. Anyway, using the rejecting mechanism as a classifier is rather unsuitable, due to the fact that the quality of rejection and classification would be potentially worse than in the case of rejection and classification realized by separated modules.

Class-Contra-Class Method in Global Rejection Architecture
In Figure 2.9, we show an intuitive discrimination method completed for native and foreign patterns. We identified regions in the feature space where either probability that patterns belong to any class is small or the difference between probabilities for different classes is small. The rejecting method class-contra-class follows the same intuition. The method applies a voting scheme to assign a class label to a pattern, that is, it assigns the class label with majority votes. Using the intuition mentioned previously, we do not assign a label: in other words, we reject a pattern, for which the number of votes for the majority class is not high enough or the difference between the number of votes for the majority and the second majority classes is not high enough. Therefore, the class-contra-class method is driven by the two parameters λ_{max} and λ_{diff}, which define minimal values of these two numbers (cf. Algorithm 3.4).

It is worth recalling that for C classes, we get at most $C-1$ votes for the majority class. Intuitively, in order to ensure correct classification, the constant λ_{max} should be equal to or slightly less than $C-1$ for an unknown pattern. On the other hand, we expect that the constant λ_{diff} is at least one, and it is desirable to have the value of this parameter greater than 1. The parameter λ_{diff} reflects, informally speaking, the granted hesitation of a rejection mechanism. The class-contra-class mechanism applied to the process a given pattern counts votes of binary classifiers. The most frequently appearing native label is the one that we assign to this pattern. However, it is interesting to check how many times the second most frequent class label appears. Intuition tells us to reject such a pattern for which there is too small difference between the counts of how many times the most frequent label and the second most frequent class label has appeared. We introduce threshold λ_{diff} to control this difference and use it as a rejection rule. The value of this parameter depends on the problem and should be adjusted experimentally based, of course, on a training set of native patterns. Also, the condition $(*)$ standing in Algorithm 3.4 could be reformulated on the basis of a problem, for example, an alternative could be turned into a conjunction.

Algorithm 3.4
A global rejecting mechanism based on the class-contra-class method
Data: $\psi_{l,i}, \quad l=1,2,\dots,C-1, \quad i=l+1,l+2,\dots,C$—a set of $C \cdot (C-1)/2$
           binary classifiers class-contra-class constituting a rejecting
           mechanism, the set of classifiers is constructed by Algorithm 3.2
           O_X—a set of unknown patterns

$\lambda_{\max}$, λ_{diff}—adjustable parameters called decision and separation thresholds

$\mathbf{V}$—vector of length C counting votes for classes

Algorithm: **for** each unknown pattern $\mathbf{X} \in O_X$ **do**

    begin

        for $i = 1$ **to** C **do** $\mathbf{V}[i] = 0$

        for each binary classifier $\psi_{l,i}$, $l = 1, 2, \ldots, C-1$,

            $i = l+1, l+2, \ldots, C$ **do** $\mathbf{V}\left[\psi_{l,i}(\mathbf{X})\right]$ + +

        $v_{\max} = \max\limits_{l=1,2,\ldots,C} \mathbf{V}[l]$

        $c_{\max} = \arg \max\limits_{l=1,2,\ldots,C} \mathbf{V}[l]$

        $v_{\max 2} = \max\limits_{\substack{l=1,2,\ldots,C \\ l \neq c_{\max}}} \mathbf{V}[l]$

(∗)   **if** $v_{\max} \geq \lambda_{\max}$ **or** $v_{\max} - v_{\max 2} \geq \lambda_{\text{diff}}$ **then**

            assume that the pattern $\mathbf{X}$ is native, with native class label $c_{\max}$

        else

            assume that the pattern $\mathbf{X}$ is foreign

    end

Result: all patterns from O_X are labeled either as native or as foreign

Let us note that Algorithm 3.4 can serve not only as a stand-alone rejection mechanism but also as a classification algorithm. We can assume that a given pattern processed with the devised method belongs to class $c_{\max}$.

3.3.3 Rejection Mechanisms in Local Rejecting Architecture

The local rejection mechanism is constructed in a way that at first we classify all patterns. As a result we obtain C subsets of patterns (corresponding to classes), and from formed subsets we reject foreign patterns. Therefore, the most suitable rejection mechanism for local architecture is the class-contra-all-other-classes method. Algorithm 3.5 outlines how to design local rejection architecture using the class-contra-all-other-classes method.

Algorithm 3.5

Local rejection employing the mechanism of the type class-contra-all-other-classes

Data: ψ—a C-class classifier constructed with the training set of native patterns

      $\psi_1, \psi_2, \ldots, \psi_C$—a set of binary classifiers constituting a rejecting mechanism, the set of classifiers is constructed by Algorithm 3.1,

      O_X—a set of unknown patterns

Algorithm: **for** each unknown pattern $\mathbf{X} \in O_X$ **do**

    begin

        using ψ determine the native class label, denote it l (class O_l)

        using ψ_l determine if $\mathbf{X}$ belongs to class O_l, or to class contra O_l^{contra}

        if ψ_l output class is *contra* **then**

            assume that the pattern $\mathbf{X}$ is foreign

        else

assume that the pattern **X** is native
end

Result: unknown patterns of the set O_X are identified either as native or as foreign, native patterns are assigned their class labels

The class-contra-all-other-classes method is a very good fit for the local rejection model. In this rejection model, once completing the classification step, we end up with separated native patterns. They are split among C groups, one per each class. Now, for patterns in each such group, we launch the rejection procedure that separates patterns belonging to the class that is supposed to be in this group from the patterns belonging to all other classes. For a group of patterns that is supposed to contain patterns coming only from class i, it is enough to launch *i-contra-all-other-classes*. The assumption that after the classification step patterns are correctly split into classes simplifies the rejection model that needs to be fitted.

Alternatively, we may apply some other rejection strategy (called class-contra-class). In Algorithm 3.6, we proceed with further elaborations.

Algorithm 3.6
Local rejection employing the mechanism of the type class-contra-class
Data: ψ—a C-class classifier constructed with the training set of native patterns
              $\psi_{l,i}$, $l = 1, 2, ..., C-1$, $i = l+1, l+2, ..., C$—a set of $C \cdot (C-1)/2$ binary classifiers class-contra-class constituting a rejecting mechanism, the set of classifiers is constructed by Algorithm 3.2,
              O_X—a set of unknown patterns
              $\lambda_{\max}$, λ_{diff}—adjustable parameters called decision and separation thresholds
Algorithm: **for** each unknown pattern $\mathbf{X} \in O_X$ **do**
              begin
                  using ψ determine the native class label, denote it l (class O_l)
                  using Algorithm 3.2 determine the native class label, denote it k (class O_k)
                  if $l = k$ **then** assume that the pattern **X** is native
                      else assume that the pattern **X** is foreign
              end
Result: unknown patterns of the set O_X are identified either as native or as foreign, native patterns are assigned their class labels

In this scheme, the first action is the classification step. Say, we process a certain pattern and the C-class classifier accounted it to native class i. Next, we employ the full class-contra-class rejection model. This means that we need to launch $C \cdot (C-1)/2$ binary classifiers, count their responses, and either accept this pattern to the class i or reject it. The same course of action needs to be repeated for all patterns that we wish to process: there is a need to launch $C \cdot (C-1)/2$ binary classifiers in order to accept or reject a pattern. In contrast, in Algorithm 3.5 (local rejection with class-contra-all-other-classes) for a given pattern, it was enough to launch a single binary classifier corresponding to class i.

3.3.4 Rejection Mechanisms in Embedded Rejecting Architecture

Embedded rejection architecture differs from the global and local architectures. It incorporates the rejection mechanism in a tree that at the same time classifies and rejects foreign patterns. By consecutive splitting, we divide a set of patterns so that at the end, in leaves, we obtain patterns coming from one class only. After each split we employ a rejection mechanism. Rejection mechanisms accompany (follow) all nodes of the embedded architecture tree.

We can employ both the class-contra-all-other-classes and the class-contra-class rejection methods. However, the classification tree in the embedded architecture changes the order in which we process native patterns. In particular, the formed tree implies that a dataset of patterns is consecutively split into two subsets. Therefore, there is a need to adjust rejection mechanisms so that they fit to reject patterns from subsets of patterns from the full dataset. Adjustments rely on redefinition of the set of classifiers, as produced by Algorithms 3.1 and 3.2, shown for each internal node. A direct consequence of a redefinition is that a set of classes assigned to each node is different.

We can interpret each node in the embedded rejection architecture as a stand-alone dataset that needs to be fitted with its own dedicated rejection mechanism (class-contra-all-other-classes and class-contra-class). In light of this, each node in embedded architecture is a stand-alone global rejection architecture, or, in other words, embedded rejection architecture is a composition of specific global rejection architectures. But such interpretations are very far reaching. In order to set up a complete embedded rejection model, we need to train as many binary classifiers as there are nodes in the tree. In practice, implementations of embedded architecture can be simplified in order to reduce processing overload associated with this quite complex architecture. There is usually no need to place rejecting mechanisms at each internal node. Lastly, let us recall that building rejecting mechanisms for a few classes requires methods different than those described in Algorithms 3.1 and 3.2.

3.4 REJECTION OPTION IN THE DATASET OF NATIVE PATTERNS: A CASE STUDY

The objective of this section is to investigate properties of classification with rejection. We are interested in two aspects of this procedure. The first aspect concerns the feature space. In particular, we are interested in investigating the quality of processing when this space is small. In such case, classification is prone to errors. The problem of a small-size feature space occurs in a few real-world situations, for instance, when producing features from signals is a costly procedure. Hence, we study the impact of dimensionality of a feature space on ensuing processing.

The second aspect of interest is the influence of the rejection mechanism on misclassification rate. As mentioned, it is a desirable property of a rejection mechanism to improve the recognition process by rejecting misclassified native patterns. In numerous applications, misclassification of patterns is much more harmful than the lack of classification decision. A standard pattern recognition problem assumes that every pattern belongs to one class. In terms of native/foreign patterns, foreign patterns are absent in the standard scheme, and, therefore, every processed pattern will get classified to some class of native patterns. In practice, classification is rarely perfect, that is, some (native) patterns are misclassified. Therefore, rejecting such misclassified patterns would improve recognizer performance; at least it would improve performance in terms of reducing the rate of misclassified patterns.

3.4.1 Dataset

Empirical evaluation of the studied methods, including the aspect of rejecting misclassified patterns, is performed by using the two datasets of native patterns:

1. Dataset of images of handwritten digits
2. Dataset of images of symbols of printed music notation

The handwritten digits dataset is available online as the MNIST dataset published by Yann LeCun *et al.* (1998). It contains images of Arabic numerals: 0, 1, …, 9. It is a popular dataset commonly used in visual pattern recognition studies, because handwritten digits are an ideal example of balanced and moderately advanced data. Obviously, this dataset used here is a 10-class dataset with 10,000 samples, approximately 1,000 samples in each class. The number of patterns in each class differs by approximately 10%. Figure 3.8 presents selected samples.

The second dataset was prepared as part of our research project, and we published this one and our other datasets in an online repository under W. Homenda *et al.* (2017). Images were cut manually from musical scores provided by various composers and present in various music genres. The dataset of musical symbols contains 20 classes. In total, there were 27,326 musical symbols to be processed. One should mention that while for handwritten digits samples are not distorted, in the presented musical symbols dataset, symbols are contaminated with other elements of music notation such as staff lines. It is also important to mention that a variety of shapes and sizes concerns not only different classes but also patterns belonging to the same class: musical symbols are imbalanced data both with respect to the quantity of patterns in classes and their properties such as shape and size. The examples shown in Figure 3.9 present patterns coming from all 20 classes, their names, and quantity. In addition, in Figure 3.10 we present an excerpt of printed music notation in order to show real sizes of symbols of music notation.

Since the dataset of printed music notation symbols is heavily unbalanced, with regard to cardinality of classes, in addition to processing the dataset as a whole, we formed two datasets with 10,000 patterns in each dataset using standard oversampling

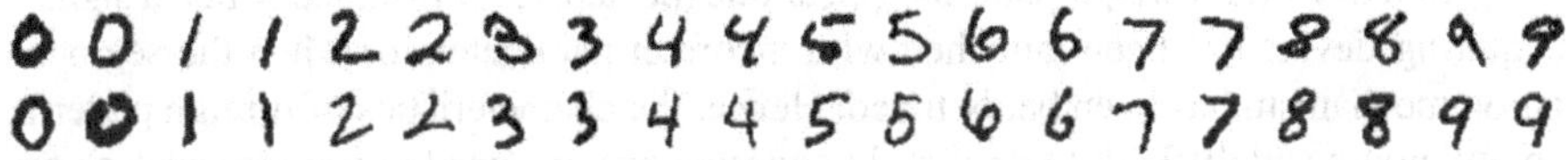

Figure 3.8   Native patterns—handwritten digits.

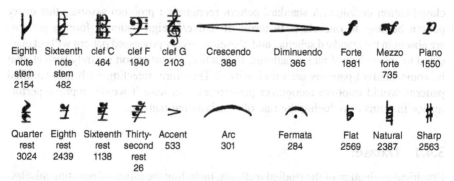

Figure 3.9 Native patterns—symbols of printed music notation symbols dataset. Below name of each symbol, we put number of samples in this class in the dataset.

Figure 3.10 An excerpt of printed music notation—visualization of imbalances in the dataset of printed music notation symbols with regard to shapes and sizes.

and undersampling techniques. Both formed datasets contained 500 samples in each class. In the case when in a given class there were more than 500 samples, we randomly selected only 500 of them. In the case when there were less than 500 available samples, we generated new data. The sampling procedures were already discussed in Chapter 1 (cf. Algorithms 1.1 and 1.2).

These two datasets, namely, handwritten digits and printed music notation symbols, were treated as native patterns. Using these patterns, we could evaluate and investigate mechanisms for standard pattern recognition, in particular various classification methods. However, these datasets are not enough if we want to evaluate rejection performance. The task of pattern rejection is analyzed in detail in this chapter and in Chapter 5.

In real world, foreign patterns appear unexpectedly, for instance, when a signal-acquiring device has been launched with incorrect parameters or when the segmentation mechanism has been badly tuned. Hence, the characteristics of foreign patterns are unknown and difficult to predict. In consequence, we need to propose such rejection models that perform well on a wide spectrum of patterns. In the study presented here, in order to validate the quality of the proposed rejection mechanisms, we use several datasets, which are here to play the role of foreign patterns. In particular, we have the following foreign datasets:

- Dataset with images of distorted handwritten digits. Digits were crossed out with digits 1 rotated by ±45°. We produced this dataset using samples coming from the MNIST database.

- Dataset of handwritten lowercase Latin alphabet letters: the set includes 32,220 patterns in total. There are 26 distinct letters, each letter represented with 1238–1241 patterns. This dataset is available at the web page (Homenda *et al.*, 2017).

- Foreign patterns originating from the dataset of printed music notation symbols; these are all nonnative patterns cut from music scores. This dataset includes 710 samples. This dataset is also available at the web page as well (Homenda *et al.*, 2017).

The samples originating from all three foreign datasets are shown in Figure 3.11.

Patterns from all sets were described by means of 171 numerical features. The set of features was reduced by removing 12 features with constant value, so then 159 features were used in further processing. They are presented in Appendix 1.A. In the case of handwritten digits, 59 correlated features were removed (with correlation coefficients assuming values >0.7) (cf. Chapter 1). Finally, the set of 100 acceptable features was used to characterize handwritten digits—this was the basic set of features for the handwritten digits dataset. Of course foreign patterns paired together with handwritten digits, distorted digits, and handwritten Latin letters were characterized with the same set of 100 features.

In the case of symbols of printed music notation, 47 features were strongly correlated. So then the set of 112 features was used to characterize symbols of printed music notation. Naturally, a foreign set taken out from musical scores was also

Figure 3.11 Samples of foreign patterns, two rows with samples of each dataset. From top to bottom: handwritten digits crossed out, handwritten Latin alphabet letters, garbage patterns of music notation.

characterized by the same set of features. Features were determined using binary (black-and-white) images (cf. Section 1.2). We performed experiments using original features, values normalized to the unit interval [0,1] and standardized features values. For details on features processing, consult Section 1.3.

We did not investigate the quality of features for foreign datasets (their correlation). This is consistent with our fundamental assumption that foreign patterns are not known at the stage of the construction of a classifier with rejection. Let us reiterate that in all cases features used to describe foreign patterns were the same as features used to describe native patterns.

In all pattern recognition procedures, we made use of two subsets of datasets of patterns: the training dataset and the test dataset. The training dataset was used for model building. The test dataset was holdout data that did not play any role in model construction; it was only used for model quality evaluation. In a few cases we used cross-validation techniques to reinforce model construction; typically it was needed in classifier training. For cross-validation model training, we also used only the training dataset.

In the case of handwritten digits, the set was split into two sets with the ratio approximately equal to $7:3$ for each class. In the training set there were 6999 samples and in the test set there were 3001 samples. These slightly unusual values appeared because the number of patterns in each class is not exactly the same; they differ by approximately 10%. We would like to make sure that each class is divided in the same way, and hence we obtained these particular values. In the case of this dataset, patterns represent handwritten decimal digits (0–9), and their class labels correspond directly to represented digits. Alike, for the case of music notation symbols, the training and test datasets were made of 70 and 30% patterns of each class, respectively.

3.4.2 Forming a Tree of Binary Classifiers

Let us now present a procedure that is completed in order to construct a binary tree of binary classifiers that altogether form a C-class classification model. The proposed idea is based on an assumption that each class of native patterns creates a regular *cloud* of points in a space of features. The default procedure described in this section is *supervised* tree construction. Alternatively, we may deal with a training set split to clusters creating regular clouds of points. In this scenario information about cluster belongingness replaces class labels. However, cluster belongingness can be considered something *weaker* than class labels, as class labels represent the *true* partition of a dataset. The variant in which we use cluster membership instead of class labels can be interpreted as *unsupervised* tree construction. In order to focus attention on methodology, we discuss the former case, that is, that patterns form known classes, and we know true class labels. Following this assumption, we propose to perform consecutive splits of the full set of native patterns until we reach subsets made of patterns belonging to a single class.

We can intuitively visualize the idea behind this procedure as follows. Each native class forms a compact cloud of points in the space of features; we can encompass such cloud using some geometrical figure, for example, an ellipsoid or a hyperrectangle (cf. Section 5.2). At first, we form a geometric region that encloses all patterns. Then, we recursively divide the set of patterns into two subsets based on an assumed dissimilarity measure; one can use a clustering procedure of choice to split patterns. The clustering procedure must be able to produce the desired number of clusters (which is always two for our purpose). Not all clustering procedures behave in this way. For example, k-means does, but DBSCAN does not. For more details on clustering, see Chapter 8. If the assumption that patterns in each class are similar one to another but dissimilar between classes, then each class forms a distinct cloud of points. The procedure of splitting is continued recursively until we reach a cloud of points designated for one class only. If the splitting procedure is performed on the cloud of all data points, it is hardly realistic to expect *pure* separation, meaning that it will rarely happen that in the end splitting will produce leaves that contain all patterns from a single class and nothing else. To mitigate this drawback, we would require to strengthen the procedure so that it is clear which cloud of points represents which class, even if diversity of patterns made clouds impure.

Alternatively, we can perform the procedure on class representatives made of single data points, instead of using clouds of points. Such representatives are, for instance, class centers. In this way we slightly simplify computations needed to complete the procedure. Algorithm 3.7 describes the details of the method used to construct a binary tree architecture for classification for a C-class dataset based on class centers representing native classes.

Algorithm 3.7
Determining architecture of a binary tree-based classifier
Data: $CC = \{\bar{\mathbf{X}}_1, \bar{\mathbf{X}}_2, \ldots, \bar{\mathbf{X}}_k\}$—a set of clusters' centers
                splitting method, for example clustering algorithm
Algorithm: **create** a tree node and label it with CC
                apply the splitting method to split CC to two groups CC_l and CC_r
                begin
                    let $CC_l = \{\bar{\mathbf{X}}_{i_1}, \bar{\mathbf{X}}_{i_2}, \ldots, \bar{\mathbf{X}}_{i_l}\}$ and $CC_r = \{\bar{\mathbf{X}}_{j_1}, \bar{\mathbf{X}}_{j_2}, \ldots, \bar{\mathbf{X}}_{j_r}\}$
                    if $|CC_l| = 1$ **then create** a tree node and label it with CC_l
                        else call recursively this algorithm with CC_l
                        being input data
                    if $|CC_r| = 1$ **then create** a tree node and label it with CC_r
                        else call recursively this algorithm with CC_r
                        being input data
                end
Result: a binary tree architecture with the root labeled by set clusters' centers CC
                and leaves labeled by single cluster's center

As a result of application of Algorithm 3.7, we obtain a tree that tells us what binary splits are needed to form a C-class classifier for a given dataset. The second step is to train component binary classifiers to implement this tree. The described mechanism is able to classify patterns, but it is still unable to reject patterns. Hence, at this point in each node of the tree, we add a rejection mechanism.

Finally, such a recognizer (classifier) is capable to classify native patterns and reject foreign ones.

3.4.3 Constructing a Tree of Binary Classifiers for the Dataset of Handwritten Digits

In this section, we discuss a particular realization of a binary tree-based classifier for the dataset of handwritten digits.

Let us recall that the experiment is based on the MNIST database (LeCun et al., 1998) (see also Section 3.4.1). Patterns were characterized by a set of features described in Chapter 1. In experiments we used 24 features picked up from the set of 100 features (cf. Section 3.4.1) according to greedy forward selection described in Section 1.4.

A quasi-balanced binary tree was obtained by running Algorithm 3.7 implemented with the use of k-means. It was the core of the classifier architecture. We split 10 classes represented with their class centers. For convenience, classes are indexed from 0 to 9 and denoted as $O_0, O_1, ..., O_9$. Classes are labeled with their indices $0, 1, ..., 9$. Centers $\bar{X}_0, \bar{X}_1, ..., \bar{X}_9$ were computed for each one of the 10 classes using the training dataset:

$$\bar{X}_l = \frac{1}{n_l} \cdot \sum_{X_i \in O_l} X_i \quad l = 0, 1, ..., 9$$

where $O_0, O_1, ..., O_9$ are classes of patterns in the training dataset and $n_0, n_1, ..., n_9$ stand for cardinalities of respective classes.

The intuition behind the proposed method is that class centers are good representatives of their classes. At the same time, clouds of patterns of different classes do not overlap or overlap very little, and class centers are different for each pair of classes. Hence, we propose to introduce a rule where similarity or dissimilarity of two class centers induces similarity or dissimilarity of two clouds of patterns. Based on this assumption, we separate one class from another. This separation is performed using a binary tree that was constructed on the basis of the knowledge about similarity/dissimilarity of cluster centers.

The tree structure discovered by Algorithm 3.7 for the dataset of handwritten digits is shown in Figure 3.12. This tree is balanced in a sense that height of any binary tree, which represents such a partitioning, cannot be smaller. It is important to underline that different clustering methods require different inputs and lead to different results. Therefore, switching from k-means clustering to another clustering method, for instance, to spectral clustering, may bring different results.

The binary tree shown in Figure 3.12 forms the core of our classifier. In the implementation each internal node of this tree, including the root, represents a binary classifier: either an SVM or an RF. Naturally, any other binary classifier can be used instead of these two. Each classifier in the tree is trained to separate patterns from classes represented by its left child and patterns from classes represented by its right child. The final architecture of the recognizer furnished with SVM and with RF is shown in Figure 3.13.

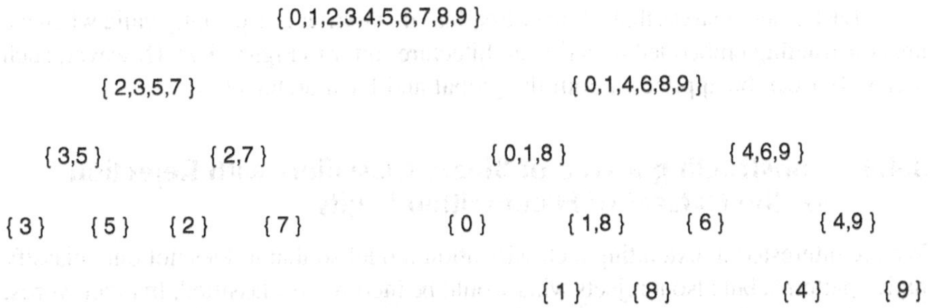

{ 0,1,2,3,4,5,6,7,8,9 }
{ 2,3,5,7 }
{ 0,1,4,6,8,9 }
{ 3,5 }
{ 2,7 }
{ 0,1,8 }
{ 4,6,9 }
{ 3 }
{ 5 }
{ 2 }
{ 7 }
{ 0 }
{ 1,8 }
{ 6 }
{ 4,9 }
{ 1 }
{ 8 }
{ 4 }
{ 9 }

Figure 3.12 The binary tree structure of the classifier built for the set of 10 classes of the MNIST database (cf. Yann LeCun *et al.*, 1998). The set $\{0, 1, 2, 3, 4, 5, 6, 7, 8, 9\}$ of 10 classes is split into two subsets, $\{2, 3, 5, 7\}$ and $\{0, 1, 4, 6, 8, 9\}$, and then the set $\{2, 3, 5, 7\}$ of four classes is split into two subsets, $\{3, 5\}$ and $\{2, 7\}$, and so on.

{ 0,1,2,3,4,5,6,7,8,9 }
{ 2,3,5,7 }
{ 0,1,4,6,8,9 }
{ 3,5 }
{ 2,7 }
{ 0,1,8 }
{ 4,6,9 }
{ 1,8 }
{ 4,9 }
{ 3 }
{ 5 }
{ 2 }
{ 7 }
{ 0 }
{ 6 }
{ 1 }
{ 8 }
{ 4 }
{ 9 }

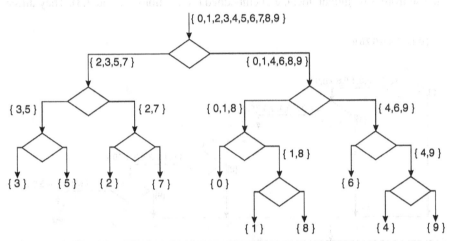

Figure 3.13 Architecture of the example tree of classifiers. In this study, random forests and SVMs were used as binary classifiers placed in nodes of this tree.

In the experiments all SVM classifiers were trained with the Gaussian kernel function. For all nodes of the tree, the values of gamma and regularization parameters were set as $\gamma = 0.0625$ and $C = 1$ and with 10-fold cross-validation applied.

The parameters of RF were defined largely following a study by Breiman (2002). The trees were trained with 500 trees and five randomly sampled features (square root of the number of 24 features being used). We have been using a voting scheme to determine class belongingness (membership), where each classifier in the ensemble votes for a class.

Some illustrative classification results are shown later in the upper part of Table 5.1. We used classifier coming from Figure 3.13 with binary SVMs regarded as local classifiers. The columns of the confusion matrix show classification results (predicted classes), while the rows present actual classes of patterns. Zeros were omitted for brevity.

Lastly, let us stress that a binary tree-based classifier is especially valid when we are constructing embedded rejection architecture (refer to Figure 3.7). However, such a classifier can be applied in both the global and local architecture.

3.4.4 Constructing a Tree of Binary Classifiers with Rejection for the Dataset of Handwritten Digits

We are interested in extending a classification model so that it does not only classify native patterns, but also it rejects what would be incorrectly classified. In other words, we wish that such a model rejects misclassified native patterns and of course rejects foreign patterns. Say, we have constructed a classification mechanism, any C-class classifier, as, for instance, a tree of binary classifiers performing the task of C-class classifier shown in Figure 3.13. At this point, we are supplementing the constructed classifier with a rejection mechanism. Let us recall that there are three schemes to choose from, say, global, local, and embedded (cf. Sections 3.2 and 3.3). They differ

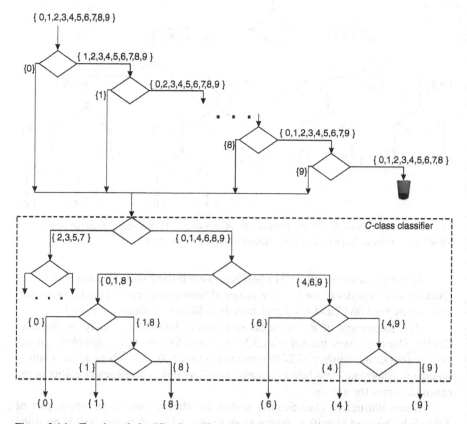

Figure 3.14 Tree-based classifier from Figure 3.13 furnished with the global rejection option. The illustration concerns the right branch of the tree (the left branch is skipped). Rejecting is performed with the same binary classifiers as in the basic C-class classifier, that is, random forests or SVMs.

in the order in which the actions of classification and rejection occur. In the global scheme, first we reject, and then we classify. In the local scheme, first we classify, and then we reject. In the embedded architecture, the actions of classification and rejection occur simultaneously; they are intertwined in the binary tree. A rejection mechanism itself can be realized with, for instance, specifically trained and setup two-class SVMs or RF.

Global rejecting architecture is set up in such a way that the rejecting mechanism is launched prior to classification. The rejection mechanism does not interfere with the classifier structure. Hence, the tree-based classifier can be replaced with any other suitable classifier. The global rejecting architecture with the basic classifier from Figure 3.13 is presented in Figure 3.14. The rejecting procedure is realized with the use of individually trained binary classifiers. There are exactly 10 such binary classifiers, one for each class. Each binary classifier is trained according to the *class-contra-all-other-classes* method, presented in detail in Algorithm 3.3, that is, each such SVM is trained to discriminate two classes: one class contains patterns from the class corresponding to a particular digit, and the other class represents all other patterns.

The rejection mechanism works as follows. We launch all 10 binary classifiers trained according to the *class-contra-all-other-classes* method, one by one. If all those classifiers classify it to *all-other-classes*, then it is rejected. In other words, if any *class-contra-all-other-classes* classifier assigns it to the *class* group, it is accepted as native.

In our experiment we implemented rejecting binary classifiers with RF and SVMs. All parameters were assumed as outlined in Section 3.4.3, that is, $\gamma = 0.0625$, $C = 1$, and with 10-fold cross-validation were assumed for SVMs. With regard to RF, five features and 500 trees were set up.

In order to realize local rejection architecture, we attach rejecting binary classifiers in the leaves of our classifier tree. Rejecting classifiers are constructed like in the case of global architecture. The classification with the local rejection option is shown in Figure 3.15. Like in the case of global rejecting, the rejection mechanism does not interfere with the classifier structure. Hence, the tree-based classifier can be replaced with any other suitable classifier.

Finally, the embedded rejecting architecture is shown in Figure 3.16. Note that the rejecting mechanisms constructed with aid of the native patterns are not suitable to be placed after the root of the tree. This is because the proposed method would not be able to discriminate a foreign group in these particular special cases, but it would just repeat the same task as the root.

3.4.5 Rejecting Misclassified Native Patterns: The Idea

We employ a rejecting mechanism to improve the recognition process by removing misclassified native patterns. In order to verify the impact of the rejection mechanism on the classification of native patterns, we need to undertake the following actions:

1. Build a classifier using the training set of native patterns.
2. Supplement the classifier with a rejecting mechanism based on the training set of native patterns.

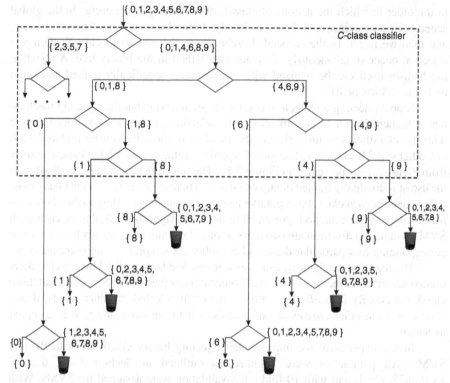

Figure 3.15   The tree-based classifier from Figure 3.13 furnished with the local rejection option. The illustration concerns the right branch of the tree (the left branch is skipped). Rejecting is performed with random forests or SVM binary classifiers attached to each leaf of the tree.

3. Subject the training and test sets of native patterns to classification without rejection (with the rejecting mechanism switched off). Create a confusion matrix for the training and test sets, that is, for each class find:

   a. The number of patterns correctly classified to this class

   b. The numbers of patterns of this class incorrectly classified to other classes

4. Subject the training and test set of native patterns to classification with rejection (with the rejecting mechanism switched on). Create a confusion matrix for the training and test set. In particular, for each class attain:

   a. The number of patterns correctly classified to this class

   b. The numbers of patterns incorrectly classified to other classes

   c. The number of rejected patterns

5. Compare results obtained in (3) and (4), in particular,

   a. Decrement in numbers of patterns classified to correct classes (3a–4a).

   b. Compare reduction in numbers of misclassified patterns (3b–4b).

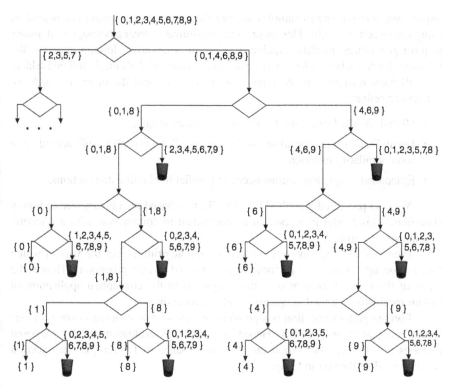

Figure 3.16 Tree-based classifier from Figure 3.13 furnished with the embedded rejection option. The illustration concerns the right branch of the tree. Rejecting is performed with SVM classifiers attached to internal nodes and leaves of the tree.

We have listed important values that we need to compute in order to evaluate the impact of a rejection mechanism on native pattern classification. Naturally, we expect that the model rejected only those native patterns that would be misclassified.

Let us stress that at this point we process only native patterns and there are no foreign patterns in the dataset. The desired situation is when misclassified patterns are rejected. The unwanted situation is when a rejection mechanism rejects native patterns that would be correctly classified. Without doubt, supplementing a classification procedure with a rejection mechanism has an impact on the modeling outcome. Further on, cf. Chapter 4, we inspect the influence of the rejection mechanism on native pattern classification quality.

3.5 CONCLUSIONS

In this chapter, we elaborated on an extended pattern recognition problem, in which, apart from the task of native pattern classification, we perform the task of foreign pattern rejection. We outlined several concepts on how to employ ensembles of binary classifiers to perform the rejection task. Noteworthy, the designed approaches do not require any foreign pattern samples to train the model, as the model is constructed using native patterns only. This, in practice, constitutes a great advantage as it allows to form generalized models, capable to reject truly unknown foreign patterns. We introduced three schemes for foreign pattern rejection: global, local, and embedded. We call them rejecting architectures. The names represent the order in which we perform rejection:

- Global: At the beginning, for the entire dataset at once.

- Local: Fine-grained for subsets of the datasets. Subsets shall be split according to some similarity criterion.

- Embedded: Rejection actions occur in parallel to classification actions.

We also presented a method for classification based on a binary tree of binary classifiers. This model is a good fit to implement in conjunction with a rejection mechanism of choice.

Presented mechanisms rely on well-known algorithms, but the way in which they are set up makes them perform an unusual and unique processing. Hence, the value of the research presented in this chapter is in the conceptual application of certain mechanisms aimed at specific kind of processing.

Foreign pattern rejection is a problem, which we encountered in our previous studies on pattern recognition. Hence, an empirical evaluation of the proposed methods and further elaboration on other approaches to foreign pattern rejection are discussed further on in Chapter 5.

REFERENCES

L. Breiman, *Manual on setting up, using, and understanding random forests V3.1*, 2002, https://www.stat.berkeley.edu/~breiman/Using_random_forests_V3.1.pdf (accessed October 9, 2017).

C. K. Chow, On optimum recognition error and reject tradeoff, *IEEE Transactions on Information Theory* 16(1), 1970, 41–46.

C. De Stefano, C. Sansone, and M. Vento, To reject or not to reject: that is the question-an answer in case of neural classifiers, *IEEE Transactions on Systems, Man, and Cybernetics, Part C: Applications and Reviews* 30(1), 2000, 84–94.

T. G. Dietterich and G. Bakiri, Solving multiclass learning problems via error-correcting output codes, *Journal of Artificial Intelligence Research* 2, 1995, 263–286.

M. Elad, Y. Hel-Or, and R. Keshet, Pattern detection using maximal rejection classifier. In: *Proceedings of International Workshop on Visual Form*, C. Arcelli *et al.* (eds.), Lecture Notes on Computer Science 2059, 2001, 514–524.

G. Fumera and F. Roli, Analysis of error-reject trade-off in linearly combined multiple classifiers, *Pattern Recognition* 37, 2004, 1245–1265.

T. M. Ha, Optimum tradeoff between class-selective rejection error and average number of classes, *Engineering Applications of Artificial Intelligence* 10(6), 1997, 525–529.

W. Homenda and A. Jastrzebska, Global, local and embedded architectures for multiclass classification with foreign elements rejection: An overview. In: *Proceedings of the Seventh International Conference of Soft Computing and Pattern Recognition (SoCPaR 2015)*, Kyushu University, Fukuoka, Japan, November 13–15, 2015, 89–94.

W. Homenda, A. Jastrzebska, and W. Pedrycz, *The web page of the classification with rejection project*, 2017, http://classificationwithrejection.ibspan.waw.pl (accessed October 9, 2017).

W. Homenda, A. Jastrzebska, W. Pedrycz, and R. Piliszek, Classification with a limited space of features: Improving quality by rejecting misclassifications. In: *Proceedings of the 4th World Congress on Information and Communication Technologies (WICT 2014)*, Malacca, Malaysia, December 8–11, 2014, 164–169.

H. Ishibuchi, T. Nakashima, and T. Morisawa, Voting in fuzzy rule-based systems for pattern classification problems, *Fuzzy Sets and Systems* 103, 1999, 223–238.

A. Koerich, Rejection strategies for handwritten word recognition. In: *9th International Workshop on Frontiers in Handwriting Recognition (IWFHR'04)*, Kokubunji, Tokyo, Japan, October 26–29, 2004, 2004.

Y. LeCun, C. Cortes, and C. J. C. Burges, *The MNIST database of handwritten digits*, 1998, http://yann.lecun.com/exdb/mnist/ (accessed October 5, 2017).

M. Li and I. K. Sethi, Confidence-based classifier design, *Pattern Recognition* 39(7), 2006, 1230–1240.

Z. Lou, K. Liu, J. Y. Yang, and C. Y. Suen, Rejection criteria and pairwise discrimination of handwritten numerals based on structural features, *Pattern Analysis and Applications* 2(3), 1999, 228–238.

L. Mascarilla and C. Frélicot, Reject strategies driven combination of pattern classifiers, *Pattern Analysis and Applications* 5(2), 2002, 234–243.

A. Meel, A. N. Venkat, and R. D. Gudi, Disturbance classification and rejection using pattern recognition methods, *Industrial & Engineering Chemistry Research* 42(14), 2003, 3321–3333.

I. Pillai, G. Fumera, and F. Roli, A classification approach with a reject option for multi-label problems, In ICIAP (1), G. Maino, G. L. Foresti, (eds.), Lecture Notes in Computer Science 6978, 2011, 98–107.

B. Scholkopf, J. C. Platt, J. Shawe-Taylor, A. J. Smol, and R. C. Williamson, Estimating the support of a high-dimensional distribution, *Neural Computation* 13(7), 2001, 1443–1471.

P. Simeone, C. Marrocco, and F. Tortorella, Design of reject rules for ECOC classification systems, *Pattern Recognition* 45(2), 2012, 863–875.

J. Xie, Z. Qiu, and J. Wu, Bootstrap methods for reject rules of fisher lda. *18th International Conference on Pattern Recognition (ICPR 2006). IEEE Computer Society*, Hong Kong, August 20–24, 2006, 425–428.

CHAPTER 4

EVALUATING PATTERN RECOGNITION PROBLEM

In the previous chapter, we have presented fundamental algorithms for pattern recognition. In this chapter, various factors and measures are defined in order to provide ways to reliably evaluate the quality of recognition without and with rejection mechanisms. These are necessary vehicles required to form and assess a good classification model. Presented measures concern both a *standard* approach to pattern recognition, in which we aim at a correct partitioning of a C-class dataset into C subsets, and also a case when we extend the problem of pattern recognition with rejecting option.

The chapter is structured as follows. First, we define generic measures that describe the results of pattern processing. We present a confusion matrix, which serves as a way to visually present the results. Then, we formulate compound (aggregate) measures to evaluate the quality of the classifier. The chapter is illustrated with several experimental studies in which we present how to evaluate the quality of native pattern recognition with foreign pattern rejection.

4.1 EVALUATING RECOGNITION WITH REJECTION: BASIC CONCEPTS

Let us recall that *native* patterns are all the proper patterns that are considered to be included in one of designed sets. Each native pattern should be paired with a correct class label. Class label unequivocally determines the corresponding class. *Foreign* patterns are everything else—the garbage. We would like to design classifiers that are capable of the following:

- Identifying all native patterns as native ones (i.e., *accept* them) and all foreign patterns as foreign ones (i.e., *reject* them)
- Classifying native patterns belonging to respective classes

Pattern Recognition: A Quality of Data Perspective, First Edition. Władysław Homenda and Witold Pedrycz.

The pertinent definitions of native and foreign patterns, along with a formulation of the task of native pattern recognition with foreign pattern rejection, are given in Chapter 3.

Quality evaluation of classification with rejection requires nonstandard measures. Intuitively, it is important to measure how exact rejection procedure is, namely, how many foreign patterns are rejected (i.e., identified as foreign ones) and how many native patterns are accepted (i.e., identified as native ones). Of course, measuring classification's quality regarded as assigning native patterns to proper classes is still of great importance (cf. Homenda and Lesinski, 2014) and we will discuss this issue later on.

4.1.1 Evaluating Effectiveness of Rejection

For better understanding of how quality of classification with rejection should be measured, we adapt parameters and quality measures used in signal detection theory and in statistics. We adapt those measures to complete discrimination between native and foreign patterns and also to classification of native patterns into respective classes. We count the number of correctly and incorrectly identified patterns and place them in a so-called confusion matrix, which is a standard way of assessing performance in pattern recognition. An example of a confusion matrix for a binary problem, here for native and foreign patterns, is presented in Table 4.1. In this table, we introduce the following notation:

- TP (true positives) – The number of native patterns classified as native ones (no matter, if assigned to a correct class or not)
- FN (false negatives) – The number of native patterns incorrectly classified as foreign ones
- FP (false positives) – The number of foreign patterns incorrectly classified as native ones
- TN (true negatives) – The number of foreign patterns correctly classified as foreign ones

The rows shown in Table 4.1 correspond to the numbers of patterns and their predicted belongingness (to the group of either native or foreign sets). Columns correspond to true belongingness (actual class) as we know of, because the dataset

TABLE 4.1 Confusion matrix for rejecting option of pattern recognition applied to two sets of native and foreign patterns without splitting the set of native patterns into classes

		True Set	
		Native	Foreign
Predicted set	Native	TP	FP
	Foreign	FN	TN

that we process is labeled. For instance, in the first column we have actual native patterns correctly identified as native (TP) and actual native patterns incorrectly identified as foreign (FN). In contrast, in the first row, we have counts of native and foreign patterns identified as native. Native patterns identified as native are, as we mentioned, denoted as TP, while foreign patterns identified as native are denoted as FP.

In addition, later on we use the following two notions:

- Positives (P), samples that truly belong to native class: $P = TP + FN$
- Negatives (N), samples that truly belong to foreign class: $N = FP + TN$

4.1.2 Imbalanced Native Versus Foreign Sets

Evaluating a single factor cannot expose classification quality. This is true in general and in the two-class problem. For instance, not only important is the proportion of the number of correctly recognized patterns of a class to the number of all patterns of this class. Obviously, the number of patterns falsely classified as belonging to this class affects intuitive meaning of quality. Especially, when we consider a class with a small number of patterns, falsely classified pattern significantly decreases intuitive evaluation of quality. Therefore, we should look for formal evaluations being in agreement with this intuition. Let us recall that in the case of imbalanced two-class problem, the minority class is called the positive one, while the majority class is the negative one.

Let us consider the notion of accuracy. It is defined as the ratio of correctly classified patterns to the number of all patterns. However, this ratio alone is too general (built at a high level of abstraction) as it does not discriminate between classification qualities with regard to particular classes present in a dataset. This limited understanding of the quality evaluation is especially risky when we are faced with imbalanced datasets. Say, we have a medical dataset with 9900 samples coming from healthy patients and only 100 samples from ill patients. We constructed a classification model that always predicts that the person is healthy. According to the meaning of accuracy defined earlier, the classifier was accurate in $9,900/10,000 = 0.99 = 99\%$. The accuracy rate of 99% sounds amazing; however the model is obviously truly useless. Statistics can be deceiving. We need a careful consideration and analysis of obtained results from different perspectives. This simple example is convincing enough to notice that we need to study not only recognition methods themselves but also objective methods for their evaluation.

Therefore, in this chapter we investigate formal evaluation methods that allow objective evaluation of recognition results. Let us recall that we deal with identification (separation) of native and foreign patterns. We consider this problem as an imbalanced problem where cardinalities of sets of native and foreign patterns are significantly different. Hence, in our case we encounter the described problem of imbalanced data in which many intuitive measures are invalid and we have to focus on prudently constructed alternative measures.

In order to compensate differences between cardinalities of sets of native and foreign patterns, we propose to balance cardinalities of both sets by using an

TABLE 4.2 Balanced confusion matrix from Table 4.1 with equalizing parameter applied

		True Belongingness	
		Native	Foreign
Predicted belongingness	Native	$\lambda \cdot \text{TP}$	$(1-\lambda) \cdot \text{FP}$
	Foreign	$\lambda \cdot \text{FN}$	$(1-\lambda) \cdot \text{TN}$

equalizing parameter λ in the confusion matrix shown in Table 4.1. Referring to quantities of this confusion matrix, the equalizing parameter is defined as follows:

$$\lambda = \frac{\text{FP} + \text{TN}}{\text{TP} + \text{FN} + \text{FP} + \text{TN}} \in [0, 1] \tag{4.1}$$

Then we come up with the balanced confusion matrix shown in Table 4.2.

4.1.3 Measuring Effectiveness of Rejecting Quality

The above quantities are adopted to evaluate quality of classification and rejection. The following general measures are used and adopted to evaluate performance of classification and rejection mechanisms:

- Accuracy articulates joined capability of correct identification of native and foreign patterns.
- Sensitivity articulates ability to identify correctly patterns from a given set; that is, it expresses the correct identification of native patterns as native ones or foreign patterns as foreign ones.
- Precision articulates ability to separate patterns coming from different sets; that is, it measures separation quality of native patterns from foreign ones and foreign patterns from native ones.

On the basis of these characteristics, we introduce the following measures

$$\text{Accuracy} = \frac{\text{TP} + \text{TN}}{\text{TP} + \text{FN} + \text{FP} + \text{TN}} = \frac{\text{TP} + \text{TN}}{\text{P} + \text{N}}$$

$$\text{Native sensitivity} = \frac{\text{TP}}{\text{TP} + \text{FN}} = \frac{\text{TP}}{\text{P}}$$

$$\text{Native precision} = \frac{\text{TP}}{\text{TP} + \text{FP}}$$

$$\text{Foreign sensitivity} = \frac{\text{TN}}{\text{TN} + \text{FP}} = \frac{\text{TN}}{\text{N}} \tag{4.2}$$

$$\text{Foreign precision} = \frac{\text{TN}}{\text{TN} + \text{FN}}$$

$$\text{F-measure} = 2 \cdot \frac{\text{Precision} \cdot \text{Sensitivity}}{\text{Precision} + \text{Sensitivity}}$$

The following brief comments may help understand the meaning of the previously mentioned measures:

- *Native precision* is the ratio of the number of patterns identified as native to the number of all patterns identified as native. Native precision evaluates the ability of the classifier to distinguish foreign patterns from native ones. The higher the value of this measure, the better ability to distinguish foreign patterns from the native ones. Native precision does not evaluate how effective the identification of native patterns is; that is, it rewards correct identification of foreign patterns. This measure is also called *positive predictive value*.

- *Native sensitivity* is the ratio of the number of patterns *correctly* identified as native to the number of all patterns that *should be* identified as native, that is, all that are actually native. This measure evaluates the ability of the classifier to identify native patterns. The higher the value of native sensitivity, the more effective identification of native patterns becomes. Unlike the native precision, this measure does not evaluate the effectiveness of separation between native and foreign patterns. This measure is also called *true positive rate*.

- *Foreign precision* corresponds to native precision. It is also called *negative predictive value*.

- *Foreign sensitivity* corresponds to native sensitivity. It is also called *specificity*.

- *Accuracy* is the ratio of the number of correctly identified native and foreign patterns to the number of all patterns. This measure describes the ability to distinguish between native and foreign patterns. Of course, the higher the value of this measure, the better the identification result.

- Precision and sensitivity are complementary, and there exists yet another characteristic that combines them, known as the *F-measure*. It is there to express the balance between precision and sensitivity since, in practice, these two affect each other. Increasing sensitivity can cause a drop in precision since, along with correctly classified patterns, there might be more incorrectly classified terms.

- The F-measure can be employed to combine native precision and native sensitivity as well as foreign precision and foreign sensitivity.

4.1.4 Separating Native and Foreign Patterns

Separation of the sets of native and foreign patterns is the most desired feature of rejecting option. Perfect rejecting mechanism should reject all foreign patterns and no native one. In other words, in the native–foreign confusion matrix, both false quantities, that is, FP and FN, should vanish (be equal to zero). Therefore, besides measures from Section 4.1.3, we introduce another four measures that directly characterize rejecting option; two of them characterize rejecting option from a perspective of native patterns and the other two, from a perspective of foreign patterns. They all have the term *separability* in their names with other adjectives defining their specifics. These measures are quite similar to sensitivity and precision. They differ from these measures in such a way that they depend on counterpart of false quantity, while sensitivity and precision do not. For instance, native precision depends on FP and not on FN, while native separability depends on both FP and FN. This dependence makes substantial difference while both FP and FN are large. In such case separability measures reflect well this problem.

$$
\begin{aligned}
\text{Strict native separability} &= \frac{TP}{TP + FN + FP} \\
\text{Native separability} &= \frac{TP + FN}{TP + FN + FP} = \frac{P}{P + FP} \\
\text{Foreign separability} &= \frac{TN + FP}{TN + FP + FN} = \frac{N}{N + FN} \\
\text{Strict foreign separability} &= \frac{TN}{FN + FP + TN}
\end{aligned}
\tag{4.3}
$$

Strict native separability measure evaluates rejection quality from a perspective of reducing incorrectly identified patterns; that is, it rewards decreasing numbers of rejected native patterns and accepted foreign patterns. For a perfect rejecting, with zero rejected native patterns and zero accepted foreign patterns, this measure produces 1, while for more realistic cases, rejections of native patterns and acceptances of foreign patterns are equally punished, and they reduce value of this measure. Interpretation of *strict foreign separability* is very similar.

Strict native separability and *strict foreign separability* could not always be reliable measures, because they do not differentiate between incorrectly identified native and foreign patterns. They are not measures fitted for certain specific real-world domains, in which we require giving higher priority to not to reject native patterns rather than to accept foreign ones (or oppositely). Examples of such domains are when there is a certain cost or risk attached to a decision and the cost differs among classes, like the risk of false medical misdiagnosis. Therefore, we define two specific forms of separability measures, namely, *native separability* and *foreign separability* that are suitable for the situations outlined earlier.

Native separability and *foreign separability* depend on false quantities FP or FN; that is, they decrease their values if FP or FN increases, respectively. However, they are not perceptive to opposite quantities. Native separability will issue value 1 if all foreign patterns are rejected (the case when FP = 0), and it does not take into account falsely rejected native patterns, so then even if most of native patterns are rejected and no foreign one is accepted, the measure issues value 1. Alike, foreign separability issues value 1 if all native patterns are accepted, that is, FN = 0, and does not depend on falsely accepted foreign patterns.

The previous measures have *positive character* in the sense that the greater their values, the better quality of the classifier being evaluated. Of course, there are complementary measures to these in the sense that simple subtraction from 1 would be the complement. In such case, we get measures of *negative character*, measures evaluating errors. For instance, false discovery rate is complementary to the positive predictive value (native precision). In this book, we refer mostly to measures of the positive character. In some cases we also evaluate errors.

4.1.5 Adaptation to Multiclass Native Patterns

The discussion presented so far was focused on a binary native/foreign identification problem. We were distinguishing between the two groups of patterns: native and foreign ones. However, an appreciable share of pattern recognition problems is of multiclass character, meaning that native patterns are distributed across some number of classes. Let us now extend the discussion to cover such processing scenarios.

Let us take a closer look at confusion matrix shown in Table 4.1. We expand this matrix by splitting native patterns into classes. Therefore, instead of a single number TP, we consider a square matrix of size C (according to general assumption that we have C native classes) to count the number of native patterns correctly identified as being native:

$$\text{TP} \rightarrow [\text{TP}_{k,l}] \quad k, l = 1, 2, \ldots, C \tag{4.4}$$

where $\text{TP}_{k,l}$ denotes the number of native patterns coming from the class l (truly belonging to this class) and classified (predicted) to class k.

In a similar way we expand the aggregative quantities FN and FP to vectors of size C in order to count native patters incorrectly identified as foreign ones and foreign patterns incorrectly identified as native ones:

$$\text{FN} \rightarrow [\text{FN}_{C+1,l}], \text{FP} \rightarrow [\text{FP}_{k,C+1}], \quad k, l = 1, 2, \ldots, C \tag{4.5}$$

where $\text{FN}_{C+1,l}$ is the number of native patterns coming from the class l falsely identified as foreign and $\text{FP}_{k,C+1}$ is the number of foreign patterns falsely identified as native and classified to the class k. Finally, we have the following composite confusion matrix of size $(C+1) \times (C+1)$ characterizing both identification and classification aspects:

$$\begin{bmatrix} \text{TP}_{k,l} & \text{FP}_{k,C+1} \\ \text{FN}_{C+1,l} & \text{TN} \end{bmatrix} \quad k, l = 1, 2, \ldots, C \tag{4.6}$$

where $\text{FN}_{C+1,l}$ and $\text{FP}_{k,C+1}$ are horizontal and vertical vectors (row and column) of the matrix, respectively.

Naturally, based on the confusion matrix representation as in (4.6), we can compute representation in the aggregative binary form, as in Table 4.1. The following straightforward formulas allow computation of elements of the binary matrix

$$\text{TP} = \sum_{k=1}^{C} \sum_{l=1}^{C} \text{TP}_{k,l}, \quad \text{FN} = \sum_{l=1}^{C} \text{FN}_{C+1,l}, \quad \text{FP} = \sum_{k=1}^{C} \text{FP}_{k,C+1} \tag{4.7}$$

4.1.6 Evaluating Multiclass Classification with Rejection Option

In order to take into account information about predicted native class belongingness, we need to expand further on collected quantities and formulate new measures. Evaluation of multiclass classification with rejection option requires adding one more quantity: CC (*correctly classified*). It denotes the number of correctly classified native patterns. This parameter supplements the set of previously defined parameters for a standard two-set *native–foreign* problem with an evaluation tool for a multiclass dataset. Since each correctly classified pattern is also identified as a native one, that is, as a TP, then of course $CC \leq TP$. CC can be obtained from the multiclass case confusion matrix characterizing classification and rejection, which is shown in (4.6). It is the sum of elements on the main diagonal of the matrix $[TP_{k,l}]$, where $k, l = 1, 2, \ldots, C$ defined by (4.4)

$$CC = \sum_{k=1}^{C} TP_{k,k} \qquad (4.8)$$

The measures of precision, sensitivity, and accuracy introduced in the previous section and used to characterize the quality of rejecting are specialized here in order to characterize both rejecting and classification. The following formulas come as their specialized forms:

$$\text{Strict accuracy} = \frac{CC + TN}{TP + FP + FN + TN}$$

$$\text{Strict native sensitivity} = \frac{CC}{TP + FN} = \frac{CC}{P} = \text{Strict native accuracy}$$

$$\text{Strict native precision} = \frac{CC}{TP + FP} \qquad (4.9)$$

$$\text{Fine accuracy} = \frac{CC}{TP}$$

The following concise comments may enhance understanding of the previous measures:

- *Strict accuracy* is absolute measure of classifier's performance joined with quality of rejecting. It is the ratio of the number of all *correctly* classified patterns, that is, foreign identified as foreign (rejected) and native identified as native (accepted) *and* classified to their respective classes, to the number of all patterns being classified. Note that accuracy is a characteristic derived from strict accuracy by ignoring the need to classify native patterns to their respective classes. In other words, unlike for strict accuracy, for accuracy it is sufficient to correctly identify whether a pattern is native or foreign.

- *Strict native sensitivity*, which is identical to strict native accuracy, is an absolute measure of classifier's performance reported on the set of native patterns with foreign patterns ignored. It is the ratio of the number of all *correctly* classified native patterns, that is, native patterns identified as native ones (accepted) *and* classified to their respective classes, to the number of all native patterns being classified.

- *Strict native precision* is the ratio of the number of native patterns identified as native ones and *correctly* classified to their classes to the number of all patterns (native and foreign) classified as native. Strict native precision evaluates the ability of the classifier to distinguish native patterns correctly classified from all native and foreign patterns incorrectly classified. The higher the value of this measure, the better ability to distinguish foreign patterns from native ones and to classify correctly native patterns. Alike native precision, strict native precision does not evaluate how effective the identification mechanism of native patterns is.

- *Fine accuracy* is the absolute measure of the classifier's performance on the set of native patterns identified as native ones. It is the ratio of the number of all *correctly* classified patterns, that is, native identified as native (accepted) *and* classified to their respective classes, to the number of native patterns identified as native. Note that fine accuracy is a characteristic derived from strict native accuracy by ignoring native patterns identified incorrectly as foreign ones. In other words, unlike strict native accuracy, fine accuracy takes into account only those native patterns that are identified correctly as native.

The following relation holds, and in the case of only a single class of native patterns, it obviously translates into equality:

$$\text{Strict accuracy} \leq \text{Accuracy} \qquad (4.10)$$

4.1.7 Illustrative Example

Let us look at an application of identification and classification measures in a practical example. For the sake of the discussion, let us at this point neglect few important issues: dataset description, classifier construction, and so on. Instead we just focus on quality measures. In later parts of this book, we address elements of the methodology that are missing here.

In the experiment, we used a native dataset of handwritten digits shown in Figure 3.6. The handwritten digits dataset is available as the MNIST repository (LeCun *et al.*, 1998). The role of foreign patterns was played by a dataset with handwritten crossed-out digits and handwritten Latin alphabet letters shown in Figure 3.11. This dataset is available at the web page (Homenda *et al.*, 2017). The native dataset is a 10-class problem with consecutive digits (0, 1, ..., 9) forming the corresponding class labels. Recognition with rejection was performed with the use of a classifier with a local rejection method (the description of this approach is presented in Section 3.4.4, while the model is illustrated in Figure 3.15). The support vector machine (SVM) parameters applied in this experiment are outlined in Section 3.4.4. The results of classification with rejection are shown in Table 4.3. Notice that native patterns are well identified; that is, only 1 out of 30 (326 out of 10,000) native patterns is identified incorrectly as a foreign one. In contrast, identification of foreign patterns is far worse for crossed-out digits: one out of five (2,098 out of 10,000) is incorrectly identified as a native one. When we consider handwritten letters as foreign patterns, their identification is superior in comparison to the crossed-out digits: only 1 out of 20 (1,598 out of 32,220) is falsely accepted as native.

Quality measures for recognition with rejection are given in Table 4.4. It is worth to draw attention that the sets of handwritten digits (native) and letters (foreign) are imbalanced in the sense of cardinalities of both sets. Therefore, we apply equalizing parameter in order to analyze quality measures of balanced model. In Table 4.4,

TABLE 4.3 Classification with rejection applied to handwritten digits (native set, 10 classes) and crossed-out digits and handwritten Latin alphabet letters (foreign sets)

					True Class								
		Native										Foreign	
Predicted class	0	1	2	3	4	5	6	7	8	9	Crossed	Letters
0	943											96
1		1114			1		1	1	7	3		703
2		3	1003	2			1	1	2		530	211
3			2	961					1	3		1
4					949	8		4	3	4	789	295
5				3		850			1	3		99
6	3	1			2	2	925		1		5	106
7			1	8				992	0	9	1	16
8	9		1	2	8	4			897	6	773	68
9								4	1	924		3
Foreign	25	17	25	34	22	28	31	26	61	57	7902	30622

The classifier with the local rejection shown in Figure 3.15 was employed. Results are outlined for the whole learning set (the training and test sets joined together). Zeroes were omitted for the sake of clarity.

TABLE 4.4 Results of classification with rejection from Table 4.3 in terms of identification of native and foreign patterns, according to (4.6) and (4.7), while CC = 9558 according to (4.8)

		True Class	
		Native	Foreign
Handwritten digits (native), crossed-out digits (foreign)			
Predicted class	Native	TP = 9674	FP = 2098
	Foreign	FN = 326	TN = 7902
Handwritten digits (native), handwritten letters (foreign)			
Predicted class	Native	TP = 9674	FP = 1598
	Foreign	FN = 326	TN = 30622
Handwritten digits (native) and letters (foreign), balanced model			
Predicted class	Native	TP = 7382	FP = 378
	Foreign	FN = 249	TN = 7253

TABLE 4.5 Characteristics of classification with rejection delineated by accuracy, sensitivity, and precision measures obtained from Tables 4.3 and 4.4

	Digits Crossed	Letters	Letters Balanced
Accuracy	87.88	95.44	95.89
Native sensitivity	96.74	96.74	96.74
Native precision	82.18	85.82	95.12
Foreign sensitivity	79.02	95.04	95.04
Foreign precision	96.04	98.95	96.68
Native separability	82.66	86.22	95.27
Foreign separability	96.84	99.00	96.84
Native F-measure	88.87	90.96	95.92
Foreign F-measure	86.70	96.95	95.85

Values of measures are given as percentages.

for comparative purposes, we present measures in both cases that are with and without balancing parameter.

The introductory observations are well characterized by the measures introduced earlier; we collected their values in Table 4.5. High score of native sensitivity indicates that native patterns are well identified. High value of foreign precision suggests that native patterns are well separated from the foreign ones. Contrasting low values of native precision and foreign sensitivity highlight the much weaker ability to identify foreign patterns, and indeed many of them were not identified correctly. Alike, low value of native precision points that foreign patterns are not well separated from the native ones. Moderate value of accuracy reflects scores of precision and sensitivity for both joined native and foreign patterns. F-measures provide more details about qualities of joined precision and sensitivity with one balanced measure. In the case of the described experiment, values of F-measures prove that native patterns were processed more successfully than foreign ones (native F-measure is higher than foreign F-measure).

Let us make a note that in the experiment presented here, we process the set of native patterns of size 10,000 with approximately 1,000 samples in each class. The set of foreign patterns is of the same size as the entire native set (10,000 samples). Hence, in this case there is no ground to apply equalizing parameter (defined by (4.1)). On the other hand, we consider a balanced model of handwritten digits (native) and handwritten letters (foreign). In Table 4.5, we present results for this case in the two variants as imbalanced and balanced. If we compare these two, it becomes apparent that imbalanced results do not reflect true quality.

Classification with rejection is aimed at the maximization of all previously mentioned measures. However, increasing one measure usually causes a drop in another one. In practice, depending on real application, some measures might be more important than others. For instance, if importance is given to minimization of contamination of the set of native patterns by foreign patterns, then native precision and native separation get higher importance than other measures, and it should be maximized. On the other hand, if the highest priority is given to minimization of the loss of native patterns, then focus should be on native sensitivity and foreign separation. If both aims are important, then native F-measure should get the highest priority.

There are two important aspects concerning interpretation of these measures. The first aspect is related to the cardinalities of both sets of native and foreign patterns or, more precisely, a proportion of the numbers of patterns located in these sets. This is, for instance, a case of imbalanced classification problems of pattern recognition. The case discussed here is well balanced: the cardinalities of native classes are close to 1000; likewise the cardinalities of the sets of native and foreign patterns are equal.

The second aspect concerns priorities given to wrong identification of a certain type of patterns. For instance, we can consider records of patients suffering from a disease as native patterns and records of healthy ones as foreign ones. In such a case, incorrect identification of native patterns as foreign ones becomes much more severe than the error made in an opposite direction. This phenomenon is the case of cost-sensitive learning. The idea is that we can attach various priorities to different classes. This can be done at the stage of model construction, as it is, for instance, discussed in Provost and Fawcett (2001), Provost and Domingos (2003), Sun et al. (2007), Garcia et al. (2007), He and Garcia (2009), and Tang et al. (2009). In our study, we do not interfere with the training of the model (so we do not engage cost-sensitive learning), but we have constructed specific measures that are capable of quantifying disparities on the processing of native and foreign patterns.

Now, let us have a look at measures provided by formulas (4.9), which supplement measures expressed in terms of (4.2). Table 4.6 contains computed values of these measures for the experiment under evaluation in this section. It turns out that values of precision, sensitivity, accuracy, and F-measure are slightly smaller than those present in Table 4.5, which is obvious due to very similar values of TP and

TABLE 4.6 Characteristics of multiclass classification with rejection delineated by precision, sensitivity, accuracy, and separability measures drawn from Tables 4.3 and 4.4

	Digits Crossed	Letters	Letters Balanced
Strict accuracy	87.30	95.17	95.31
Strict native sensitivity	95.58	95.58	95.58
Strict native precision	81.19	84.79	93.98
Fine accuracy	98.80	98.80	98.80
Strict native F-measure	87.80	89.86	94.77
Strict native separability	79.96	83.41	92.17
Strict foreign separability	76.53	94.09	92.04

Values of measures are given as percentages.

CC, but CC is slightly smaller. It is worth to draw attention to fine accuracy, which was missing in the analysis of a rejection mechanism alone. Fine accuracy (independent on the results on foreign patterns) reaches 98.80%, which means that most of patterns identified as native ones are correctly classified to respective classes. This observation may be important in some applications of classification problems with rejection where false classification would be costly.

We draw attention to the balanced case of handwritten digits and letters. Let us reiterate that balancing outlines more objective evaluation of the model; in this case almost all values are significantly greater than those in the imbalanced case.

4.2 CLASSIFICATION WITH REJECTION WITH NO FOREIGN PATTERNS

It should be noted that there are two scenarios in which we may employ rejection. First, an apparent one is when we are at a true risk of encountering foreign patterns. However, we may also apply a rejection mechanism to a scenario in which we are certain that it is based on native patterns only (with no foreign ones). In this case, rejecting should be directed toward falsely classified patterns. Rejecting all falsely classified patterns and accepting all patterns classified to correct classes is the ideal situation in such case. Unfortunately, in real-life problems this ideal state is unattainable. Therefore, a satisfying situation is when we are minimizing the number of rejected correctly classified patterns, while the number of rejected incorrectly classified ones is favorably high.

In order to depict such case, the set of measures defined in Section 4.1.2 is reduced due to vanishing parameters: $FP = 0$ (there are no foreign patterns that could be incorrectly accepted as native one) and $TN = 0$ (there are no foreign patterns that could be correctly rejected as foreign ones). As a consequence, (4.2) and (4.9) are reduced to parameters related to the set of native patterns only. Finally, the three measures remain valid: accuracy, strict accuracy, and fine accuracy. Let us have a closer look at the shape of these measures in the case when there are no foreign patterns:

$$\text{Accuracy} = \frac{TP + TN}{TP + FN + FP + TN} = \frac{TP}{TP + FN} = \text{Native sensitivity}$$

$$\text{Strict accuracy} = \frac{CC + TN}{TP + FP + FN + TN} = \frac{CC}{TP + FN}$$

$$\text{Fine accuracy} = \frac{CC}{TP} = \frac{\text{Strict accuracy}}{\text{Accuracy}} = \text{Native precision} \tag{4.11}$$

$$(\text{Native})\text{F-measure} = 2 \cdot \frac{\text{Precision} \cdot \text{Sensitivity}}{\text{Precision} + \text{Sensitivity}} = 2 \cdot \frac{\text{Accuracy}}{1 + \text{Accuracy}}$$

Accuracy measure evaluates the performance of the rejecting mechanism by counting (native) patterns, which are rejected and not classified. Therefore, from this perspective, the best rejecting mechanism is just the rule to not reject.

Strict accuracy measure evaluates the quality of joined rejecting mechanism and classifier. However, this measure does not offer a clear evaluation of the rejecting mechanism. On the one hand, if none of the patterns are rejected, then strict accuracy achieves its maximum. On the other hand, rejecting patterns classified to incorrect classes is highly desirable despite the fact that the value of strict accuracy becomes reduced. Gathering these arguments we may conclude that the ideal recognizer furnished with a rejection option should reject only incorrectly classified patterns, which will maximize strict accuracy measure. Yet, real-world recognizers with a rejection option are not ideal, and the strict accuracy measure does not clearly answer the question whether a rejection mechanism performance is desired or not. Therefore, we refer to another quality measure for recognition with rejection employed for native patterns only—fine accuracy.

Fine accuracy directly evaluates the quality of recognition with rejection for the scenario, in which incorrect classification is much more harmful than a lack of classification. In other words, the maximization of fine accuracy forces classification with high success rate and removes patterns difficult (or impossible) to classify correctly. Unfortunately, maximization of this measure may certainly lead to increasing the number of rejected patterns correctly classified to respective classes, which is a highly undesirable outcome. Hence, an F-measure like a combination of accuracy and fine accuracy would produce a required compromise. This type of combined measure would engage some priority mechanisms (factors) to favor either accuracy or fine accuracy:

$$(\text{Accuracy})\text{F-measure} = 2 \cdot \frac{\xi \cdot \text{Accuracy} \cdot (1 - \xi) \cdot \text{Fine accuracy}}{\xi \cdot \text{Accuracy} + (1 - \xi) \cdot \text{Fine accuracy}} \tag{4.12}$$

where $\xi \in (0, 1)$ defines a priority index (weight) of component measures.

Now, let us present an evaluation for two processing schemes: pure classification (without rejection) and classification with rejection applied to native patterns only. In both tasks the same C-class classifier is employed for classification. In the scenario with rejection, the local rejection mechanism (as in the previous part of this chapter) is employed. In order to compare both schemes, we look at two confusion matrices. The first one, obtained for classification with no rejection, is given in Table 4.7. Note that the row corresponding to the rejected native patterns includes all entries equal to zero as there is no rejection mechanism. Furthermore, the column corresponding to true foreign patterns includes zeroes as well since this experiment is based on the native patterns only. The second confusion matrix is given in Table 4.8.

TABLE 4.7 Confusion matrix for classification without rejection applied to the set of native patterns only

<table>
<tr><th rowspan="3" colspan="2"></th><th colspan="12">True Class</th></tr>
<tr><th colspan="10">Native</th><th rowspan="2">Foreign</th></tr>
<tr><th>0</th><th>1</th><th>2</th><th>3</th><th>4</th><th>5</th><th>6</th><th>7</th><th>8</th><th>9</th></tr>
<tr><th rowspan="10">Predicted class / Native</th><th>0</th><td>953</td><td></td><td>1</td><td></td><td></td><td>1</td><td>3</td><td></td><td>11</td><td></td><td></td></tr>
<tr><th>1</th><td>3</td><td>1121</td><td></td><td></td><td>1</td><td></td><td>6</td><td>1</td><td>2</td><td>4</td><td></td></tr>
<tr><th>2</th><td>2</td><td>5</td><td>1019</td><td>5</td><td></td><td></td><td>1</td><td>3</td><td>6</td><td>1</td><td></td></tr>
<tr><th>3</th><td>1</td><td>1</td><td>4</td><td>985</td><td></td><td></td><td></td><td></td><td>3</td><td>11</td><td></td></tr>
<tr><th>4</th><td></td><td></td><td>1</td><td></td><td>965</td><td>9</td><td>3</td><td>6</td><td>4</td><td>5</td><td></td></tr>
<tr><th>5</th><td></td><td>1</td><td></td><td>4</td><td></td><td>865</td><td>2</td><td></td><td>2</td><td>8</td><td></td></tr>
<tr><th>6</th><td>4</td><td>2</td><td></td><td></td><td>3</td><td>2</td><td>939</td><td></td><td>6</td><td></td><td></td></tr>
<tr><th>7</th><td></td><td>1</td><td>1</td><td>10</td><td></td><td>2</td><td></td><td>1007</td><td>2</td><td>15</td><td></td></tr>
<tr><th>8</th><td>15</td><td></td><td>4</td><td>5</td><td></td><td>8</td><td>4</td><td></td><td>932</td><td>13</td><td></td></tr>
<tr><th>9</th><td>2</td><td>4</td><td></td><td>1</td><td>13</td><td>5</td><td></td><td>11</td><td>6</td><td>952</td><td></td></tr>
<tr><th colspan="2">Foreign</th><td></td><td></td><td></td><td></td><td></td><td></td><td></td><td></td><td></td><td></td><td></td></tr>
</table>

Quantities FN, FP, and TN are not available in this case. CC = 9738 (the sum numbers on the main diagonal). Zeroes were omitted for the sake of clarity.

TABLE 4.8 Confusion matrix for classification with local rejection for the native set of patterns only, CC = 9558

| | | True Class | | | | | | | | | | |
| | | Native | | | | | | | | | | Foreign |
		0	1	2	3	4	5	6	7	8	9	
Predicted class	Native 0	943										
	1		1114			1		1	1	7	3	
	2		3	1003	2		0	1	1	2		
	3			2	961					1	3	
	4					949	8		4	3	4	
	5				3		850			1	3	
	6	3	1	1		2	2	925		1		
	7				8				992		9	
	8	9		1	2		4			897	6	
	9					8			4	1	924	
	Foreign	25	17	25	34	22	28	31	26	61	57	

Zeroes were omitted for the sake of clarity.

TABLE 4.9   Summary of recognition results without rejection and with rejection

	Without Rejection	With Rejection
TP	10,000	9674
CC	9738	9558
TP – CC	262	116
FN	0	326

TABLE 4.10   Characteristics of classification with and without rejection delineated by accuracy measures drawn from Tables 4.7 and 4.8

	Without Rejection	With Rejection
Accuracy	100.00	96.74
Strict accuracy	97.38	95.58
Fine accuracy	97.38	98.80
Accuracy F-measure	98.67	97.76

Values of measures are given as percentages.

Note that Table 4.8 is mostly a repetition of Table 4.3—with the difference that there are no foreign patterns. Importantly, the last row in Table 4.8 contains information on how many native patterns of respective classes were rejected.

Notice that the rejected patterns (the last row in Table 4.8) originate from the two categories: native patterns that would be correctly classified if they had not been rejected and native patterns that would be misclassified if they had not been rejected. We visually check if the rejection mechanism saved more from the first or from the second category by comparing the number on the main diagonal in Table 4.7 with numbers on the main diagonal in Table 4.8. The larger drop in those values we observe, the more harmful effect the rejection mechanism exerts on the recognition process. In contrast, if we compare content (numbers) outside the main diagonal (Table 4.7) with numbers outside the main diagonal in Table 4.8, we are satisfied if the drop is substantial as this would signify that the rejection mechanism rejected misclassified patterns. Ideally, we wish to have nonzero values only on the main diagonal.

Table 4.9 contains a summary of the results, of which values were obtained by summing appropriate cells from Tables 4.7 (without rejection) and 8 (with rejection). Table 4.10 contains quality evaluation measures needed to compare both recognition schemes. Adding the rejection mechanism helped reduce the number of misclassification by 146, that is, from 262 to 116. Comparing the values shown in Table 4.10, we observe that the rejection mechanism increased fine accuracy and decreased the remaining measures.

4.3   CLASSIFICATION WITH REJECTION: LOCAL CHARACTERIZATION

In this section, we discuss a local evaluation of classification methods. First of all, the classification quality from the perspective of a single class is discussed. Local evaluation technique is fine grained, and it is in opposition to the aggregative (global) evaluation technique, in which we use a single numerical value to describe some quantity. Local evaluation is extended to a multiclass problem. Here, we take into account quality of classification in terms of one class versus other classes. Then, we evaluate classifier's performance in such frames. Furthermore, we characterize classification from the perspective of single elements from outside of the main diagonal of confusion matrix, which exhibit numbers of incorrectly classified native patterns—in other words, local errors. Finally, we propose aggregative parameters to alternatively characterize global properties of classification.

4.3.1 Characterization of a Multiclass Problem

In the case of a multiclass classification problem, there is no single universal quality parameter. The variety of relevant quantities in a multiclass problem requires consideration of different viewpoints on classification's efficiency. The obvious adaptation of a local approach to evaluation of classification relies on considering measures individually for each class. In formulas (4.13), we define an adaptation of the accuracy measures to the local case. The adaptation of other measures is completed in a similar manner.

$$\text{Class accuracy}_l = \frac{\sum_{1 \le i \le C} \text{TP}_{i,l}}{\sum_{1 \le i \le C} \text{TP}_{i,l} + \text{FN}_{C+1,l}} = \frac{\text{TP}_l}{P_l} = \text{Class sensitivity}_l$$

$$\text{Strict class accuracy}_l = \frac{\text{TP}_{l,l}}{\sum_{1 \le i \le C} \text{TP}_{i,l} + \text{FN}_{C+1,l}} = \frac{\text{TP}_{l,l}}{P_l} = \text{Strict class sensitivity}_l$$

$$\text{Fine class accuracy}_l = \frac{\text{TP}_{l,l}}{\sum_{1 \le i \le C} \text{TP}_{i,l}} = \frac{\text{TP}_{l,l}}{\text{TP}_l} = \frac{\text{Strict class accuracy}_l}{\text{Class accuracy}_l}$$

$$(4.13)$$

for $l = 1, 2, \ldots, C$.

Let us reiterate the meaning of abbreviations used in (4.13):

- l—the class for which we compute the score
- $\text{TP}_{i,l}$—an element in i-th row and l-th column in confusion matrix, informing how many patterns from class l were classified to class i
- P_l—the number of all patterns in class l
- $\text{FN}_{C+1,l}$—the number of rejected patterns from the native class l

Class accuracy$_l$ is expressed as a proportion of the number of native patterns coming from class l being identified as native, that is, classified to any native class, to the number of all patterns from class l. The interpretation of the previous two measures (strict class accuracy and fine class accuracy) is similar.

Besides the measures discussed earlier, we use *Local error*$_{k,l}$ measure. This measure is viewed as a proportion of the number of native patterns coming from class l and wrongly classified to class k and cardinality of the class l:

$$\text{Local error}_{k,l} = \frac{\text{TP}_{k,l}}{\sum_{1 \le i \le C} \text{TP}_{i,l} + \text{FN}_{C+1,l}} = \frac{\text{TP}_{k,l}}{\text{P}_l}, \quad k \ne l \tag{4.14}$$

for $k, l = 1, 2, \ldots, C$.

However, there is a need for a combined evaluation of a solution with respect to all classes treated en block. Modified variations characterize different aspects of classification and rejecting quality. The variations of accuracy measure defined by (4.15) and (4.16) explain the adaptation of measures that express global characteristics of multiclass problem. Adaptation of other measures is straightforward. With regard to the local error, this adaptation is very similar with an exception that it concerns two classes.

The first modification comes as the mean value of class measures:

$$\text{Mean strict class accuracy} = \frac{1}{C} \sum_{1 \le l \le C} \text{Strict class accuracy}_l$$

$$\text{Mean local error} = \frac{1}{C^2 - C} \sum_{1 \le k \le C} \sum_{1 \le l \le C, l \ne k} \text{Local error}_{k,l} \tag{4.15}$$

The next modification concerns the maximal and minimal values of class local measures. Examples of such measures are expressed in the form

$$\text{Min strict class accuracy} = \min_{1 \le l \le C} \{ \text{Strict class accuracy}_l \}$$

$$\text{Max local error} = \max_{1 \le l \le C, 1 \le k \le C, l \ne k} \{ \text{Local error}_{k,l} \} \tag{4.16}$$

Variants of accuracy and error measures defined in (4.15) and (4.16) express global characteristics of a multiclass problem by aggregating information on a local level.

Let us notice that accuracy (as in (4.11)) has similar characteristics. Let us reiterate that accuracy favors classes of bigger cardinalities, while influence of infrequent classes is imperceptible. Its application is justified when numerous classes are of similar cardinalities and detection of patterns of small ones is less important. Alternatively one may apply weighting factors to mediate for this property.

The mean class accuracy is suitable to be applied when a dataset is imbalanced and no weighting factors are needed. It equalizes numerous classes with small ones; that is, it makes them uniform. In practice, it favors small classes. For instance, when there are more infrequent classes than frequent ones, a good classifier's performance observed on small classes increases this factor too much.

Conversely, the adaptations in (4.16) reflect the worst case of local class measures. Min class accuracy informs about classification quality of the class that achieved the worst local accuracy. Max local error informs about the worst case of native patterns from one class, which are falsely classified to another class. By analogy we can construct other measures, for instance, max class accuracy and mean local error.

4.3.2 Illustrative Example

Let us illustrate the performance of the local evaluation measures in an experiment dealing with a multiclass classification problem. The experiment aimed at processing

TABLE 4.11   References to class names, which are used in Tables 4.12 and 4.13

0	1	2	3	4	5	6	7	8	9
Accent	Clef C	Clef G	Fermata	Flat	Natural	Thirty-second rest	Quarter rest	Eighth rest	Sharp

symbols of music notation. The dataset has been introduced and described in Section 3.4.1. For the sake of clarity, in this experiment we used only 10 selected representative classes of this dataset. So then, in this problem, we consider 10 native classes. In particular, we used four classes that are infrequent, namely, accent, clef C, fermata, and thirty-second rest. The most rare class is thirty-second rest with only 26 representatives. We also took six regular classes: clef G, flat, natural, quarter rest, eighth rest, and sharp. The most frequent class is quarter rest, which contains 3024 samples.

Note that instead of true class names, we used numbers in order to fit the forthcoming tables on the screen. The reference numbers are shown in Table 4.11.

Having in mind that the dataset is challenging to process, we decided to form the model on a modified dataset. We oversampled rare classes and undersampled frequent classes so that the actual processed dataset was made of 500 samples of each class, which in total gave us a dataset comprising 5000 patterns. Pattern generation technique was according to the *on intervals* scheme presented in Algorithm 1.1. In addition, we used standardized feature values. We split patterns into the training and test sets; the training set was made of 350 samples of each class (70% of all samples). We selected first 8 features out of an ordered list of 20 best features. The list of 20 best features is in Appendix 1.B. Altogether the number of data used for classification was modest and quite heavily preprocessed.

We used a binary tree classifier constructed with SVMs as a classification mechanism. The tree was developed using k-means on class centers and is structured as shown in Figure 4.1. SVMs forming the tree were trained with gamma parameter set to 0.01 and cost parameter to 1; 10-fold cross-validation was applied. We supplemented the binary tree with local rejection option shown in Figure 3.15. Rejecting mechanisms were trained according to the *one-versus-all-other-classes* method presented in Algorithm 3.1. We used binary SVMs with gamma and cost parameters being the same as for classifying models. Rejecting mechanisms, like classification model, were trained on the balanced dataset.

In the experiment, we compare native pattern rates of classifications without and with rejection. However, testing is performed on the entire dataset of standardized 10 class patterns, not on the set of balanced patterns. As a result, the testing set is made of 17,102 samples. A limited number of these samples were used in the training procedure (they were used in the balanced set); we estimate that it was around 3000 samples. The remaining were not used in the training process.

In Tables 4.12–4.14, we present a range of characteristics describing two experiments: musical symbol classification with and without rejection. Tables 4.12 and 4.13 contain confusion matrices and three selected local quality measures: class accuracy, strict class accuracy, and fine class accuracy. Table 4.12 concerns the

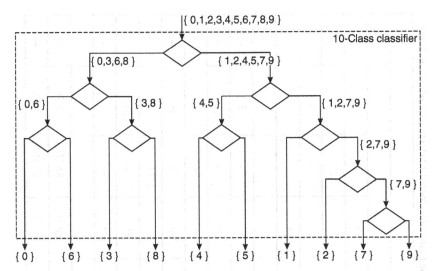

Figure 4.1   The structure of binary tree-based classifiers used in musical symbol classification.

experiment on classification without rejection performed on the set of native patterns only. Table 4.13 illustrates results for classification with rejection applied to native and foreign patterns. Comparing local measures in both tables, it is clear that rejecting option increases fine accuracy. However, this gain is sometimes quite costly. For instance, 602 patterns correctly classified to class no. 5 (naturals) and 106 incorrectly classified naturals were rejected. On the other hand, only 12 out of 118 incorrectly classified naturals were not rejected; that is, only 1 out of 10 naturals classified to incorrect class was not rejected (!). These two contrasting observations are reflected in local measures: class accuracy and strict class accuracy. The values of these measures were reduced from 100% and 96% to circa 70%, while fine class accuracy reached almost 100%. In contrast, in the case of class no. 4 (flats), the cost of losing 57 correctly classified flats was compensated by rejecting 136 wrongly classified flats. Moreover, only 25 incorrectly classified patterns were not rejected. As in the previous case, these observations are clearly reflected in local measures: fine class accuracy was increased from 93.73% to 98.95%, while strict class accuracy was reduced from 93.73% to 91.51%.

Table 4.14 contains statistics on minimal and mean values of different accuracies. We also see maximal and mean local error. Fine accuracy and local error reflect positive aspects of the supplemented rejection mechanism—rejection is a method to remove native patterns that would have been misclassified otherwise. Even if the comprehensive evaluation points us toward the conclusion that the rejection mechanism improved recognition for the dataset under investigation, we have to denote that it affects (in a smaller, but still noticeable degree) classification of native patterns that would be correctly classified as well. This is reflected in Table 4.14 in the deterioration of class accuracy and strict accuracy.

At the first sight, one may see quite alarming rates of minimum class accuracy and minimum strict accuracy, both equal to 53.85%. However, if we look in detail in

TABLE 4.12 Confusion matrix for classification without rejection applied to the set of symbols of music notation (selected 10 classes) standardized features

| Predicted class | | True Class | | | | | | | | | | |
| | | Native | | | | | | | | | | Foreign |
		0	1	2	3	4	5	6	7	8	9	
Native	0	516										—
	1		460		11	8	1			22	9	—
	2		1	1932			12		14	7	7	—
	3	15			273							—
	4					2408	23		84	6	8	—
	5		1	3		56	2299		78	3	10	—
	6			105		40	13	26	11	74	2	—
	7			35		11	27		2743	2	68	—
	8	2		3		20	1		1	2321	4	—
	9		2	25		26	11		93	4	2455	—
	Foreign	—	—	—	—	—	—	—	—	—	—	—
Class accuracy												100.00
Strict class accuracy		96.81	99.14	91.87	96.13	93.73	96.31	100.0	90.71	95.16	95.79	
Fine class accuracy		96.81	99.14	91.87	96.13	93.73	96.31	100.0	90.71	95.16	95.79	

Empty cells contain zero patterns; we skipped zeros for the sake of clarity.

TABLE 4.13 Confusion matrix for classification with local rejection for selected 10 classes from the dataset of music notation symbols, standardized features

| | | True Class | | | | | | | | | | |
| | | Native | | | | | | | | | | Foreign |
Predicted class		0	1	2	3	4	5	6	7	8	9	Foreign
Native	0	516			4	2				5		18
	1		454								1	62
	2			1755			4		6		1	
	3				268							57
	4					2351	2		8	4	3	47
	5			2		1	1667				4	3
	6			9		6	1	14	2	2		27
	7	1		4		2			2129		41	14
	8	1		13		1				2137		51
	9					13	5		37		1914	20
	Foreign	15	10	320	12	193	708	12	842	291	599	411
	Class accuracy	97.19	97.84	84.78	95.77	92.49	70.34	53.85	72.16	88.07	76.63	
	Strict class acc	96.81	97.84	83.45	94.37	91.51	69.84	53.85	70.40	87.62	74.68	
	Fine class acc	99.61	100.0	98.43	98.53	98.95	99.29	100.0	97.57	99.49	97.45	

TABLE 4.14 Characteristics of classification with and without rejection delineated by accuracy measures drawn from Tables 4.12 and 4.13

	Without Rejection	With Rejection
Min class accuracy	100.00	53.85
Mean class accuracy	100.00	82.91
Min strict accuracy	90.71	53.85
Mean strict accuracy	95.56	82.04
Min fine accuracy	90.71	97.45
Mean fine accuracy	95.56	98.93
Max local error	4.99	1.60
Mean local error	0.49	0.10

Values of measures are given as percentages.

Table 4.12, we note that these results concern the most infrequent class in our data-set—thirty-second rest. This class contains just 26 samples. Results were not scaled, so these, perhaps alarming, rates are in fact an effect of a lack of proper scaling, rather than some drastic underperformance of the model. If we would have had removed the thirty-second rest class, the second worst results are equal to 70.34% (class accuracy) and 69.84 (strict class accuracy), which are quite satisfying.

A formulation of further measures that may be directed to enhance the evaluation of some selective aspects of a model is straightforward. It is a common practice to develop problem-oriented quality evaluation methods, as we presented in this chapter. Let us mention that there is a substantial volume of literature, where various problem-specific quality measures are discussed, for instance, Foody (2005) for thematic classification, Okeke and Karnieli (2006) for fuzzy classification in remote sensing, and Crum et al. (2006) and Gegundez-Arias et al. (2012) for medical image analysis and others. Naturally, one may also encounter a great deal of studies about general use quality evaluation techniques, for instance, in Bradley (1997) concerning ROC curves and in Kautz et al. (2017) focused on the evaluation of a generic multiclass classification problem. In addition, one may consult Sokolova and Lapalme (2009) for a review of various performance measures.

4.4 CONCLUSIONS

The presented methodology for evaluating quality of native pattern recognition with foreign pattern rejection is applied in other chapters of this book. We believe that the presented range of measures is beneficial to flexibly illustrate the performance of the classifier. Naturally, these measures are of a very generic character and can be easily adapted to a broad spectrum of problems.

REFERENCES

A. P. Bradley, The use of the area under the roc curve in the evaluation of machine learning algorithms, *Pattern Recognition* 30(7), 1997, 1145–1159.

W. R. Crum, O. Camara, and D. L. G. Hill, Generalized overlap measures for evaluation and validation in medical image analysis, *IEEE Transactions on Medical Imaging* 25(11), 2006, 1451–1461.

G. M. Foody, Local characterization of thematic classification accuracy through spatially constrained confusion matrices, *International Journal of Remote Sensing* 26(6), 2005, 1217–1228.

V. Garcia, J. S. Sanchez, R. A. Mollineda, R. Alejo, and J. M. Sotoca, The class imbalance problem in pattern recognition and learning. In: *II Congreso Espanol de Informatica*, Zaragoza, Spain, September 11–14, 2007, 283–291.

M. E. Gegundez-Arias, A. Aquino, J. M. Bravo, and M. Diego, A function for quality evaluation of retinal vessel segmentations, *IEEE Transactions on Medical Imaging* 31(2), 2012, 231–239.

H. He and E. A. Garcia, Learning from imbalanced data, *IEEE Transactions on Knowledge and Data Engineering* 21(9), 2009, 1263–1284.

W. Homenda, A. Jastrzebska, and W. Pedrycz, *The web page of the classification with rejection project*, 2017, http://classificationwithrejection.ibspan.waw.pl (accessed October 5, 2017).

W. Homenda and W. Lesinski, Imbalanced pattern recognition: Concepts and evaluations. In: *Proceedings of the 2014 International Joint Conference on Neural Networks*, Beijing, China, July 6–11, 2014, 3488–3495.

T. Kautz, B. M. Eskofier, and C. F. Pasluosta, Generic performance measure for multiclass-classifiers, *Pattern Recognition* 68, 2017, 111–125.

Y. LeCun, C. Cortes, and C. J. C. Burges, *The MNIST database of handwritten digits*, 1998, http://yann.lecun.com/exdb/mnist/ (accessed October 5, 2017).

F. Okeke and A. Karnieli, Methods for fuzzy classification and accuracy assessment of historical aerial photographs for vegetation change analyses. Part I: Algorithm development, *International Journal of Remote Sensing* 27(1), 2006, 153–176.

F. Provost and P. Domingos, Tree induction for probability-based ranking, *Machine Learning* 52(3), 2003, 199–215.

F. Provost and T. Fawcett, Robust classification for imprecise environments, *Machine Learning* 42(3), 2001, 203–231.

M. Sokolova and G. Lapalme, A systematic analysis of performance measures for classification tasks, *Information Processing & Management* 45(4), 2009, 427–437.

Y. Sun, M. S. Kamel, A. K. C. Wong, and Y. Wang, Cost-sensitive boosting for classification of imbalanced data, *Pattern Recognition* 40(12), 2007, 3358–3378.

Y. Tang, Y. Q. Zhang, N. V. Chawla, and S. Krasser, SVMs modeling for highly imbalanced classification, *IEEE Transactions on Systems Man and Cybernetics Part B-Cybernetics* 39(1), 2009, 281–288.

CHAPTER 5

RECOGNITION WITH REJECTION: EMPIRICAL ANALYSIS

The problem of classification of native patterns with foreign pattern rejection is first and foremost driven by practical applications, in which we need to process contaminated datasets. In this chapter, we apply the methods introduced and discussed in the previous chapters.

First, we present an empirical evaluation of the three rejecting architectures: global, local, and embedded. Let us recall that these are ensemble models that could be constructed, for example, with the use of binary classifiers. We process handwritten digits and symbols of printed music notation.

Next, we propose another kind of rejecting method, based on geometrical figures. We study how elementary geometrical figures such as hyperrectangles and ellipsoids play an essential role of rejecting mechanisms. Geometrical figures enclose native patterns, and by the rule of exclusion, we reject patterns located outside the area occupied by such figures. An empirical evaluation of this approach is also based on the datasets of handwritten digits and symbols of printed music notation.

There are two essential aspects of recognition enhanced by rejection option, namely, a quality of foreign pattern rejection and an influence of a rejection mechanism on native pattern classification. Ideally, we wish to remove as many foreign patterns as possible, but in practice increasing the rate of foreign pattern rejection increases also the rate of native pattern rejection. These two conflicting criteria of the rate of acceptance of native patterns and the rate of rejection of foreign patterns should be balanced. Geometrical regions-based rejection arises as an intuitive example of a model that allows a flexible adjustment of these two criteria. Manipulating the volume of such geometrical regions enclosing native patterns helps us control the balance between rejection and acceptance rates. Another important aspect is an influence of the rejection mechanism on native pattern classification. We demonstrate that a rejection mechanism can improve classification quality by rejecting patterns that

Pattern Recognition: A Quality of Data Perspective, First Edition. Władysław Homenda and Witold Pedrycz.
© 2018 John Wiley & Sons, Inc. Published 2018 by John Wiley & Sons, Inc.

would be incorrectly classified. However, in an empirical study, we show that in real-world applications, there is always some trade-off.

5.1   EXPERIMENTAL RESULTS

In the following sections we investigate properties of the three rejection models: global, local, and embedded. The evaluation is based on a series of experiments on handwritten digits and printed music notation recognition. In order to implement the three architectures, we use either support vector machines (SVMs) or random forest (RF) algorithm. These are two well-performing classification methods, both described in detail in Chapter 2. We use and compare results for the two different algorithms to show that key properties of the proposed models are, in a nutshell, independent on the choice of the algorithm.

The role of foreign patterns in the experiments is played by crossed-out handwritten digits and Latin alphabet letters for the set of handwritten digits and garbage patterns from music notation for the set of symbols of printed music notation. It shall be mentioned that foreign patterns are very realistic, which means that some of them are similar to actual native patterns, for instance, digit 0 and letter o, digit 6 and letter b, and so on. Also, few foreign patterns of music notation are similar to native patterns of this set, but the difference is that foreign patterns were those cut out imprecisely and are not completely correct (for instance, only two thirds of some pattern was cut out).

5.1.1   Comparison of Rejecting Architectures

We conducted an experiment in which all four architectures shown in Figures 3.13–3.16, were employed to process the training set and the test set of handwritten digits as native patterns. Detailed results of classification without rejection (Figure 3.13; SVMs were used as binary classifiers) and classification with rejection in local architecture (Figure 3.15; also SVMs were used as binary classifiers) are presented in the confusion matrix shown in Table 5.1. Both classifiers are displayed in Figures 3.13 and 3.15, respectively. Binary classifiers in both architectures are realized as binary SVM with parameters set up as outlined previously, that is, $\gamma = 0.0625$, $C = 1$; furthermore 10-fold cross-validation was considered.

In Table 5.1, the columns represent prediction results, that is, we have there the number of patterns from corresponding classes, which are classified to classes corresponding to the rows. And vice versa, the rows represent actual results, that is, in the rows we have the number of patterns corresponding to the columns (cf. Section 4.1). Therefore, on the main diagonal, we have the numbers of patterns correctly classified to corresponding classes. In the case of classification with rejection, in the last row we have the number of rejected patterns from native classes $FN_{C+1,l}$, $l = 1, 2, …, C$, and, in the last element $FN_{C+1,C+1}$ of this row, which is denoted TN, we have the number of rejected foreign patterns. When classification with rejection is applied only for native patterns, we skip the last column since it has only zeroes (there are no foreign patterns).

TABLE 5.1   Confusion matrices for the classifiers without and with rejection shown in Figures 3.13 and 3.15, respectively

			0	1	2	3	4	5	6	7	8	9
Classification without rejection (in %)	The training set	0	670			1		1	2		5	
		1	1	789	1							1
		2	2	3	717	3					4	
		3	1		1	695	0	6			2	5
		4					681			5	3	2
		5				1		606	1		2	1
		6	3	1				2	666		2	
		7				5				710	1	8
		8	8		2	2		6	2		658	5
		9	1	1		1	6	3		5	5	684
	The test set	0	283						1		6	
		1	2	332	1		1		6	1	2	3
		2		2	302	2			1	3	2	1
		3		1	3	290		3			1	6
		4			1		284		3	1	1	3
		5		1		3		259	1			7
		6	1	1			3		273		4	
		7		1	1	5		2		297	1	7
		8	7		2	3		2	2		274	8
		9	1	3			7	2		6	1	268
Classification with rejection (in %)	The training set	0	662								4	
		1		788								
		2		2	707	2					1	
		3			1	684		5				2
		4				0	673			4	3	2
		5				1		595			1	1
		6	2					2	662		1	
		7				4		0		701	0	5
		8	4					4		0	640	3
		9					5			2	1	662
		−1	18	4	14	16	9	18	9	13	31	31
	The test set	0	281								3	
		1		326			1		1	1		3
		2		1	296				1	1	1	
		3			1	277		3			1	1
		4					276					2
		5				2		255				2
		6	1	1			2		263			
		7			1	4				291		4
		8	5		1	2					257	3
		9					3			2		262
		−1	7	13	11	18	13	10	22	13	30	26

Zeroes were omitted for the sake of clarity. −1 stands for the group of rejected patterns.

Looking at Table 5.1, we see no drastic discrepancies in classification rates for a case of classification with rejection in comparison with classification with no rejection. It indicates that adding the rejection mechanism did not hinder the quality of classification. More importantly, the rejection mechanism is able to remove native patterns falsely classified—these are those native patterns that without rejecting would be incorrectly classified otherwise. Let us draw attention to elements beside the main diagonals in Table 5.1. In confusion matrices for the model including the rejection mechanism, there are far more blank spots (that stand for zero) than for confusion matrices for the model with no rejection. It has to be stressed, though, that when we added the rejection mechanism, the number of correctly classified elements dropped a bit as well. However, the number of rejected correctly classified patterns (loss) is smaller than the number of rejected misclassified patterns (gain). The following quantities in terms of maximal and average values for all 10 classes confirm the aforementioned observations. For the training set, we have:

- 1.45% points—Loss (decrease) of strict accuracy (from 98.24 to 96.79%)
- 0.83% points—Gain (increase) of fine accuracy (from 98.24 to 99.09%)
- Maximal and average gains (decrease) on errors are:
  - ∘ 1.68% points—Maximal gain on predictive class error, that is, for patterns of a given class incorrectly classified to other classes (from 3.52 to 1.84%)
  - ∘ 0.89% points—Gain on the average of predictive class errors (reduction from 1.79 to 0.90%)

  For the test set corresponding values are:

- 2.70% points—Loss (decrease) of strict accuracy (from 95.37 to 92.77%)
- 2.73% points—Gain (increase) of fine accuracy (from 95.37 to 98.10%)
- Maximal and average gains (decrease) on errors are:
  - ∘ 6.60% points—Maximal gain on predictive class error (from 11.55 to 4.95%)
  - ∘ 2.85% points—Gain on the average of predictive class errors (from 4.65 to 1.80%)

Hence, we may conclude that the studied rejection mechanism improved the overall recognition quality.

In Table 5.2, we present generic measures evaluating the influence of various rejection schemes on the classification outcome. Measures presented in this table were discussed in Chapter 4. The upper part of the table concerns the training set, and the lower part, the test set (confusion matrices of classification without rejection and with local rejection are presented in Table 5.1). We compare global, local, and embedded architectures based on the RF and on the SVM binary classifiers. In addition, we present classification quality in a case when we do not perform rejection. We show results for the C-class classifier from Figure 3.13, used in the local and global architectures and results produced for the embedded architecture implemented as illustrated in Figure 3.15. The results were reported for the features space made of 24 best features computed with the best-first search with SVM as a feature evaluator (for details on features selection, consult Chapter 1).

TABLE 5.2   Comparison of classification results with rejection (global, local, and embedded architectures) on the train and test sets of native patterns of handwritten digits with classification results without a rejection mechanism

	No Rejection		Global		Local		Embedded	
Rejecting Architecture								
Base Classifier	RF	SVM	RF	SVM	RF	SVM	RF	SVM
Dataset			*Native patterns, training set*					
Accuracy, sensitivity	100.00	100.00	100.00	99.69	100.00	97.67	100.00	97.19
Strict acc., strict sens.	100.00	98.24	100.00	98.09	100.00	96.79	100.00	96.43
Fine acc., strict precision	100.00	98.24	100.00	98.39	100.00	99.09	100.00	99.22
Max strict class accuracy	100.00	99.37	100.00	99.37	100.00	99.24	100.00	98.74
Min strict class accuracy	100.00	96.48	100.00	95.75	100.00	93.77	100.00	92.96
Maximal local error	0.00	1.17	0.00	1.42	0.00	0.80	0.00	0.80
Mean local error	0.00	0.20	0.00	0.21	0.00	0.10	0.00	0.05
Dataset			*Native patterns, test set*					
Accuracy, sensitivity	100.00	100.00	89.47	96.70	88.90	94.57	88.17	94.04
Strict Acc., strict sens.	94.04	95.37	87.37	93.70	87.34	92.77	86.70	92.34
Fine Acc., strict precision	94.04	95.37	97.65	96.90	98.24	98.10	98.34	98.19
Max strict class accuracy	96.77	97.42	94.43	96.45	94.43	95.60	94.13	95.58
Min strict class accuracy	87.79	88.45	75.00	86.47	75.00	86.47	72.95	85.81
Maximal local error	4.29	2.64	2.05	2.64	1.65	1.70	1.65	1.70
Mean local error	0.66	0.52	0.23	0.35	0.17	0.20	0.16	0.18

RF, results for random forests; SVM, results for support vector machines.

As mentioned before, all SVMs in the experiment have been constructed with a 10-fold cross-validation in order to avoid overfitting. For RF cross-validation was not applied since the RF construction procedure itself can produce a classifier with reduced variance. RF aggregate and average multiple decision trees built on different parts of the same training set. We have elaborated on details of RF in Section 2.5.3. Nonetheless, results on the test set are noticeably poorer than on the training set.

On the training set, for classifiers employing binary RF, all positive measures (accuracy, sensitivity, precision) are equal to 100%, while negative measures (errors) are equal to 0. However, on the test set, positive measures are significantly smaller, while negative measures substantially worse (increased). Classifiers employing binary SVMs perform slightly worse on the training set, but much better on the test set compared with RF-based classifiers.

The results presented in Table 5.2 show that local architecture leads to the lowest deterioration in the quality of native pattern classification, while global architecture performs slightly worse than the embedded architecture. On the other hand, individual measures indicate that the quality ranking may be different.

5.1.2   Reducing the Set of Features

Classification is a data-driven procedure, in which data are represented with some measurable characteristics. As we have mentioned and described in Chapter 1, these characteristics are features. The classification outcome is strictly related to the quality of features space in which we perform processing in order to discriminate between different classes of patterns. Poor quality of data manifests itself in two aspects. First, it is the cardinality of the computed set of features that corresponds to dimensionality of features space. The second aspect is the quality of the produced features. We have elaborated on both issues in Chapter 1. In this section let us briefly revisit the first aspect: dimensionality of features space. Let us illustrate it with a practical example of a handwritten digits recognition problem and the influence of the number of features on the recognition outcome.

The issue with the number of features is that there is an apparent property that the more features we have, the more we are prone to overfit a classifier during the training procedure. On the other hand, a small number of features often do not suffice to make a good model. The right solution is to find a balance: we need not too many and not too few features to train a model. An extended discussion on this topic is in Section 1.4. In this section we empirically verify the effect of limiting the space of features on the performance of classifiers and how the quality of classification was improved by rejecting misclassified elements in different feature spaces. We continue with the discussion initiated in Section 3.4.5.

The experiment presented in this study relies on reducing the set of features by four and repeating the procedure outlined in Section 3.4.5, for the reduced set of features. Let us recall that up to this point we employed the set of 24 features to construct

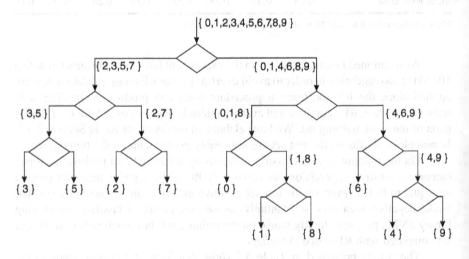

Figure 5.1 Structure of binary tree-based classifiers and its textual representation for the handwritten digits dataset.

the full classifier and rejecting mechanism studied in that section. This set of features was selected with greedy forward selection with the limited expansion method defined in Algorithm 1.8. Features were then sorted according to decreasing values of the SVM-based evaluator, and the four of them with the lowest values of this test were dropped at each reduction step. Finally, models were built for sets of features with cardinalities 24, 20, 16, 12, 8, and 4, each time the best features were used. For each set of features, the classifier structure (binary tree) is determined with k-means clustering. Parameters of all SVM components are as follows: Gaussian kernel function, cost equal to 1, γ equal to 0.0625, and 10-fold cross-validation were utilized (cf. Section 3.4.3). At the final stage, rejecting SVM classifiers were also trained on the basis of these values of parameters. Let us mention that each SVM classifier could be individually tuned in order to provide optimal values of gamma and regularization parameters.

In Figure 5.1, we include a textual representation of the classifier coming from Figure 3.13. Notice that each node of the tree is represented with the sequence of digits forming the corresponding class labels. For instance, the root is denoted by the sequence of all classes 0123456789, because all classes are present at the input of the root. In contrast, leaves are denoted with single digits. By analogy we represent other nodes. In Figure 5.2, we present classifiers constructed for full and for reduced sets of features. Numerical results are shown in Table 5.3.

In Figure 5.2, the classifier for the complete set of 24 features is fully represented in the textual form in the upper part of the diagram. Classifiers for the consecutive sets of features are represented mostly as strings of dots: dots replaced digits not changed compared with the corresponding ones in the classifier for a larger features set. Note that the first split of all 10 classes into two clusters is exactly the same for all sets of features, except for models constructed for the smallest sets of features with just 8 and 4 features. We observe that for smaller sets of features, there were more significant changes in the structure of the classifier.

The results displayed in Figure 5.2 and Table 5.3 lead to the following conclusions:

1. For recognition without rejection, the accuracy is equal to 100% because there are no rejected patterns (no false negatives). For the same reason strict accuracy and fine accuracy are equal.

2. Of course, performance quality deteriorates as we decrease the number of features.

3. Models with 20, 16, and 12 features are slightly worse in comparison with the model with the full set of 24 features.

4. Models with 8 and 4 features (less than half of the full set) perform much worse. Class distribution in the classifier tree changed considerably, which is manifested in significant changes in similarity/dissimilarity measures between classes that are in the background of method forming the binary tree classifier. For these feature spaces, there is an additional level of the binary classifier tree. The drop in the performance is huge, especially for the smallest model with 4 features: 20–30% points lower accuracy measures and 10–20% points increased

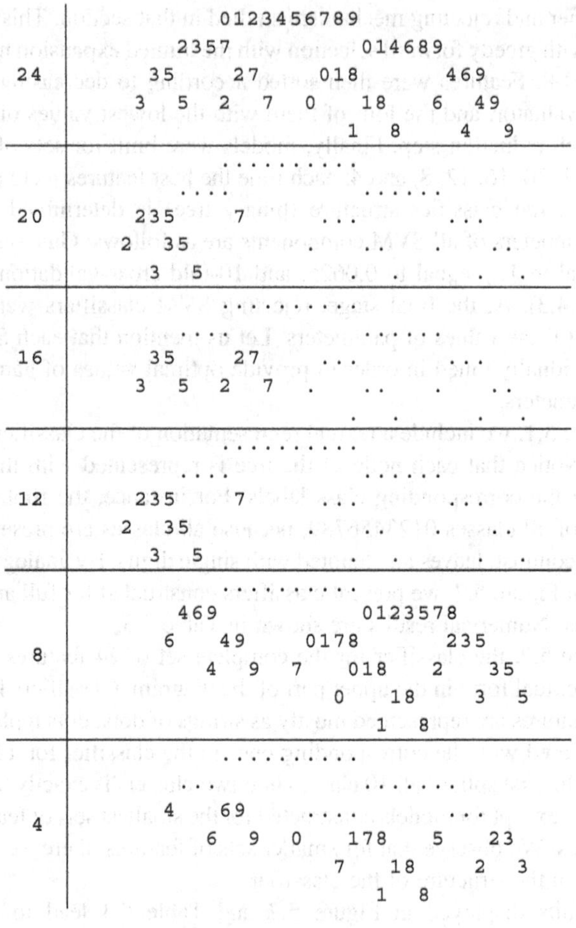

```
                          0123456789
                2357                014689
   24       35      27      018           469
            3  5  2  7    0    18     6   49
                          1    8       4    9
          ─────────────────────────────────────
                     . . . . . . . . .
                   . . . .           . . . . . .
   20       235     7      ...            ...
            2  35           .    ..    .  ...
            3  5  5         .    ..    .   .
          ─────────────────────────────────────
                     . . . . . . . . .
                   . . . .           . . . . . .
   16       35      27      ...            ...
            3  5  2  7      .    ..    .   ..
                           .   .   .      .   .
          ─────────────────────────────────────
                     . . . . . . . . .
                   . . . .           . . . . . .
   12       235     7      ...            ...
            2  35           .    . .    . ...
            3  5            .    . .    .   . ..
          ─────────────────────────────────────
                     . . . . . . . . .
                469                0123578
             6      49      0178         235
                4  9  7      018     2    35
                          0    18     3    5
                          1    8
          ─────────────────────────────────────
                     . . . . . . . . .
                   . .               . . .
    4           4     69     ....         ...
                6   9  0    178     5    23
                          7   18     2    3
                          1    8
```

Figure 5.2 Structure of binary tree-based classifiers for various sets of features: of size 24, 20, 16, 12, 8, and 4. The classifier for the full set of 24 features is fully represented in the textual form (it is the first shown structure). Dots represent digits (class labels), which are not changed in comparison with the classifier above.

    maximal error in comparison with the performance reported for the model with 12 features model.

5. As we reduce the number of features, we observe the deterioration of global measures (accuracy, strict accuracy, fine accuracy, and errors).

6. It is especially worth to draw attention to the fact that classification with rejection keeps fine accuracy and errors almost stable.

5.1.3 Quality of Classifier Versus Rejecting Performance

Another experiment is aimed at comparing rejecting performance for basic classifiers of different quality. In this experiment we operate again in the features space formed

TABLE 5.3 Classification performance of classifiers constructed in features spaces of different dimensionalities: ranging from 24 to 4 features

Num. Features	24		20		16		12		8		4	
Dataset	Train	Test	Train	Test	Train	Test	Train	Test	Train	Test	Train	Test
Classification without rejection												
Accuracy	100.00	100.00	100.00	100.00	100.00	100.00	100.00	100.00	100.00	100.00	100.00	100.00
Strict accuracy	98.24	95.37	97.66	94.97	96.47	94.50	94.80	93.00	88.37	87.64	75.18	75.21
Fine accuracy	98.24	95.37	97.66	94.97	96.47	94.50	94.80	93.00	88.37	87.64	75.18	75.21
Max St. Cl. Acc.	99.37	97.42	99.27	97.07	99.12	97.65	98.49	96.77	94.32	94.92	90.39	92.88
Min St. Cl. Acc.	96.48	88.45	95.75	88.78	94.05	88.45	92.21	87.46	82.53	77.23	55.96	55.78
Max local error	1.17	2.64	1.60	3.63	2.50	4.29	3.35	6.12	12.66	14.52	21.61	25.48
Mean local err.	0.20	0.52	0.26	0.56	0.40	0.61	0.58	0.78	1.31	1.39	2.76	2.76
Classification with rejection												
Accuracy	97.67	94.57	97.97	94.94	96.51	95.20	95.16	94.24	90.68	90.60	73.88	74.08
Strict accuracy	96.79	92.77	97.39	93.04	95.33	92.17	92.81	90.57	84.61	83.54	64.61	64.85
Fine accuracy	99.09	98.10	99.40	98.00	98.77	96.81	97.54	96.11	93.30	92.20	87.45	87.54
Max St. Cl. Acc.	99.24	95.60	98.99	96.13	98.36	96.94	98.24	95.60	90.68	91.56	88.06	88.81
Min St. Cl. Acc.	93.77	86.47	95.60	87.13	91.50	86.14	87.54	84.49	75.80	71.29	35.93	32.01
Max local error	0.80	1.70	0.64	1.98	1.28	4.22	2.62	3.30	9.13	11.22	11.46	14.85
Mean local error	0.10	0.20	0.07	0.21	0.13	0.34	0.26	0.41	0.69	0.80	1.03	1.03

Results concern classifying models alone (upper part) and classifiers with rejection (bottom part). Results are expressed as percent. Reported are the following measures: accuracy, strict accuracy, fine accuracy, class maximal accuracy, class minimal accuracy, maximal local error, and mean local error.

of the best 24 features. We tested and compared global and local rejection mechanisms added to various 10-class classifiers. We compared the selected results. In particular, we wanted to collate results achieved by:

- A rejecting/classifying model that exhibits high classification rates on training and tests sets (with strict accuracy equal to 99% and 96% on the training set and the test set, respectively)

- An overfitted rejecting/classifying model with strict accuracy equal to 100% and 82% on the training/test sets

- A model of an average quality with strict accuracy equal to 95% and 94% on the training/test sets

- A poor model with strict accuracy equal to 92% and 92% on the training and test set, respectively

We discussed the tree-based classifier shown in Figure 3.13, employed in the global and in the local architecture as shown in Figures 3.14 and 3.15. Based on the architecture shown in Figure 3.13, we constructed four different classifiers of different quality: good quality, overfitted, average quality, and poor quality ones. We applied SVMs to form all four models. All kernels were Gaussian: in all cases there was 10-fold cross-validation employed, but we intentionally manipulated with the values of γ and cost parameters in order to achieve the desired classification properties. Detailed results for the poor quality classifier supplemented with the local rejecting architecture are presented in the form of confusion matrices shown in Table 5.4. A summary of the results obtained from all configurations is outlined in Table 5.5.

Basic C-class classifier of good quality, overfitted, average quality, and poor one was then equipped with rejecting mechanism in global and local architecture as shown in Figures 3.14 and 3.15. The parameters of rejecting SVMs were set up as described in Sections 3.4.3 and 3.4.5, that is, $\gamma = 0.0625$, $C = 1$, and with 10-fold cross-validation. The results were reported for the handwritten digits in the features space made of 24 best features computed using best-first search with SVM as feature evaluator (for details on features selection, consult Chapter 1).

Detailed results for poor quality classifier employed in local architecture are presented in the form of the confusion matrices in Table 5.4. Summarized results obtained from all configurations are reported in Table 5.5.

Comparing Table 5.1 and Table 5.4, we can draw several general conclusions. The following indicators reported in terms of maximal and average values for all 10 classes confirm the aforementioned observations. For the training set, we have:

- 0.09% points—Loss (decrease) of strict accuracy (from 92.53 to 92.44%)

- 7.02% points—Gain (increase) of fine accuracy (from 92.53 to 99.55%)

- Maximal and average gains (decrease) on errors are:
  - 10.87% points—Maximal gain on predictive class error, that is, for patterns of a given class incorrectly classified to other classes (from 11.90 to 1.03%)
  - 9.12% points—Gain on the average of predictive class errors (reduction from 7.54 to 0.42%)

TABLE 5.4   Confusion matrix for the low quality basic classifier

			0	1	2	3	4	5	6	7	8	9
Classification without rejection (in %)	The training set	0	649		1			1	3		18	2
		1	2	777	1	1	1		4	2	6	4
		2	9	6	651	5	2	1	13	7	6	
		3	3	4	41	676		35	1	3	7	17
		4		4	10		637	3	4	14	5	28
		5	2	1	1	3		565	1	2	4	6
		6	9	1			3	1	630		6	
		7			7	14		2		649	1	12
		8	10		6	6	5	13	15	1	620	15
		9	2	1	4	2	39	3		42	9	622
	The test set	0	281					1	1		9	1
		1	1	328				1	8	1	1	5
		2	2	3	286	2	2	1	5	2	2	2
		3		2	16	278		8		1	4	9
		4		1			265	1		3	1	8
		5		3		2		254	4	1		3
		6	3	2	1		3		266		4	
		7		1	4	16	1			283		6
		8	6		3	4	1	1	3		270	6
		9	1	1		1	22	2		17	1	263
Classification with rejection (in %)	The training set	0	647								3	
		1		777								
		2		2	651	1					1	
		3				675		5				1
		4					637			1	2	
		5				1		565				1
		6	1						630		1	
		7								649		
		8	2			1	1				619	1
		9					3			1		620
		−1	36	15	71	29	46	54	41	69	56	83
	The test set	0	278								3	
		1		323				1	4	1		3
		2			283			1		1	2	1
		3			1	270		3			1	1
		4					264				1	1
		5				2		249				3
		6	2	1					260			
		7		1	1	2				277		2
		8	4			2	1		1		254	1
		9					4			2		254
		−1	10	16	23	28	25	15	22	26	32	38

Patterns are characterized by the full set of 24 features. Score is computed for both training and test sets. Zeroes were omitted for the sake of clarity.

TABLE 5.5   Results of the considered models without rejection (upper part) and with rejection (bottom part)

Classifier Dataset	Good Quality (99–96%)		Overfitted (100–82%)		Average Quality (95–94%)		Poor (92–92%)	
	Train	Test	Train	Test	Train	Test	Train	Test
No rejection								
Accuracy	100.00	100.00	100.00	100.00	100.00	100.00	100.00	100.00
Strict accuracy	99.73	96.50	100.00	82.34	95.49	94.50	92.53	92.44
Fine accuracy	99.73	96.50	100.00	82.34	95.49	94.50	92.53	92.44
Max class accuracy	100.00	98.71	100.00	99.35	98.49	97.07	97.86	96.19
Min class accuracy	99.27	92.08	100.00	75.37	91.50	89.11	88.10	86.80
Max local error	0.48	2.64	0.00	24.34	3.37	4.41	5.83	7.46
Mean local error	0.03	0.39	0.00	1.97	0.51	0.61	0.84	0.84
Global rejection								
Accuracy	99.69	96.70	99.69	96.70	99.69	96.70	99.69	96.70
Strict accuracy	99.49	94.57	99.69	81.54	95.41	92.80	92.46	90.70
Fine accuracy	99.80	97.79	100.00	84.32	95.71	95.97	92.75	93.80
Max class accuracy	99.87	97.42	100.00	98.06	98.49	96.26	97.86	95.24
Min class accuracy	98.83	89.44	99.12	75.07	91.22	86.80	87.82	84.49
Max local error	0.48	1.70	0.00	21.70	3.37	3.39	5.83	6.78
Mean local error	0.02	0.24	0.00	1.69	0.48	0.44	0.81	0.67
Local rejection								
Accuracy	99.66	96.30	99.50	82.27	95.89	94.54	92.86	92.17
Strict accuracy	99.46	94.47	99.50	81.24	95.41	92.64	92.44	90.37
Fine accuracy	99.80	98.10	100.00	98.74	99.51	97.99	99.55	98.05
Max class accuracy	99.87	97.10	100.00	97.10	98.49	95.60	97.86	94.72
Min class accuracy	98.83	89.11	98.83	75.07	91.22	86.47	87.82	83.83
Max local error	0.48	1.70	0.00	1.36	0.80	1.69	0.80	1.39
Mean local error	0.02	0.20	0.00	0.12	0.05	0.21	0.05	0.20

Results are expressed in percentages.

For the test set, the corresponding values are as follows:

- 2.07% points—Loss (decrease) of strict accuracy (from 92.44 to 90.37%)
- 5.61% points—Gain (increase) of fine accuracy (from 92.44 to 98.05%)
- Maximal and average gains (decrease) on errors are:
  - 10.57% points—Maximal gain on predictive class error (from 13.20 to 3.63%)
  - 5.77% points—Gain on the average of predictive class errors (from 7.58 to 1.81%)

Hence, we may conclude that the studied rejection mechanism improved the overall recognition quality much higher than for the case of results shown in Table 5.1. The analysis of the synthesized results presented in Table 5.5 leads to the conclusion that the poorer the basic classifier is, the smaller losses and greater gains of the quality measures.

5.1.4   Classification with Rejection for Imbalanced Datasets

In this section, we take a closer look at a range of various quality measures being used to evaluate the quality of classification with rejection completed with the aid of the three architectures discussed so far. As it was already indicated, an evaluation of recognition outcome must be performed with regard to the character of the processed data. Imbalanced datasets skew quality rates if they are not appropriately adjusted. Hence, we compare quality rates in two cases: when computations include scaling completed for imbalanced data and when no scaling was involved.

The experiment was primarily completed for the datasets of handwritten digits and 10 representative classes of symbols of printed music notation (refer to the description of the experiments presented in Section 4.3.2). The list of selected 10 classes is reported in Table 4.11. We compare the results of a series of experiments on music notation dataset with analogous experiments conducted for the dataset of handwritten digits. We evaluate classification with rejection. The comparison concerns models constructed using the three architectures: global, local, and embedded.

The datasets of handwritten digits (native patterns) and crossed-out handwritten digits and handwritten letters of Latin alphabet (foreign patterns) used in this experiment are described in Section 3.4. Let us recall that all classifying and rejecting models were constructed using 70% of data (original values of features, not processed in any way). The training set contained approximately 700 patterns of each native class. The datasets of handwritten digits and crossed-out digits include 10,000 patterns each, while the dataset of handwritten letters included 32,220 patterns.

The underlying classifier used for symbols of printed music notation dataset was a binary tree, displayed in Figure 5.3. The dataset contains 20 best features selected via greedy search procedure, as described in Chapter 1. The tree structure was determined using k-means applied to class centers. In order to instantiate a complete architecture, we trained binary classifiers. All classifiers (performing classification or rejection) were constructed using a balanced training set with standardized features values. The reason for using balanced set is that we observe substantial differences between class cardinalities in the dataset of music notation symbols. The least frequent class in the selected set contained only 26 patterns, while the most frequent class involved 3024 samples (cf. Table 4.11 and Figure 3.9). The balanced training set contained 350 samples belonging to each class; the method used to oversample data followed

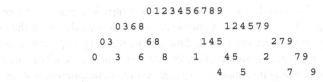

Figure 5.3   Textual representation of the structure of binary tree-based classifiers applied in classification of music notation symbols.

the *on the intervals* scheme described in Algorithm 1.1. Altogether, the training set of music notation symbols contained 3500 patterns. Even though we used processed dataset for model training, model evaluation was performed on original patterns from these classes exhibiting their own cardinality, that is, without oversampling or undersampling, although the features values were standardized. The full dataset used in this experiment contained 16,392 samples distributed unevenly across 10 classes. Rejecting mechanisms, similarly as classifying mechanisms, were constructed based on a balanced training set. We applied the *class-contra-all-other-classes* method (cf. Section 3.3). All binary classifiers (in classifying tree and in rejecting mechanisms) were trained using the SVM algorithm. All training procedures used RBF kernels with gamma parameter fixed at 0.0625 and cost equal 1. We used a 10-fold cross-validation.

In order to verify the quality of the models, we used symbols of music notation pairing native patterns with music notation garbage symbols (cf. Figures 3.9 and 3.11). The size of the set of foreign patterns (710 patterns) is approximately 0.04 of the size of the set of native patterns (16,392 symbols). These disproportional cardinalities entail the need for scaling the evaluation measures.

The results of the experiment are displayed in Table 5.6. We computed a range of global quality criteria, and we compare results across different rejecting models: global, local, and embedded. We compare results for the two datasets: handwritten digits and music notation symbols. In addition, for pairs of imbalanced native/foreign sets, we also display the rates in two cases: without and with scaling.

The results indicate that the global architecture is the least invasive when we evaluate its interference with acceptance of native patterns—this architecture removed the least number of native patterns. It manifests itself in high scores of native sensitivity obtained for the set of handwritten digits, namely, 98.79% for the global architecture, and with this score global architecture overheads local and embedded architectures more than a couple of percent points. For the set of music notation symbols, this difference is rather small, namely, 99.19% for the global architecture and slightly less for the two other architectures.

Even though the global architecture *retains* many native patterns, it is not necessarily preserving *good* native patterns. By this, we mean that it does not reject native patterns that would be incorrectly classified. In contrast, the other two architectures, local and embedded, reject more native patterns, but they also reject more misclassified native patterns than the global architecture. This is visible by looking at the strict accuracy rates, which are lower for the global architecture than for the other two architectures for the set of crossed-out handwritten digits.

Obviously, it is easier to accept a native pattern than to reject a foreign pattern. We stated this because all examined models were constructed with the use of native patterns only. This conclusion can be drawn by comparing native and foreign F-measures. The native F-measures are higher in all the studied models with the only exception when the values were not scaled. This confirms a very important remark, already addressed in the previous studies, that set scaling is a necessity when we process imbalanced data. Skipping this important step affects the description of real dependencies in the data. Note that for data not being scaled with the majority of foreign patterns (the set of handwritten Latin letters, three times as many as native digits),

TABLE 5.6 Comparison of experiments on classification with rejection (global, local, and embedded architectures) for two native datasets: handwritten digits and printed music notation

Dataset of Native Patterns	Digits			Musical Symbols	
Scaling	No		Yes	No	Yes
Dataset of Foreign Patterns	Crossed	Letters	Letters	Garbage	Garbage
Architecture	*Global rejecting*				
Accuracy	81.71	95.74	96.79	98.12	86.28
Strict accuracy	80.70	95.26	95.78	97.97	86.21
Native sensitivity	98.79			99.19	
Foreign sensitivity	64.62	94.79		73.38	
Native separability	73.87	85.63	95.05	98.86	78.98
Strict native separability	72.97	84.59	93.90	98.06	78.34
Foreign separability	98.80	99.63	98.80	84.22	99.20
Strict foreign separability	63.85	94.44	93.66	61.80	72.79
Native F-measure	84.37	91.65	96.85	99.02	87.85
Foreign F-measure	77.94	97.14	96.73	76.39	84.25
Architecture	*Local rejecting*				
Accuracy	87.88	95.44	95.89	98.04	86.24
Strict accuracy	87.30	95.17	95.31	97.94	86.19
Native sensitivity	96.74			99.11	
Foreign sensitivity	79.02	95.04		73.38	
Native separability	82.66	86.22	95.27	98.86	78.98
Strict native separability	79.96	83.41	92.17	97.98	78.27
Foreign separability	96.84	99.00	96.84	82.94	99.12
Strict foreign separability	76.53	94.09	92.04	60.86	72.73
Native F-measure	88.87	90.96	95.92	98.98	87.81
Foreign F-measure	86.70	96.95	95.85	75.67	84.21
Architecture	*Embedded rejecting*				
Accuracy	88.74	95.73	95.91	98.09	89.03
Strict accuracy	88.22	95.49	95.39	98.00	88.99
Native sensitivity	96.24			98.91	
Foreign sensitivity	81.23	95.57		79.15	
Native separability	84.20	87.52	95.76	99.11	82.75
Strict native separability	81.03	84.23	92.16	98.02	81.85
Foreign separability	96.38	98.85	96.38	79.87	98.92
Strict foreign separability	78.29	94.47	92.11	63.22	78.30
Native F-measure	89.52	91.44	95.92	99.00	90.02
Foreign F-measure	87.82	97.16	95.89	77.46	87.83

Whenever applicable we compute rates in two variants: original and scaled (balanced). Scaling concerns a case when we test native/foreign sets that have disproportional cardinalities. Results are expressed as percentages.

we have substantially higher foreign F-measure (97.14%) than native F-measure (91.65%). At the same time, for not scaled data with vast majority of native patterns (the set of music notation symbols vs. small set of music notation garbage patterns), we have a drastic supremacy of native F-measure (99%) over foreign F-measure (77.46%).

In other words, not scaled measures vote in favor of identification of a set with the greater number of patterns. The difference can be substantial. Let us compare perhaps the most extreme case: strict native separability for symbols of music notation. In the case when the rate is computed without scaling, it is equal to 98.86%. In contrast, when we consider unequal cardinalities (we have a very infrequent foreign set) and scale the rates, strict native separability is equal to 78.98%. This high dispersion (19.88% points) is an evident argument for a careful construction of quality measures.

The greatest capability to reject foreign patterns was manifested by the embedded architecture. This comes as no surprise, since this model is made of more rejecting components than the other two architectures.

All in all, the constructed models perform very well. Investigating numerical results in Table 5.6, we may conclude the experiment by saying that we recommend the local architecture. It performs satisfyingly, and at the same it is computationally not that demanding as the embedded architecture.

Comparing results for the set of symbols of music notation and the set of hand-written digits, we see that it was easier to separate music notation garbage patterns, that is, to reject them from the set of music notation symbols, than to reject crossed-out digits and Latin alphabet letters from the set of handwritten digits (cf. strict foreign separability).

If we compare the scores for the set of handwritten digits paired with the set of crossed-out digits and the set of handwritten digits paired with the set of letters of the Latin alphabet, we see that the latter pair was easier to distinguish. All measures are higher for the pair with Latin letters; the average difference is about 13% points. Disproportions are the highest for the global architecture (about 18% points), the lowest for the embedded architecture (about 9% points).

If we compare and summarize the results for native/foreign handwritten digits (native) and Latin letters (foreign) and music notation symbols (native) and garbage patterns of music notation (foreign), we see that the former pair (digits/letters) was the easiest to recognize. The average difference between measures for the two pairs of native/foreign datasets with handwritten digits being the native set is about 11% points (in favor of digits/letters). In contrast, comparing results for handwritten digits (native) and foreign crossed-out digits (foreign) with music notation symbols (native) and garbage patterns of music notation patterns (foreign) leads us to a conclusion that the recognition rates for this pair of native/foreign datasets is very similar (the difference is ca. 1% point in favor of music notation sets). In all cases, differences are the most substantial for the global architecture and the smallest for the embedded architecture.

Last but not least, let us state that the proposed three architectures perform very well in the task of native pattern classification with foreign pattern rejection. In the pursuit of yet more powerful models, let us now present a group of geometrical models. We forego the need for pattern classification, and we focus on the simple discrimination between native and foreign patterns.

5.2   GEOMETRICAL APPROACH

The geometrical approach to foreign pattern rejection is based on an intuition that the set of native patterns occupies a regular area in an M-dimensional real coordinate space of features R^M and that this area can be easily enclosed in basic geometrical figures. The second intuitive note is that foreign patterns fall out of the area occupied by native patterns or, at least, the majority of foreign patterns can be separated from the majority of native patterns in such way. It is desirable to define an area occupied by native patterns using simple geometrical figures, which would be easy for numerical processing. For convenience we assume that figures used to describe such area are convex and compact and that such area is a union of a family of such figures.

Let us denote geometrical figures $Z_1, Z_2, \ldots, Z_C$ and assume that Z_l includes all patterns from the set Tr_l—the set of training patterns from the class Tr_l, $l = 1, 2, \ldots, C$. Hence

$$Z = Z_1 \cup Z_2 \cup \ldots \cup Z_C \qquad (5.1)$$

represents a region in R^M, which includes all patterns coming from the training set. For convenience, we assume that all figures Z_l are convex and compact. Such a region will be used to determine if a given features vector from R^M represents a native or foreign pattern. We write $\mathbf{x} \in Z$ to indicate the fact that $\mathbf{x} \in R^M$ belongs to one or more figures Z_l.

The direct mechanism developed to deal with foreign patterns assumes that each native pattern from the lth class is positioned inside a geometrical figure Z_l. This holds for all native patterns of all classes, and therefore we assume that all native patterns are enclosed in the union Z of figures Z_l. In other words, location inside or outside of the designed union of geometrical figures discriminates native patterns from foreign patterns, of course with an assumption that native patterns are located inside the figures. This particular strategy, when adapted to the overall classification with rejection, could be adjusted to deal with specific data processing problems.

Forming the region Z in the features space R^M, we guarantee that no native pattern of the training set lies outside of this area. In practice, an area constructed in this way, despite its optimality with respect to native patterns of the training set, may be too wide for rejecting purposes. On the other hand, there may be native patterns that are distant from the *core* area of a class. Therefore, smaller figures Z_l spanned not over the entire class, but on all patterns except distant ones, may significantly reduce the size of the area occupied by native patterns. In this way, the area for foreign patterns, which complements native patterns space, is increased, which improves the efficiency of rejecting.

There is an undeniable relationship between a depth contour of a dataset and a geometrical figure enclosing a dataset. Depth contours are, in fact, geometrical figures. In addition, their construction method causes that they tightly enclose data points in a dataset. Hence, they are very well adjusted to actual training data points and perform well in the task of border-patterns (outliers) identification. However, procedures for discovering a convex hull are computationally very costly, and there is not much of a use of this method if dimensionality of a features space is greater than three.

In contrast, procedures for computing elementary geometrical figures are more efficient in high dimensional spaces.

There is still a need to specify which geometrical figures we will use to construct the rejecting mechanism, that is, to construct the region Z. We decided to apply M-dimensional hyperrectangles and ellipsoids. Hence, we must define how to establish membership to each of them. The choice of the figures has been dictated first and foremost by their simplicity (hyperrectangles are very straightforward to compute) and intuition (ellipsoidal figures are in agreement with the expectation that features vectors will be normally distributed in each class).

5.2.1   Hyperrectangles

The simplest type of a geometrical figure is a hyperrectangle defined as the Cartesian product of intervals defined in the real line. It is a direct conclusion from this assumption that edges of a hyperrectangle are parallel to coordinate system's axes. Therefore, it is easy to find a hyperrectangle enclosing a given set of native patterns in terms of enclosing vectors of features representing the patterns. Also, it is straightforward to test inclusion of a pattern into a hyperrectangle.

A hyperrectangle in the space of features R^M is just the Cartesian product of intervals. We assume that hyperrectangles are finite and closed:

$$H = I_1 \times I_2 \times \cdots \times I_M = [l_1, r_1] \times [l_2, r_2] \times \cdots \times [l_M, r_M] \qquad (5.2)$$

where $I_i = [l_i, r_i]$ are finite (and closed) intervals, that is, $-\infty < l_i < r_i < \infty$, $i = 1, 2, \ldots, M$. Hence, generic geometrical figures denoted previously $Z, Z_1, Z_2, \ldots,$ Z_C realized with hyperrectangles will be denoted as $H, H_1, H_2, \ldots, H_C$.

To continue with the discussion, we need to recall the formula for calculating the volume of a hyperrectangle, which is the product of lengths of intervals:

$$V_H = |I_1| \cdot |I_2| \cdot \ldots \cdot |I_M| = \prod_{i=1}^{M} (r_i - l_i) \qquad (5.3)$$

A hyperrectangle H, defined as in (5.2), enclosing a set of patterns $O = \{o_1, o_2, \ldots, o_n\}$ described with M-dimensional features vectors in R^M is easily constructed by setting endpoints of the intervals $I_1, I_2, \ldots, I_M$:

$$l_k = \min_{o_i \in O} \{x_{k,i}\} \quad r_k = \max_{o_i \in O} \{x_{k,i}\}, \quad k = 1, 2, \ldots, M \qquad (5.4)$$

where $(x_{1,k}, x_{2,k}, \ldots, x_{M,k})^T$ is the features vector of the pattern $o_k \in O$.

In contrast, a given pattern represented by a vector $(x_{1,k}, x_{2,k}, \ldots, x_{M,k})^T$ is included in the hyperrectangle defined by (5.2) if and only if

$$l_l \leq x_{l,k} \leq r_l \text{ for all } l = 1, 2, \ldots, M$$

If we have each class Tr_l of training patterns enclosed in one dedicated hyperrectangle H_l, $l = 1, 2, \ldots, C$, then we can turn (5.1) to the union of these hyperrectangles:

$$H = H_1 \cup H_2 \cup \ldots \cup H_C \qquad (5.5)$$

which entails that H encloses the whole training set of patterns Tr.

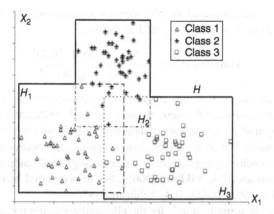

Figure 5.4 Construction of a hyperrectangles-based area enclosing the training set of native patterns. The area is the union of hyperrectangles built on classes of patterns from the training set.

An example with three classes and two features ($C = 3$ and $M = 2$) is shown in Figure 5.4. The three classes of patterns are marked with triangles, crosses, and squares. Patterns are characterized by two features (X_1 and X_2). Hyperrectangles $H = H_1 \cup H_2 \cup H_3$ enclose all native patterns from the training set.

5.2.2 Ellipsoids

Like in the case of hyperrectangles, let us now construct minimal volume ellipsoids enclosing a given set of patterns. Unlike in the case of hyperrectangles, construction of such ellipsoids is not so direct. Even though ellipsoids are not as easily defined and computed as hyperrectangles, they are intuitive, and, in addition, it is relatively easy to determine if a pattern is included in an ellipsoid.

An ellipsoid defined in the features space R^M can be defined as follows:

$$E(\mathbf{x}_0, A) = \left\{ \mathbf{x} \in R^M : (\mathbf{x} - \mathbf{x}_0)^T A (\mathbf{x} - \mathbf{x}_0) \leq 1 \right\} \tag{5.6}$$

where $\mathbf{x}_0 \in R^M$ is the center of the ellipsoid and A is a positive-definite matrix (Kumar and Yildirim, 2005). It is clear that the center $\mathbf{x}_0$ and matrix A uniquely define an ellipsoid.

From now on generic geometrical figures denoted $Z, Z_1, Z_2, \ldots, Z_C$ when realized with ellipsoids are denoted $E, E_1, E_2, \ldots, E_C$, where E_l is an ellipsoid enclosing a train set of native patterns from the lth class. As in the case of the hyperrectangles-based model, if each class Tr_l of native training patterns is enclosed in a dedicated ellipsoid E_l, $l = 1, 2, \ldots, C$, then the region defined by (5.1) is realized with a union of these ellipsoids, which enclose the whole training set of patterns Tr:

$$E = E_1 \cup E_2 \cup \ldots \cup E_C \tag{5.7}$$

We assume that for each class of native patterns, construction of an ellipsoid E_l enclosing class l relies on the minimization of its volume. In order to satisfy this assumption, ellipsoids enclosing classes of native patterns need to be fitted. Let us notice that for a given volume of the M-dimensional unit hypersphere V_0, the volume of the ellipsoid $E(\mathbf{x}_0, A)$ can be computed as follows:

$$\text{Vol}(E(\mathbf{x}_0, A)) = \frac{V_0}{\sqrt{\det(A)}} = \frac{\text{Vol}(E(0, I))}{\sqrt{\det(A)}} \tag{5.8}$$

where I is the unit matrix, 0 is the origin of the coordinate system, and $\det(A)$ is the determinant of the matrix A. Of course, the unit hypersphere V_0 is the ellipsoid with its center in the origin of the coordinate system and with the unit matrix I defining this special-case ellipsoid. We do not discuss here techniques of constructing of a minimal volume ellipsoid enclosing a given set of patterns. Suggested reading on this subject includes P. Kumar and E.A. Yildirim (2005). For tests reported here, we used algorithms provided by M.J. Todda, E.A. Yildirim (2007).

The following formula stands for the ellipsoid enclosing the whole lth class:

$$E_l = E(\mathbf{x}_l, A_l) = \left\{ \mathbf{x} \in R^M : (\mathbf{x} - \mathbf{x}_l)^T A_l (\mathbf{x} - \mathbf{x}_l) \le 1 \right\} \quad l = 1, 2, \dots, C \tag{5.9}$$

where $\mathbf{x}_l \in R^M$ is the center of the lth ellipsoid and the matrix A_l defines it.

Finally, a given pattern $\mathbf{x} \in R^M$ is identified as a foreign one if and only if

$$(\mathbf{x} - \mathbf{x}_l)^T A_l (\mathbf{x} - \mathbf{x}_l) > 1 \quad \text{for each} \quad l = 1, 2, \dots, C$$

If each class Tr_l of the training set is enclosed in a dedicated ellipsoid E_l, $l = 1, 2, \dots, C$, then a union of these ellipsoids

$$E = E_1 \cup E_2 \cup \dots \cup E_C \tag{5.10}$$

encloses the whole training set of patterns Tr. An example with three classes and two features ($C = 3$ and $M = 2$) is shown in Figure 5.5. By analogy to Figure 5.4, we have three classes of patterns marked with triangles, crosses, and squares and two features (X_1 and X_2). Three ellipsoids, E_1, E_2, E_3, enclose all patterns coming from the training set.

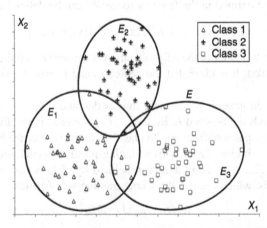

Figure 5.5 Construction of an area enclosing the training set of patterns. The area is the union (E) of three ellipsoids (E_1, E_2, E_3) built on classes of patterns from the training set.

5.2.3   Limiting the Area Reserved for Native Patterns in a Geometrical Model

The region Z defined in (5.1) guarantees that all native patterns from the training set are enclosed in it. Of course, there are no means to assure that all native patterns from a test set or, more general, all native patterns outside of the training set will fall into Z. However, we can expect that if a training set provides a good representation of native patterns, then most newly coming native patterns will be located in this region. On the other hand, we expect that most foreign patterns will fall out of this region. In other words, we expect that we will correctly reject most foreign patterns based on their location outside of the designed geometrical model. Intuitively, the smaller the region Z, the fewer foreign patterns will fall into Z, that is, will not be rejected. Therefore, in order to increase the chance for rejection, we should reduce the region Z to be as compact as possible. At the same time, we may expect that unskillful reduction of the space Z will cause not only the increase of the rejection rate for foreign patterns but also the rejection rate of native patterns. Of course, the latter outcome is undesired. However, we shall state that there are two objectives that are in conflict: to increase the acceptance rate of native patterns and to reduce the acceptance rate of foreign patterns. In the following sections we discuss methods for reducing the region reserved for native patterns so that we can produce a model that provides an acceptable balance for the two aforementioned conflicting criteria.

The most intuitive and direct attempt to reduce the region Z is simply to reduce the volume of components Z_l that together form Z. Let us recall that each component figure Z_l corresponds to the lth class of native patterns. Reduction of Z_l's volume relies on removing patterns that are located at their boundaries. In order to achieve a tight control over the degree to which we compress the volume of Z_l, we realize the process in an iterative fashion. It means that in each iteration of such a procedure, only one or only a few border patterns are removed from the training set, and then the new region enclosing remaining patterns is defined. To describe this process, let us return to (5.1) and rewrite it in the following form:

$$Z^0 = Z_1^0 \cup Z_2^0 \cup \ldots \cup Z_C^0 \qquad (5.11)$$

where the superscripts set to 0 indicate the fact that no turn of such iterative process has been done yet. Therefore, all patterns from the training set fall into the region Z^0. We use the name k-spans to denote shrunken areas of native patterns:

$$Z^k = Z_1^k \cup Z_2^k \cup \ldots \cup Z_C^k \qquad (5.12)$$

k-Spans are defined by creating the following sequence of thinned training set classes:

$$Tr^k = Tr_1^k \cup Tr_2^k \cup \ldots \cup Tr_C^k \qquad (5.13)$$

for $k = 1, 2, \ldots, s$; here s stands for the iteration index. Of course, Z^0 is based on the original training set Tr^0. Then, each k-span Z^k is created on a reduced training set Tr^k, which, in turn, consists of thinned classes of native patterns.

There are questions how to identify border patterns and what to do if we have a number of them. It is clear that for a given class of patterns, it is desirable to remove from the training set at first those patterns that are located far from the area occupied by the majority of patterns from this class. However, to make this assumption valid, it is necessary that patterns that belong to one class form a joint (perhaps even dense) area where we can find the majority of patterns. A minor share of patterns is located outside of this *majority* area, and we proceed with eliminating these *loose* patterns. Let us call such patterns *outermost* patterns. In the successive sections we discuss these two questions in more detail.

Hyperrectangles

The simplest strategy for reducing the area of native patterns is just to remove patterns from the training set that constitute endings of intervals for each feature. In the case where there is more than one such pattern for a given feature, we assume that only one of them is removed in one turn of this strategy. On the other hand, one pattern can serve as an ending point for more than one feature. Therefore, up to $2M$ patterns in one step would be removed from each class. If all patterns, which constitute endings of intervals, are removed, then more than $2M$ patterns may be removed from each class.

This direct and trivial method in practice can be inefficient, because it does not distinguish native patterns, which are close to the area of concentration of native patterns from patterns located far from this area. It is clear that removing a pattern close to the concentration area results in a small reduction of the hyperrectangle volume, while removing loosely scattered distant patterns is a much more effective strategy. Therefore, we discuss here methods of removing the *outermost* patterns, that is, the methods for identification of loosely scattered patterns, positioned relatively far from the majority of patterns in a given class. Removing *outermost* patterns allows us to substantially reduce the volume of a hyperrectangle enclosing remaining patterns in a given class.

Let us consider native patterns belonging to a given class present in the training set. Let us assume that the hyperrectangle enclosing them is defined as in (5.2). An answer to the question how to identify the outermost patterns is a matter of computing reductions of hyperrectangle's volume defined in (5.3) related to retracting points of intervals $[l_i, r_i]$ for each feature $i = 1, 2, ..., M$, that is, computing $2M$ reductions, and then removing this pattern (or more than one pattern if there is more than one pattern at the end of the selected interval), which gives the maximal volume reduction. Let us look in detail at this procedure. Having determined the interval $[l_i, r_i]$ for the ith feature, we need to find the smallest value of this feature, which is greater than l_i, say, l_i', and the greatest value, which is smaller than r_i, say, r_i'. There is an additional requirement, in practice quite easy to satisfy, that the feature i takes on more than 3 distinct values. Otherwise, there are too few features values to carry on with the proposed computations. For the sake of the discussion, let us assume that there were enough distinct values of feature i, and we have identified l_i' and r_i'. Next, referring to (5.3), the reduced volumes are $V_H \cdot (r_i - l_i') / (r_i - l_i)$ and $V_H \cdot (r_i' - l_i) / (r_i - l_i)$, respectively. In other words, corresponding volume reductions are $V_H \cdot (l_i' - l_i) / (r_i - l_i)$ and $V_H \cdot (r_i - r_i') / (r_i - l_i)$. It is clear that volume reductions depend on relative length reduction of intervals and do not depend on the volume of the hyperrectangle V_H. Therefore, instead of computing volumes, we can simply compute ratios of relative interval length reductions $\left(l_i'-l_i\right)/(r_i-l_i)$ and $\left(r_i-r_i'\right)/(r_i-l_i)$ for $i=1,2,\ldots,M$ and then remove this endpoint (out of $2M$ endpoints) for which the ratio is the greatest. For simplicity, let us call these ratios *shrinking factors*.

In the aforementioned discussion, we considered removing endpoints of intervals while, in fact, an equivalent action is to remove appropriate patterns from the training set, which in consequence reduces respective endpoints at the stage of hyperrectangle formation. If we can assume that values of features are unique for all patterns, that is, there is no such pair of patterns that has the same value of any feature, then removing any interval endpoint is equivalent to removing a single pattern. For the sake of simplicity, let us at first assume in Algorithms 5.1 and 5.2 that all patterns in the training set take on unique values of the feature. Later, we discuss a more general case without this assumption.

Algorithm 5.1
Reducing the set of patterns by removing the outermost pattern

Data: $O=\{o_1,o_2,\ldots,o_r\}$, the set of r native patterns (in particular the training set)

          $\mathbf{x}_i=\left(x_{1,i},x_{2,i},\ldots,x_{M,i}\right)^T$ the vector of features of the pattern $o_i,\ \ i=1,2,\ldots,r$

Algorithm: **for** each (feature) $l=1$ to M **do**
            begin

              $i_{\min}=\arg\min_{o_i\in O}\{x_{l,i}\},\ \ \Delta x_{l,\min}=\min_{o_i\in O-\{o_{i_{\min}}\}}\{x_{l,i}\}-\min_{o_i\in O}\{x_{l,i}\}$

              $i_{\max}=\arg\max_{o_i\in O}\{x_{l,i}\},\ \ \Delta x_{l,\max}=\max_{o_i\in O}\{x_{l,i}\}-\max_{o_i\in O-\{o_{i_{\max}}\}}\{x_{l,i}\}$

(*)   **if** $\Delta x_{l,\min}>\Delta x_{l,\max}$ **then** $\Delta V_l\cong\dfrac{\Delta x_{l,\min}}{r_l-l_l},\ \ \text{del}_l=i_{\min}$

                      else $\Delta V_l\cong\dfrac{\Delta x_{l,\max}}{r_l-l_l},\ \ \text{del}_l=i_{\max}$

            end
(**) $k=\arg\max_{l=1,2,\ldots M}\{\Delta V_l\}$
(***)   remove o_{del_k} from the set of patterns O
Results: the set $O-\{o_{\text{del}_k}\}$ of $r-1$ patterns with the *outermost* pattern removed.

Algorithm 5.2
Construction of s-span hyperrectangle H^s
            based on the training set of patterns

Data: $Tr=Tr_1\cup Tr_2,\ldots,Tr_C$, the training set Tr being a union of C subsets, each subset Tr_i contains patterns belonging to the i-th class

Algorithm: **construct** 0-span H^0 applying (5.3) and (5.4) to the training set $Tr^0=Tr$
            for $k=1$ to s **do**
            begin
               for each (class) $l=1$ to C **do**
                  remove outermost pattern(s) from Tr_l^{k-1} in order to form Tr_l^k
               construct the training set $Tr^k=Tr_1^k\cup Tr_2^k,\ldots,Tr_C^k$
               construct k-span H^k applying (5.3) and (5.4) to the training set Tr^k
            end
Result: the s-span hyperrectangle H^s

An example method of removing the outermost pattern from a given set of patterns is given in Algorithm 5.1. In each iteration of the "for k" loop in Algorithm 5.2, we are reducing the set of training patterns. At the beginning we start with the full training set Tr^0. After we finish the first run ($k = 1$), we reduce the training set and we denote it as Tr^1. After the second loop ($k = 2$), we perform further reducing, and the outcome training set is denoted Tr^k. Finally, after the procedure is finished, we obtain training set Tr^s. Let us notice that having the training set

$$Tr^{k-1} = Tr_1^{k-1} \cup Tr_2^{k-1} \cup \ldots \cup Tr_C^{k-1}$$

we construct the reduced training set

$$Tr^k = Tr_1^k \cup Tr_2^k \cup \ldots \cup Tr_C^k \qquad (5.14)$$

by applying Algorithms 5.1 and 5.2 to each class Tr_i^{k-1} of native patterns, that is, removing the outermost pattern from each class. It is worth noting that each step of Algorithm 5.2 reduces the cardinality of the training set by exactly one (outermost) pattern for each class. Of course, in this case, general figures denoted Z in (5.11) and (5.12) should be replaced by k-span hyperrectangles, so those formulas are turned to

$$H^k = H_1^k \cup H_2^k \cup \ldots \cup H_C^k \qquad (5.15)$$

An illustrative example of k-spans (up to 3-span) is shown in Figure 5.6. Here 3-span means that three outermost patterns were removed from each class.

This method of hyperrectangle shrinking was chosen for its simplicity. One may easily envision other methods for hyperrectangle reduction, which may be quantitatively more efficient for a given dataset than the method presented here. Later on we will briefly show other ideas for hyperrectangle shrinking. However, we would like to focus the discussion on the methodology without getting into detailed investigations of minor aspects of the proposed procedures. Let us mention that we can easily consider other strategies for the identification of patterns that can be removed in order to construct a shrunken geometrical model. Such methods can utilize diverse notions of distance or density, different metric systems, different definitions of class or cluster center, or patterns dispersion. For instance, we can propose to remove patterns, which are the *farthest* from the barycenter of the class; we may remove an assumed percentage of the *farthest* patterns; we may split classes of native patterns into clusters and then enclose such clusters into separate hyperrectangles; and so on. However, let us emphasize that this study aims at discussing general methodology instead of investigating the detailed algorithmic aspects.

Let us look at Algorithm 5.1. In practice, it is rather a rare situation when for each endpoint of an interval, there is exactly one corresponding pattern. It is quite likely that we will have several patterns corresponding to intervals' endpoints. For instance, let us consider datasets of handwritten digits or symbols of music notation. The values of many features in these datasets are integers from intervals of length of a few dozens, while the cardinality of each class is several hundreds. Hence, statistically, there will be several or even more patterns corresponding to each possible value of a feature. Even in such a case, Algorithm 5.1 works correctly: for the endpoint chosen in the instruction denoted with ($*$), only one corresponding pattern is removed.

(a)

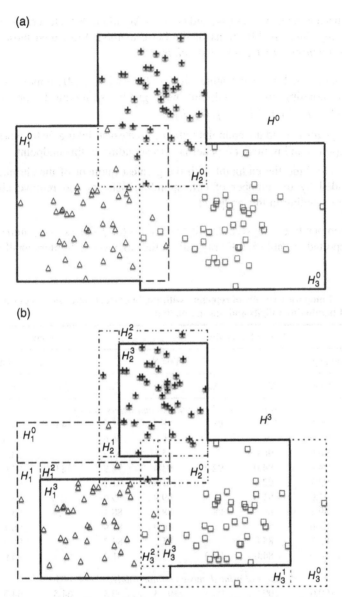

Figure 5.6   Construction of k-spans H^k for two features and three classes of native patterns: (a) H^0 span is the union of rectangles enclosing all patterns of classes. (b) H^3 span is constructed based on removed three outermost patterns from each class.

Therefore, it is sufficient to repeat it until all patterns at a given endpoint are removed. Through such repetitions of the instruction denoted with (∗), we are able to remove the given endpoint. However, the process of removing patterns one by one is not optimal. In such case it is recommended to remove all patterns corresponding to the chosen endpoint at once. In order to execute this action and remove all relevant patterns at once, we need to modify lines (∗∗) and (∗∗∗) in Algorithm 5.1. Hence, let us consider the following three modifications of the basic method. We called those modified criteria *all features*, *one feature*, and *volume*:

1. *All features*: For each feature, that is, for $i = 1, 2, \ldots, M$, remove all patterns corresponding to the endpoint with greater shrinking factor out of the $(l'_i - l_i)/(r_i - l_i)$ and $(r'_i - r_i)/(r_i - l_i)$.

2. *One feature*: Find the endpoint with the greatest shrinking factor among all $2M$ endpoints and remove all patterns corresponding to this endpoint.

3. *Volume*: Find the endpoint with the greatest quotient of the shrinking factor divided by the number of corresponding patterns and remove all patterns corresponding to this endpoint.

A comparative analysis of the three attempts (*all features*, *one feature*, and *volume*) is reported in Table 5.7. Empty cells in the table were left intentionally, because

TABLE 5.7 Comparing results of rejection with the hyperrectangles model performed for datasets of handwritten digits and music notation

Native Sensitivity						Foreign Sensitivity		
Training Set			Test Set			Foreign Patterns		
All	One	Vol.	All	One	Vol.	All	One	Vol.
Handwritten digits (native) and letters (foreign)								
100.0	100.0	100.0	98.6	98.7	98.7	74.6	74.6	74.6
	98.0	98.0		96.6	96.6		91.5	92.2
	96.0	96.0		95.1	94.7		93.4	94.5
94.0	94.0	94.0	92.7	93.4	92.9	88.9	94.0	95.6
	92.7	92.7		91.5	91.3		94.4	95.9
	91.0	91.0		89.9	89.6		94.7	96.6
89.3	89.3	89.3	88.7	88.0	88.1	93.9	95.8	96.8
	87.6	87.6		86.9	85.9		96.1	97.1
	85.8	85.8		85.2	83.7		96.7	97.4
84.0	84.0	84.0	83.2	83.3	81.2	96.4	96.9	97.6
Symbols of printed music notation (native and foreign)								
100.0	100.0	100.0	90.3	90.3	90.3	64.5	64.5	64.5
	98.0	98.0		86.9	86.8		77.0	77.5
	96.0	96.0		83.6	83.6		81.6	83.3
93.9	93.9	93.9	82.8	80.9	80.7	81.7	87.0	86.3
	91.9	91.9		78.8	78.4		88.9	88.0
	89.9	89.9		76.3	76.4		90.4	91.7
87.9	87.9	87.9	75.3	73.7	74.1	89.6	91.1	92.2
	84.8	84.8		70.3	70.7		92.1	93.8
	81.7	81.7		66.3	67.3		92.5	94.1
78.7	78.7	78.7	67.1	63.9	64.1	93.2	93.4	94.9

the *all features* method after each iteration induces larger differences to the results. This is due to the fact that it removes more elements at once than the other two methods. The upper part of the table concerns handwritten digits with features that are original, that is, they were neither normalized nor standardized. The lower part of the table concerns music notation. We processed this dataset with features that were also original, but the dataset was balanced with the *on interval* balancing method. For details on dataset preprocessing, consult Chapter 1.

We used two datasets described in Section 3.4. Now, let us reiterate information suitable for this experiment:

- Native patterns: Handwritten digits from the MNIST database (LeCun *et al.*, 1998), 10,000 patterns in total, about 1,000 patterns in each class. From each class we chose randomly 70% samples for the training set, with remaining 30% made the test set; foreign patterns: handwritten lowercase letters of the Latin alphabet from our own database, 32,220 patterns in total (Homenda *et al.*, 2017).

- Native patterns: Symbols of printed music notation from our own database (Homenda *et al.*, 2017); the original dataset is heavily imbalanced: it consists of 27,299 patterns unevenly distributed among 20 classes. Imbalancedness is observed first and foremost in class cardinalities. For instance, the most infrequent class is thirty-second rest for which we have only 26 samples. In contrast, the most frequent class is quarter rest with 3024 samples. Knowing that imbalanced datasets cause problems in classifier training, we preprocessed this dataset and produced a balanced version. In the balanced version we have 500 patterns in each one of 20 classes, so in total there were 10,000 samples. We used the *on intervals* method to generate new patterns in rare classes. For overly frequent classes, we performed undersampling, and we randomly selected desired number of patterns. Next, we randomly selected 70% patterns of each class to make a training set, with the remaining 30% made the test set. We paired this balanced native dataset with foreign patterns from the dataset of music notation symbols. Foreign patterns are garbage and not native symbols of printed music notation, 710 patterns in total.

In Table 5.7, we outline native sensitivity (acceptance rates for native patterns) at training and test sets and foreign sensitivity (rejection rates) of foreign patterns (cf. Section 4.1.3). Let us recall that native sensitivity (acceptance rate) is the percentage of native patterns falling into an area, say, model, defined by (5.15) out of all native patterns checked in this process. Similarly, the foreign sensitivity (rejection rate) is the percentage of foreign patterns falling out of the area defined in (5.15) out of all foreign patterns. Of course, the greater both values are, the better. Columns are called *all*, *one*, and *vol.*, which correspond to three volume-reducing methods: (a) *all features*, (b) *one feature*, and (c) *volume*, respectively. In each row we refer to acceptance rates for training and test sets of native patterns and rejection rate of foreign patterns. We include results after the first, the second, and the third turn of Algorithm 5.2 for (a) *all features* volume-reducing method applied in Algorithm 5.1. Also, we report corresponding and two intermediate results of (b) *one feature* and (c) *volume* volume-reducing methods applied in Algorithm 5.1.

Numbers show that the first volume-reducing method ((a) *all features*) is slightly better at the set of music notation symbols than the other two methods, because for the same level of accuracy at the native training set, it achieves higher accuracy at the native test set. However, all three methods give similar results on the set of handwritten digits. In contrast, the third method outperforms the two other methods on the set of foreign patterns, while the first method is noticeably worse on the set of foreign patterns.

The detailed characteristics of two best methods (out of the three proposed here) of removing outermost patterns are presented in Figure 5.7. These methods are called (b) *one feature* and (c) *volume*. In the figure, we present acceptance rates obtained on the training and test sets of (native) patterns and rejection rates at sets of foreign patterns as functions of the number of iterations performed in Algorithm 5.2. We observe that as we progress with iterations, the *volume* method (of removing

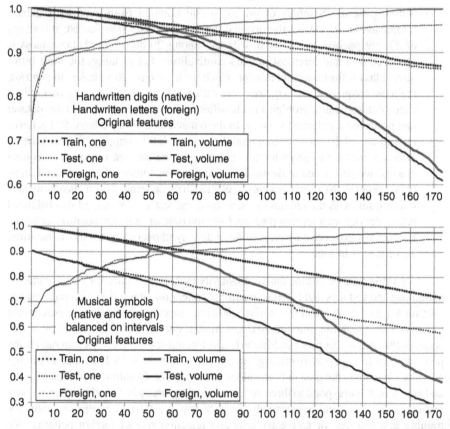

Figure 5.7   Geometrical rejection with hyperrectangles—characteristics for handwritten digits and symbols of music notation datasets. Plots outline acceptance rate (for the training and test sets of native patterns) and rejection rate (for the set of foreign patterns) as functions of iterations performed in Algorithm 5.2. Compared are two methods of removing patterns: *one feature* and *volume*.

outermost patterns) starts outperforming the *one feature* method with respect to the rejection rate for foreign patterns. However, this gain comes together with a deterioration of the acceptance rate for native patterns, which is much worse for the (c) *volume* than for the (b) *one feature* volume-reducing method. Still, if we stop comparing classification and rejection rates as functions of algorithm iterations and switch to comparing the divergence between a fixed acceptance rate at the train set with corresponding acceptance rate of the test set native patterns and foreign pattern rejection rate, the *volume* method outperforms the *one feature* method.

Ellipsoids

Let us notice that area defined in (5.7) includes all native patterns. However, it may include many foreign patterns too. Therefore, as in the case of hyperrectangles, it would be reasonable to reduce this area in order to improve the quality of rejection of native patterns. We can apply the apparent method of removing *outermost* patterns. However, identification of outermost patterns is not so direct in the case of ellipsoids as it was in the case of hyperrectangles. Computationally efficient methods produce an approximation of the minimum volume ellipsoid enclosing a set of patterns. Therefore, there might not be any patterns located exactly on the surface of the formed figure, that is, such case that inequality in (5.9) turns to equality. On the other hand, if some outermost patterns are removed, then it is necessary to rebuild an ellipsoid enclosing a reduced set of patterns, as it was in the case of hyperrectangle. However, the construction of ellipsoids is computationally more costly and more sophisticated than the construction of hyperrectangles. In addition, as we mentioned, constructed ellipsoids are approximations of the exact minimal volume ellipsoid enclosing all patterns, unlike exact hyperrectangles, construction of which is straightforward (cf. Section 5.2.1).

Let us adjust (5.11) to the case of ellipsoids:

$$E^0 = E_1^0 \cup E_2^0 \cup \dots \cup E_C^0 \tag{5.16}$$

where ellipsoids enclosing patterns belonging to a given classes are defined as follows:

$$E_l^0 = E\left(\mathbf{x}_l^0, A_l^0\right) = \left\{ \mathbf{x} \in R^M : \left(\mathbf{x} - \mathbf{x}_l^0\right)^T A_l^0 \left(\mathbf{x} - \mathbf{x}_l^0\right) \le 1 \right\} \quad l = 1, 2, \dots, C \tag{5.17}$$

Finally, a given pattern $\mathbf{x} \in R^M$ is identified as a foreign one if and only if

$$\left(\mathbf{x} - \mathbf{x}_l^0\right)^T A_l^0 \left(\mathbf{x} - \mathbf{x}_l^0\right) > 1 \quad \text{for each} \quad l = 1, 2, \dots, C \tag{5.18}$$

Then, like in the case of hyperrectangles, let us create consecutive k-spans:

$$E^k = E_1^k \cup E_2^k \cup \dots \cup E_C^k \tag{5.19}$$

k-spans are defined by creating the following sequence of reduced training sets of native patterns

$$Tr^k = Tr_1^k \cup Tr_2^k \cup \dots \cup Tr_C^k \tag{5.20}$$

for $k = 0, 1, 2, \dots, s$. Formation of spans is given in Algorithms 5.3 and 5.4. Illustration of this method is presented in Figure 5.8.

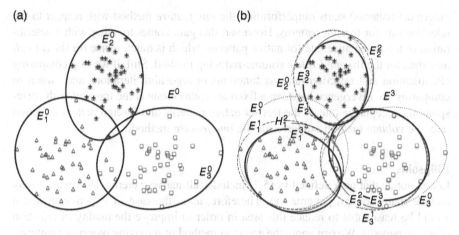

Figure 5.8   Construction of ellipsoids-based k-spans E^k for two features and three classes of native patterns: (a) E^0 span is the union of three ellipsoids; each ellipsoid encloses all patterns of one class. (b) Spans E^1, then E^2, and then E^3 are constructed by removing three (one from each class) outermost patterns from each class in each iteration.

Algorithm 5.3

Reducing the set of patterns by removing the outermost pattern in an ellipsoids-based model

Data:	$O = \{o_1, o_2, \ldots, o_r\}$, the set of patterns O
	$\mathbf{x}_i = (x_{1,i}, x_{2,i}, \ldots, x_{M,i})^T$ vector of features of patterns o_i, $i = 1, 2, \ldots, r$
Algorithm:	create the minimal volume ellipsoid $E(\mathbf{x}_0, A)$ enclosing patterns

for each pattern $o_i \in O$ **do**
    compute $r_i = (\mathbf{x}_i - \mathbf{x}_0)^T A (\mathbf{x}_i - \mathbf{x}_0)$
$k = \arg\max_{o_i \in O} \{r_i\}$
remove o_k from the set of patterns O

Result:	the set $O - \{o_k\}$ of $r - 1$ patterns with the *outermost* pattern removed.

Algorithm 5.4

Construction of the s-span E^s based on the training set of patterns

Data:	$Tr = Tr_1 \cup Tr_2, \ldots, Tr_C$, the training set and its split to classes
Algorithm:	**construct** 0-span E^0 applying (5.16) and (5.17) to the training

    set $Tr^0 = Tr$
for $k = 1$ to s **do**
begin
    for each (class) $l = 1$ to C **do**
        reduce the set of patterns Tr_l^{k-1} to Tr_l^k using Algorithm 5.3
    construct the training set $Tr^k = Tr_1^k \cup Tr_2^k, \ldots, Tr_C^k$
    construct k-span E^k applying (5.19) and (5.20) to the training set Tr^k
end

Result:	the s-span E^s

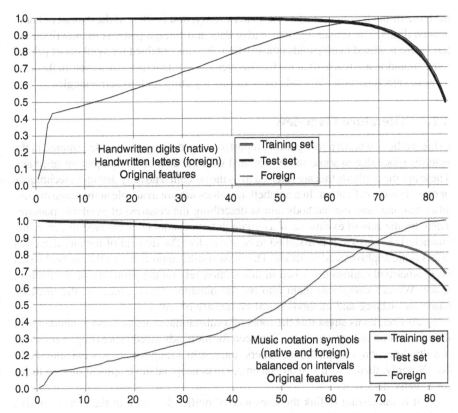

Figure 5.9   Geometrical rejection with ellipsoids—characteristics for handwritten digits and symbols of music notation datasets. Plots outline acceptance rate (for the training and test sets of native patterns) and rejection rate (for the set of foreign patterns) as functions of iterations performed in Algorithm 5.4.

Like in the case of hyperrectangles, more than one pattern may be removed in one call of Algorithm 5.3. For the case of ellipsoids, we tested the approach relying on removing a given number of patterns with the largest radius $r = (\mathbf{x} - \mathbf{x}_0)^T A (\mathbf{x} - \mathbf{x}_0)$.

The characteristics of geometrical rejection using ellipsoids are shown in Figure 5.9. We present acceptance rates for training and test sets of (native) patterns and rejection rate of the set of foreign patterns as functions of the number of iterations performed in Algorithm 5.4. These characteristics allow comparing the rejection capability of an ellipsoids-based model with a hyperrectangles-based model presented in Figure 5.7. It is clear that rejecting with ellipsoids is much more effective than with hyperrectangles. It is worth drawing attention to the dataset of handwritten digits, where the acceptance rate for the training and test sets of patterns is almost identical for the whole scale. Moreover, it is worth emphasizing the rates of native pattern acceptance and rate of foreign pattern rejection. Both are very high and both reach about 97% after only 64 iterations. Let us stress that ellipsoids were constructed on the basis of the training set of native patterns only, that is, 70% of the dataset of handwritten digits. The remaining 30% native data and foreign patterns (handwritten letters) were not used in ellipsoids construction. Results obtained for symbols of printed music notation are worse than for handwritten digits and letters. However, in this case as well, rejecting with ellipsoids outperforms rejecting with hyperrectangles.

5.2.4 Literature Overview

In the studies on pattern recognition, reported are certain operations determined in the features space that in some sense is related with the task of foreign pattern rejection. These are the methods that are referred to as the so-called novelty detection techniques or one-class classification. In a nutshell, novelties are not available at the stage of classifier training, and the methods aim at describing the class(es) of available patterns. Then, by the rule of exclusion, if a certain pattern cannot be accounted to one of the characterized classes, it is assumed to be a novelty. The domain of novelty detection offers quite sophisticated methods. The most similar approaches to the method based on geometrical regions described in this section rely on the geometry of the features space. We can categorize them into two dominating streams, namely, the methods based on distance and methods based on density of patterns.

At first, let us stress that the roots of a substantial share of these methods are in the research on outlier detection. The need for more advanced processing, which was novelty detection, rekindled researchers' interest in a range of already existing methods. Notably, there was a vast advancement in conceptual development of algorithms for novelty detection.

It is interesting to link the relevance of outlier detection in the context of this book. Let us cite Hawkins' definition of an outlier: "an outlier is an observation that deviates so much from other observations as to arouse suspicions that it was generated by a different mechanism" (Hawkins, 1980). In light of this definition, we see some analogy between a foreign pattern and an outlier. Nonetheless, there is a primary difference between these two. Let us emphasize again that outliers are native and rejecting them is a measure taken to assure that non-outliers will be processed correctly, while foreign patterns should be removed, as they do not belong to the native dataset. In addition, characteristics of foreign patterns are not known, while outliers belong to the native set and their characteristics are known. A discussion facing the problem of outliers in complex geometrical figures constructs built for native patterns can be found, for instance, in Mozharovskyi *et al.* (2015). However, we do not intend to discuss this problem here, because it is a marginal relevance vis-à-vis the main focus of this study. We only touch this issue in the context of geometric constructs built on sets of native patterns.

In our opinion, the most similar approaches to the geometrical figures-based approach are depth-based methods that have been primarily developed as a tool for outlier detection. Depth-based methods, for instance, algorithms named as ISO-DEPTH (Ruts and Rousseeuw, 1996) and FDC (Fast Depth Contour) (Johnson *et al.*, 1998), are based on the idea of a depth contour of a dataset. The notion of a depth contour itself has been proposed by J. Tukey (1975, 1977). In a nutshell, depth-based methods rely on computing convex hulls of a dataset that is being iteratively reduced. In other words, in each iteration we are able to reduce the dataset by some number of patterns, and we recalculate the convex hull. However, procedures for discovering a convex hull are computationally very costly, and there is not much of a use of depth-based method if dimensionality of a features space is greater than three. In contrast, procedures for computing elementary geometrical figures are more efficient in high dimensional spaces.

The second, less similar but very important method is one-class SVM (Tax and Duin, 2004). It is an alteration of the *regular* SVM algorithm, presented in Section 2.3. The difference is that the *regular* approach is planar, that is, it designs separating margins in a hyperplane. The one-class SVM designs a spherical boundary. The procedure aims at minimizing volume of a sphere. There are two possible choices: first that all patterns from the training set are included in the designed hypersphere and second that we have a soft margin and allow errors on the training set.

It is also worth to point out that a range of clustering techniques were rediscovered as tools for novelty detection. For instance, let us take any cluster formed in an unsupervised form for a given dataset. We can easily determine two most distant data points in a given cluster. Next, we can use this distance to construct a novelty detection rule saying that if a certain pattern is further from all patterns in a given cluster than this distance (plus some small tolerance margin), then from the perspective of this cluster, this pattern is a novelty.

Fuzzy clustering is another valid method for novelty detection. Let us recall that in standard clustering belongingness is expressed as either zero or one—an element either belongs or it does not. In contrast, in fuzzy clustering, belongingness is expressed as a number from [0,1] interval (cf. Chapter 8). The most popular fuzzy clustering method is fuzzy c-means, a fuzzy variant of k-means (Bezdek *et al.*, 1984). We can use this flexible scale as means for novelty detection. If a certain pattern does not belong to any cluster with a membership degree greater than some predefined threshold, then we can account it as a novelty.

Among other approaches we find methods that generate synthetic data to extend a one-class-classification problem to a binary classification problem (Hempstalk *et al.*, 2008), probabilistic methods that aim at determining density of a dataset with deep roots in well-known probabilistic clustering techniques (McLachlan and Basford, 1988), and many more. A comprehensive review of novelty detection techniques has been published recently in Pimentel *et al.* (2014).

Interestingly, the proposed foreign pattern rejection method based on architectures of specifically trained classifiers is unique, and there are no similar compound methods in the literature.

5.3 CONCLUSIONS

In this chapter, we have empirically evaluated the three rejecting architectures such as global, local, and embedded. It has been shown that the global architecture is the most general, as it is independent of the classification step. It constitutes a good fit not only for the presented classification task but also for clustering. While the global architecture is the most universal one, it achieves the weakest results. On the opposite side, we have the embedded architecture, which is fully integrated with the classification mechanism and heavily depends on the construction of the classifier. The embedded architecture exploits internal structure of the classifier, and as a result of this, it yields superior performance. The downside is that the model is fairly complex. The local architecture is positioned somewhere in between the global and the embedded architectures. It can be regarded as a compromise, in which we have a relatively simple model achieving satisfying results.

An alternative to the discussed ensemble methods for foreign pattern rejection are methods operating in features space, for instance, the presented method based on geometrical figures. We have shown and compared models based on hyperrectangles and on ellipsoids. The first option achieves poorer results, but is computationally cheap and intuitive to apply and verify. In contrast, ellipsoids-based method achieves better results, but it is incomparably more resource demanding. The great advantage of geometrical models in general is their flexibility. By manipulating with volume of a model, we can balance between the rate of foreign pattern rejection and native pattern acceptance. This arises a very desirable feature when we deal with real-world data for which separation between native and foreign patterns is not a trivial task.

To sum up, we have shown two paths for designing a rejection model: indirect, based on an ensemble of binary classifiers, and direct, designing a model directly in a features space. The main disadvantage of geometrical models is that it is not realistic to apply the same model for classification and rejection. In contrast, ensemble models combine classification with rejection.

REFERENCES

J. C. Bezdek, R. Ehrlich, and W. Full, FCM: The fuzzy c-means clustering algorithm, *Computers & Geosciences* 10(2–3), 1984, 191–203.

D. Hawkins, *Identification of Outliers*, London, Chapman and Hall, 1980.

K. Hempstalk, E. Frank, and I. H. Witten, One-class classification by combining density and class probability estimation. In: *Proceedings of Joint European Conference on Machine Learning and Knowledge Discovery in Databases ECML PKDD 2008: Machine Learning and Knowledge Discovery in Databases*, Antwerp, Belgium, September 15–19, 2008, 505–519.

W. Homenda, A. Jastrzebska, and W. Pedrycz, *The web page of the classification with rejection project*, 2017, http://classificationwithrejection.ibspan.waw.pl (accessed October 6, 2017).

T. Johnson, I. Kwok, and R. T. Ng, Fast computation of 2-dimensional depth contours. In: *Proceedings of the Fourth International Conference on Knowledge Discovery and Data Mining (KDD-98)*, New York, August 27–31, 1998, 224–228.

P. Kumar and E. A. Yildirim, Minimum-volume enclosing ellipsoids and core sets, *Journal of Optimization Theory and Applications* 126(1), 2005, 1–21.

Y. LeCun, C. Cortes, and C. J. C. Burges, *The MNIST database of handwritten digits*, 1998, http://yann.lecun.com/exdb/mnist/ (accessed October 6, 2017).

G. J. McLachlan and K. E. Basford, *Mixture Models: Inference and Applications to Clustering*, vol. 1, New York, Marcel Dekker, 1988.

P. Mozharovskyi, K. Mosler, and T. Lange, Classifying real-world data with the DDα-procedure, *Advances in Data Analysis and Classification*, 9(3), 2015, 287–314.

M. A. F. Pimentel, D. A. Clifton, L. Clifton, and L. Tarassenko, A review of novelty detection, *Signal Processing* 99, 2014, 215–249.

I. Ruts and P. J. Rousseeuw, Computing depth contours of bivariate point clouds, *Computational Statistics and Data Analysis*, 23, 1996, 153–168.

D. M. J. Tax and R. P. W. Duin, Support vector data description, *Machine Learning* 54(1), 2004, 45–66.

M. J. Todda and E. A. Yildirim, On Khachiyan's algorithm for the computation of minimum-volume enclosing ellipsoids, *Discrete Applied Mathematics* 155(13), 2007, 1731–1744.

J. Tukey, Mathematics and the picturing of data, *Proceedings of the 1975 International Congress of Mathematics* 2, 1975, 523–531.

J. Tukey, *Exploratory Data Analysis*, Reading, MA, Addison-Wesley, 1977.

PART II

ADVANCED TOPICS: A FRAMEWORK OF GRANULAR COMPUTING

CHAPTER 6

CONCEPTS AND NOTIONS OF INFORMATION GRANULES

In this chapter, we introduce the main concepts of information granules and discuss their roles. We elaborate on their formalizations and characterizations, leading altogether to the emergence of the area of granular computing. The concepts of information granules and information granularity offer a general perspective for information processing and generate far-reaching implications in numerous application areas. This is predominantly due to the fact that information granules help realize abstraction processes completed at various levels of hierarchy. This aspect is emphasized and exemplified through selected examples.

6.1 INFORMATION GRANULARITY AND GRANULAR COMPUTING

Information granules are intuitively appealing constructs, which play a pivotal role in human cognitive and decision-making activities (Bargiela and Pedrycz, 2003, 2005, 2008; Pedrycz and Bargiela, 2002; Zadeh, 1997, 1999, 2005). We perceive complex phenomena by organizing existing knowledge along with available experimental evidence and structuring them in a form of some meaningful, semantically sound entities, which are central to all ensuing processes of describing the world, reasoning about the environment, and supporting decision-making activities.

The term information granularity itself has emerged in different contexts and numerous areas of application. It carries various meanings. One can refer to artificial intelligence in which case information granularity is central to a way of problem solving through problem decomposition, where various subtasks could be formed and solved individually. Information granules and the area of intelligent computing revolving around them being termed granular computing are quite often presented with a direct association with pioneering studies. He coined an informal, yet highly descriptive and compelling concept of information granules by Zadeh (1997). In a general way, by information granules one regards a collection of elements drawn together

Pattern Recognition: A Quality of Data Perspective, First Edition. Władysław Homenda and Witold Pedrycz.

by their closeness (resemblance, proximity, functionality, etc.) articulated in terms of some useful spatial, temporal, or functional relationships. Subsequently, granular computing is about representing, constructing, processing, interpreting, and communicating information granules.

It is again worth emphasizing that information granules permeate almost all human endeavors. No matter which problem is taken into consideration, we usually set it up in a certain conceptual framework composed of some generic and conceptually meaningful entities—information granules, which we regard to be of relevance to the problem formulation, further problem solving, and a way in which the findings are conveyed to the community. Information granules realize a framework in which we formulate generic concepts by adopting a certain level of abstraction. Let us refer here to some areas that offer compelling evidence as to the nature of underlying processing and interpretation.

Image Processing
In spite of the continuous progress in the area, a human being assumes a dominant and very much uncontested position when it comes to understanding and interpreting images. Surely, we do not focus our attention on individual pixels and process them as such, but group them together into a hierarchy of semantically meaningful constructs—familiar objects we deal with in everyday life. Such objects involve regions that consist of pixels or categories of pixels drawn together because of their proximity in the image, similar texture, color, and brightness. This remarkable and unchallenged ability of humans dwells on our effortless ability to construct information granules, manipulate them, and arrive at sound conclusions.

Processing and Interpretation of Time Series
From our perspective we can describe them in a semi-qualitative manner by pointing at specific regions of such signals. Medical specialists can effortlessly interpret various diagnostic signals including ECG or EEG recordings (Pedrycz and Gacek, 2002). They distinguish some segments of signals and interpret their combinations. In stock market, one analyzes numerous time series by looking at amplitudes, trends, and emerging patterns. Experts can interpret temporal readings of sensors and assess a status of the monitored system. Again, in all these situations, the individual samples of the signals are not the focal point of the analysis, synthesis, and signal interpretation. We always granulate all phenomena (no matter if they are originally discrete or analog in their nature). When working with time series, information granulation occurs in time and in the features space where the data are described.

Granulation of Time
Time is another important and omnipresent variable that is subjected to granulation. We use seconds, minutes, days, months, and years. Depending upon a specific problem we have in mind, who the user is, the size of information granules (time intervals) could vary quite significantly. To the high-level management, time intervals of quarters of year or a few years could be meaningful temporal information granules on the basis of which one develops any predictive model. For those in charge of everyday operation of a dispatching center, minutes and hours could form a viable scale of time granulation.

Long-term planning is very much different from day-to-day operation. For the designer of high-speed integrated circuits and digital systems, the temporal information granules concern nanoseconds, microseconds, and perhaps milliseconds. Granularity of information (in this case time) helps us focus on the most suitable level of detail.

Data Summarization
Information granules naturally emerge when dealing with data, including those coming in the form of data streams. The ultimate objective is to describe the underlying phenomenon in an easily understood way and at a certain suitable level of abstraction. This requires that we use a vocabulary of commonly encountered terms (concepts) and discover relationships between them and possible linkages among the underlying concepts. An illustrative example is visualized in Figure 6.1. Having a collection of detailed numeric weather data concerning temperature, precipitation, and wind speed, they are transformed into a linguistic description at the higher level of abstraction. It is noticeable that information granularity emerges with regard to several variables present in the data.

Design of Software Systems
We develop software artifacts by admitting a modular structure of an overall architecture of the designed system where each module is a result of identifying essential functional closeness of some components of the overall system. Modularity (granularity) is a holy grail of the systematic software design supporting a production and maintenance of high quality software products.

Information granules are examples of artifacts. As such they naturally give rise to hierarchical structures: the same problem or system can be perceived at different levels of specificity (detail), depending on the complexity of the problem, available computing resources, and particular needs to be addressed. A hierarchy of information granules is inherently visible in processing of information granules. The level of details (which is represented in terms of the size of information granules) becomes an essential facet facilitating a way a hierarchical processing of information with different levels of hierarchy is indexed by the size of information granules.

Even such commonly encountered and simple examples presented so far are convincing enough to lead us to ascertain that (a) information granules are the key components of knowledge representation and processing, (b) the level of granularity

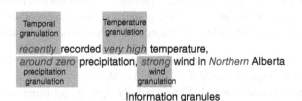

Information granules

DATA: Temperature, precipitation, and wind data recorded in
Northern Alberta

Figure 6.1   From a stream of numeric data to their granular description completed with the aid of information granules.

of information granules (their size, to be more descriptive) becomes crucial to the problem description and an overall strategy of problem solving, (c) hierarchy of information granules supports an important aspect of perception of phenomena and delivers a tangible way of dealing with complexity by focusing on the most essential facets of the problem, and (d) there is no universal level of granularity of information; commonly the size of granules is problem oriented and user dependent.

Human centricity comes as an inherent feature of intelligent systems. It is anticipated that a two-way effective human–machine communication is imperative. Human perceive the world, reason, and communicate at some level of abstraction. Abstraction comes hand in hand with non-numeric constructs, which embrace collections of entities characterized by some notions of closeness, proximity, resemblance, or similarity. These collections are referred to as information granules. Processing of information granules is a fundamental way in which people process such entities. Granular computing has emerged as a framework in which information granules are represented and manipulated by intelligent systems. The two-way communication of such intelligent systems with the users becomes substantially facilitated because of the usage of information granules.

By no means, the aforementioned quite descriptive definition of information granules is formal. It rather intends to emphasize the crux of the idea and link it to the human centricity and computing with perceptions rather than plain numbers.

What has been said so far touched upon a qualitative aspect of the problem. The visible challenge is to develop a computing framework within which all these representation and processing endeavors could be formally realized.

While the notions of information granularity and information granules themselves are convincing, they are not operational (algorithmically sound) until some formal models of information granules along with the related algorithmic framework have been introduced. In other words, to secure the algorithmic realization of granular computing, the *implicit* nature of information granules has to be translated into the constructs that are *explicit* in their nature, namely, described formally in which information granules can be efficiently computed with.

The common platform emerging within this context comes under the name of granular computing. In essence, it is an emerging paradigm of information processing. While we have already noticed a number of important conceptual and computational constructs built in the domain of system modeling, machine learning, image processing, pattern recognition, and data compression, in which various abstractions (and ensuing information granules) came into existence, granular computing becomes an innovative and intellectually proactive endeavor that manifests in several fundamental ways:

- It identifies the essential commonalities between the surprisingly diversified problems and technologies used there, which could be cast into a unified framework known as a granular world. This is a fully operational processing entity that interacts with the external world (that could be another granular or numeric world) by collecting necessary granular information and returning the outcomes of the granular computing.

- With the emergence of the unified framework of granular processing, we get a better grasp as to the role of interaction between various formalisms and visualize a way in which they communicate.

- It brings together the existing plethora of formalisms of set theory (interval analysis) under the same banner by clearly visualizing that, in spite of their visibly distinct underpinnings (and ensuing processing), they exhibit some fundamental commonalities. In this sense, granular computing establishes a stimulating environment of synergy between the individual approaches.

- By building upon the commonalities of the existing formal approaches, granular computing helps assemble heterogeneous and multifaceted models of processing of information granules by clearly recognizing the orthogonal nature of some of the existing and well-established frameworks (say, probability theory coming with its probability density functions and fuzzy sets with their membership functions)

- Granular computing fully acknowledges a notion of variable granularity, whose range could cover detailed numeric entities and very abstract and general information granules. It looks at the aspects of compatibility of such information granules and ensuing communication mechanisms of the granular worlds.

- Granular computing gives rise to processing that is less time demanding than the one required when dealing with detailed numeric processing.

- Interestingly, the inception of information granules is highly motivated. We do not form information granules without reason. Information granules arise on purpose as an evident realization of the fundamental paradigm of abstraction.

On the one hand, granular computing as an emerging area brings a great deal of original, unique ideas. On the other, it dwells substantially on the existing well-established developments that have already happened in a number of individual areas. In a synergistic fashion, granular computing brings fundamental ideas of interval analysis, fuzzy sets, and rough sets and facilitates building a unified view at them where an overarching concept is the granularity of information itself. It helps identify main problems of processing and its key features, which are common to all the formalisms being considered.

Granular computing forms a unified conceptual and computing platform. Yet, what is important, it directly benefits from the already-existing and well-established concepts of information granules formed in the setting of set theory, fuzzy sets, rough sets, and others. Reciprocally, the general investigations carried out under the rubric of Granular computing offer some interesting and stimulating thoughts to be looked at within the realm of the specific formalism of sets, fuzzy sets, shadowed set, or rough sets. There is a plethora of formal approaches toward characterization and processing information granules such as probabilistic sets (Hirota, 1981), rough sets (Pawlak, 1982, 1985, 1991; Pawlak and Skowron, 2007a, 2007b), and axiomatic set theory (Liu and Pedrycz, 2009).

6.2 FORMAL PLATFORMS OF INFORMATION GRANULARITY

A number of formal platforms exist in which information granules are conceptualized, defined, and processed.

Sets (intervals) realize a concept of abstraction by introducing a notion of dichotomy: we admit element to belong to a given information granule or to be excluded from it. Along with the set theory comes a well-developed discipline of interval analysis (Alefeld and Herzberger, 1983; Moore, 1966; Moore *et al.*, 2009). Alternatively to an enumeration of elements belonging to a given set, sets are described by characteristic functions taking on values in {0,1}. Formally, a characteristic function describing a set A is defined as follows:

$$A(x) = \begin{cases} 1, & \text{if } x \in A \\ 0, & \text{if } x \notin A \end{cases} \tag{6.1}$$

where $A(x)$ stands for a value of the characteristic function of set A at point x. With the emergence of digital technologies, interval mathematics has appeared as an important discipline encompassing a great deal of applications. A family of sets defined in a universe of discourse X is denoted by $P(X)$. Well-known set operations—union, intersection, and complement—are the three fundamental constructs supporting a manipulation on sets. In terms of the characteristic functions, they result in the following expressions:

$$(A \cap B)(x) = \min(A(x), B(x)) \quad (A \cup B)(x) = \max(A(x), B(x)) \quad \bar{A}(x) = 1 - A(x) \tag{6.2}$$

where $A(x)$ and $B(x)$ are the values of the characteristic functions of A and B at x and $\bar{A}$ denotes the complement of A.

Fuzzy sets deliver an important conceptual and algorithmic generalization of sets. By admitting partial membership of an element to a given information granule, we bring an important feature that makes the concept to be in rapport with reality. It helps working with the notions, where the principle of dichotomy is neither justified nor advantageous. The description of fuzzy sets is realized in terms of membership functions taking on values in the unit interval. Formally, a fuzzy set A is described by a membership function mapping the elements of a universe X to the unit interval [0,1]:

$$A : X \rightarrow [0, 1] \tag{6.3}$$

The membership functions are therefore synonymous of fuzzy sets. In a nutshell, membership functions generalize characteristic functions in the same way as fuzzy sets generalize sets. A family of fuzzy sets defined in X is denoted by $F(X)$. Fuzzy sets are generalizations of sets and are represented as a family of nested sets (representation theorem).

Operations on fuzzy sets are realized in the same way as already shown by (6.2), where the arguments are not 0 and −1 values assumed by characteristic functions, but membership values with their values in the unit interval. However, given the fact that we are concerned with membership grades in [0,1], there are a number of alternatives in the realization of fuzzy sets operators (logic *and* and *or* operators, respectively). Those are implemented through the so-called t-norms and t-conorms (Schweizer and Sklar, 1983, Klement *et al.*, 2000; Pedrycz and Gomide, 2007).

Shadowed sets (Pedrycz, 1998, 2005) offer an interesting description of information granules by distinguishing among three categories of elements. Those are the elements that (i) fully belong to the concept and (ii) are excluded from it and (iii) their belongingness is completely *unknown*. Formally, these information granules are described as a mapping $X: X \rightarrow \{1, 0, [0,1]\}$ where the elements with the membership quantified as the entire $[0,1]$ interval are used to describe a shadow of the construct. Given the nature of the mapping here, shadowed sets can be sought as a granular description of fuzzy sets where the shadow is used to localize unknown membership values, which in fuzzy sets are distributed over the entire universe of discourse. Note that the shadow produces non-numeric descriptors of membership grades. A family of fuzzy sets defined in X is denoted by $S(X)$.

Probability-oriented information granules are expressed in the form of some probability density functions or probability functions. They capture a collection of elements resulting from some experiment.

In virtue of the fundamental concept of probability, the granularity of information associates with a manifestation of occurrence of some elements. Probability function and probability density function are commonly encountered abstract descriptors of experimental data. The abstraction offered by probabilities is apparent: instead of coping with huge masses of data, one produces their abstract manifestation in the form of a single or a few probability functions. Histograms are examples of probabilistic information granules arising as a concise characterization of one-dimensional data; see Figure 6.2. In case of data belonging to a single class, an information granule—histogram composed of $c + 1$ bins—is described by vectors of cutoff points a, $a = [a_1, a_2, ..., a_c]$ and the corresponding vector of counts n, $n = [n_1, n_2, ..., n_c, n_{c+1}]$, that is, $H = (a, n)$. Their description could be provided in different ways. For instance, if the data come from a two-class problem, a histogram is an information granule containing probabilities (frequencies) of data belonging to a certain class and falling into a given interval (bin). In this case, a histogram H is an information granule in the form $H = (a, p)$ with p standing for a vector of the corresponding probabilities; see Figure 6.2.

When dealing with data coming from several classes, a suitable description of a histogram is composed of the vector of cutoff points and the associated entropy values, $H = (a, h)$, where h denotes a vector of entropies computed for data falling within the corresponding bins.

Rough sets emphasize a roughness of description of a given concept X when being realized in terms of the indiscernibility relation provided in advance (Pawlak, 1982). The roughness of the description of X is manifested in terms of its

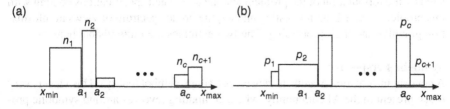

Figure 6.2   Histogram as an example of information granule with data coming from a single-class (a) and two-class problem (b).

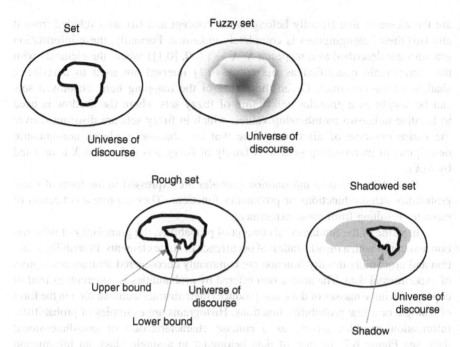

Figure 6.3   Conceptual realization of information granules—a comparative view.

lower and upper approximations of a certain rough set. A family of rough sets defined in X is denoted by $R(X)$.

Figure 6.3 visualizes the main differences present between selected sets, fuzzy sets, rough sets, and shadowed sets. As noted, the key aspect is about a representation of a notion of belongingness of elements to the concept. It is binary in case of sets (yes–no, dichotomy), partial membership (fuzzy sets), lower and upper bounds (rough sets), and uncertainty region (shadowed sets). Rough sets and shadowed sets, even though they exhibit conceptual differences that give rise to constructs that highlight some items endowed with uncertainty (undefined belongingness), are conceptually different.

Other formal models of information granules involve axiomatic sets, soft sets, and intuitionistic sets (in which raised is a notion of membership and nonmembership).

The choice of a certain formal setting of information granulation is mainly dictated by the formulation of the problem and the associated specifications coming with this problem. There is an interesting and a quite broad spectrum of views at information granules and their processing. The two extremes are quite visible here.

Symbolic Perspective

A concept information granule is viewed as a single symbol (entity). This view is very much present in the AI community, where computing revolves around symbolic processing. Symbols are subject to processing rules, giving rise to results, which are again symbols coming from the same vocabulary one has started with.

Numeric Perspective
Here information granules are associated with a detailed numeric characterization. Fuzzy sets are profound examples with this regard. We start with numeric membership functions. All ensuing processing involves *numeric* membership grades so in essence it focuses on number crunching. The results are inherently numeric. The progress present here has resulted in a diversity of numeric constructs. Because of the commonly encountered numeric treatment of fuzzy sets, the same applies to logic operators (connectives) encountered in fuzzy sets.

There are a number of alternatives of describing information that are positioned in between these two extremes, or the descriptions could be made more in a multilevel fashion. For instance, one could have an information granule described by a fuzzy set whose membership grades are symbolic (ordered terms, say, *small, medium, high*; all defined in the unit interval).

With regard to the formal settings of information granules as briefly highlighted previously, it is instructive to mention that all of them offer some operational realization (in different ways, though) of implicit concepts by endowing them by a well-defined semantics. For instance, treated implicitly an information granule *small error* is regarded just as a symbol (and could be subject to symbolic processing as usually realized in artificial intelligence); however once explicitly articulated as an information granule, it becomes associated with some semantics (becomes calibrated), thus coming with a sound operational description, say, characteristic or membership function. In the formalism of fuzzy sets, the symbol *small* comes with the membership description, which could be further processed.

6.3 INTERVALS AND CALCULUS OF INTERVALS

Sets and set theory are the fundamental notions of mathematics and science. They are in common usage when describing a wealth of concepts, quantifying relationships, and formalizing solutions. The underlying fundamental notion of set theory is that of *dichotomy*: a certain element belongs to a set or is excluded from it. A universe of discourse X over which a set or sets are formed could be very diversified depending upon the nature of the problem.

Given a certain element in a universe of discourse X, a process of dichotomization (binarization) imposes a binary, *all-or-none* classification decision: we either accept or reject the element as belonging to a given set. If we denote the acceptance decision about the belongingness of the element by 1 and the reject decision (non-belongingness) by 0, we express the classification (assignment) decision of $x \in X$ to some given set (S or T) through a characteristic function:

$$S(x) = \begin{cases} 1, & \text{if } x \in S \\ 0, & \text{if } x \notin S \end{cases} \qquad T(x) = \begin{cases} 1, & \text{if } x \in T \\ 0, & \text{if } x \notin T \end{cases} \tag{6.4}$$

The empty set $\emptyset$ has a characteristic function that is identically equal to 0, $\emptyset(x) = 0$ for all x in X. The universe X itself comes with the characteristic function that is identically equal to 1, that is, $X(x) = 1$ for all x in X. Also, a singleton

$A = \{a\}$, a set comprising only a single element, has a characteristic function such that $A(x) = 1$ if $x = a$ and $A(x) = 0$ otherwise.

Characteristic functions $A: X \rightarrow \{0, 1\}$ induce a constraint with well-defined binary boundaries imposed on the elements of the universe X that can be assigned to a set A. By looking at the characteristic function, we see that all elements belonging to the set are non-distinguishable—they come with the same value of the characteristic function, so by knowing that $A(x_1) = 1$ and $A(x_2) = 1$, we cannot tell these elements apart. The operations of union, intersection, and complement are easily expressed in terms of the characteristic functions. The characteristic function of the union comes as the maximum of the characteristic functions of the sets involved in the operation. The complement of A, denoted by $\bar{A}$, comes with a characteristic function equal to $1 - A(x)$.

Interval analysis has emerged with the inception of digital computers and was mostly motivated by the models of computations therein, which are carried out for intervals implied by the finite number of bits used to represent any number on a (digital) computer. This interval nature of the arguments (variables) implies that the results are also intervals. This raises awareness about the interval character of the results. Interval analysis is instrumental in the analysis of propagation of granularity residing within the original arguments (intervals).

Here we elaborate on the fundamentals of interval calculus. It will become apparent that they will be helpful in the development of the algorithmic fabric of other formalisms of information granules.

We briefly recall the fundamental notions of numeric intervals. Two intervals $A = [a, b]$ and $B = [c, d]$ are equal if their bounds are equal, $a = c$ and $b = d$. A degenerate interval $[a, a]$ is a single number. There are two categories of operations on intervals, namely, set-theoretic and algebraic operations.

Set-Theoretic Operations

Assuming that the intervals are not disjoint (have some common elements) they are defined as follows:

$$\text{Intersection } \{z|\, z \in A \text{ and } z \in B\} = [\max(a, c), \min(b, d)]$$
$$\text{Union } \{z|z \in A \text{ or } z \in B\} = [\min(a, c), \max(b, d)] \tag{6.5}$$

For the illustration of the operations, refer to Figure 6.4.

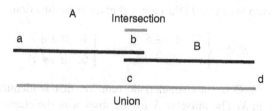

Figure 6.4 Examples of set-theoretic operations on numeric intervals.

Algebraic Operations on Intervals

The generic algebraic operations on intervals are quite intuitive. As before, let us consider the two intervals $A = [a, b]$ and $B = [c, d]$. The results of addition, subtraction, multiplication, and division are expressed as follows (Moore, 1966):

$$A + B = [a + c, b + d]$$

$$A - B = [a - d, b - c] = A + [-1 - 1] * B = [a, b] + [-d, -c] = [a - d, b - c]$$

$$A * B = [\min(ac, ad, bc, bd), \max(ac, ad, bc, bd)]$$

$$A/B = [a, b][1/d, 1/c] \text{ (it is assumed that 0 is not included in the interval } [c, d])$$

$$(6.6)$$

All these formulas result from the fact that the aforementioned functions are continuous on a compact set; as a result they take on the largest and the smallest value as well as the values in between. The intervals of the obtained values are closed—in all these formulas one computes the largest and the smallest values.

In addition to the algebraic operations, for a continuous unary operation $f(x)$ on the space of real numbers R, the mapping of the interval $A = [a, b]$ produces an interval $f(A)$:

$$f(A) = [\min f(x), \max f(x)]$$

$$(6.7)$$

where the minimum (maximum) are taken for all xs belonging to A. Examples of such mappings are x^k, $\exp(x)$, $\sin(x)$, and so on. For the monotonically increasing or decreasing functions, the aforementioned formula are significantly simplified:

- Monotonically increasing functions

$$f(A) = [f(a), f(b)]$$

$$(6.8)$$

- Monotonically decreasing functions

$$f(A) = [f(b), f(a)]$$

$$(6.9)$$

Example

Let us consider two intervals $A = [-1, 4]$ and $B = [1, 6]$. The algebraic operations applied to A and B produce the following results:

- Addition $A + B = [-1 + 1, 4 + 6] = [0, 10]$
- Subtraction $A - B = [-1 - 6, 4 - 1] = [-7, -3]$
- Multiplication $A * B = [\min(-1, -6, 4, 24), \max(-1, -6, 4, 24)] = [-6, 24]$
- Division $A/B = [-1, 4] * [1/6, 1/1] = [\min(-1/6, -1, 4/6, 4), \max(-1/6, -1, 4/6, 4)] = [-1/6, 4]$

Distance between Intervals

The distance between two intervals A and B is expressed as

$$d(A, B) = \max(|a - c|, |b - d|)$$

$$(6.10)$$

One can easily show that the properties of distances are satisfied: $d(A, B) = d(B, A)$ (symmetry), $d(A, B)$ is nonnegative with $d(A, B) = 0$ (nonnegativity) if and only if (iff) $A = B$, $d(A, B) \leq d(A, C) + d(B, C)$ (triangle inequality). For real numbers the distance is reduced to the Hamming one.

6.4   CALCULUS OF FUZZY SETS

Fuzzy sets and the corresponding membership functions form a viable and mathematically sound framework to formalize concepts with gradual boundaries. The fundamental idea of fuzzy set is to relax this requirement by admitting intermediate values of class membership (Klir and Yuan, 1995; Nguyen and Walker, 1999; Zadeh, 1965, 1975, 1978). Therefore we may assign intermediate values between 0 and 1 to quantify our perception on how compatible these values are with the class (concept) with 0 meaning incompatibility (complete exclusion) and 1 compatibility (complete membership). Membership values thus express the degrees to which each element of the universe is compatible with the properties distinctive to the class. Intermediate membership values underline that no *natural* (binary) threshold exists and that elements of the universe could be the members of a class and at the same time belong to other classes with different degrees. Allowing for gradual, hence less strict, nonbinary membership degrees is the crux of fuzzy sets.

Formally, a fuzzy set A is described by a membership function mapping the elements of a universe X to the unit interval [0,1]. The membership functions are therefore synonymous of fuzzy sets. In a nutshell, membership functions generalize characteristic functions in the same way as fuzzy sets generalize sets.

Being more descriptive, we may view fuzzy sets as elastic constraints imposed on the elements of a universe. As emphasized before, fuzzy sets deal primarily with the concept of elasticity, graduality, or absence of sharply defined boundaries. In contrast, when dealing with sets, we are concerned with rigid boundaries, lack of graded belongingness, and sharp, binary boundaries. Gradual membership means that no natural boundary exists and that some elements of the universe of discourse can, contrary to sets, coexist (belong) with different fuzzy sets with different degrees of membership.

6.4.1   Membership Functions and Classes of Fuzzy Sets

Formally speaking, any function $A: X \rightarrow [0, 1]$ could be qualified to serve as a membership function describing the corresponding fuzzy set (Dubois and Prade, 1979, 1997, 1998). In practice, the form of the membership functions should be reflective of the problem at hand for which we construct fuzzy sets. They should reflect our perception (semantics) of the concept to be represented and further used in problem solving, the level of detail we intend to capture, and a context, in which the fuzzy set are going to be used. It is also essential to assess the type of a fuzzy set from the standpoint of its suitability when handling the ensuing optimization procedures. It also needs to accommodate some additional requirements arising as a result of further needs for optimization procedures such as differentiability of membership functions. Given these criteria in mind, we elaborate on the most commonly used categories of membership functions. All of them are defined in the universe of real numbers, that is, $X = R$.

Triangular Membership Functions

The fuzzy sets are expressed by their piecewise linear segments described in the form

$$A(x, a, m, b) = \begin{cases} 0 & \text{if } x \le a \\ \dfrac{x-a}{m-a} & \text{if } x \in [a, m] \\ \dfrac{b-x}{b-m} & \text{if } x \in [m, b] \\ 0 & \text{if } x \ge b \end{cases} \quad (6.11)$$

Using more concise notation, the aforementioned expression can be written down in the form $A(x, a, m, b) = \max\{\min[(x-a)/(m-a), (b-x)/(b-m)], 0\}$. The meaning of the parameters is straightforward: m denotes a modal (typical) value of the fuzzy set, while a and b are the lower and upper bounds, respectively. They could be sought as the extreme elements of the universe of discourse that delineate the elements belonging to A with nonzero membership degrees. Triangular fuzzy sets (membership functions) are the simplest possible models of membership functions. They are fully defined by only three parameters. As mentioned, the semantics is evident as the fuzzy sets are expressed on a basis of knowledge of the spreads of the concepts and their typical values. The linear change in the membership grades is the simplest possible model of membership one could think of.

Trapezoidal Membership Functions

They are piecewise linear function characterized by four parameters, a, m, n, b, each of which defines one of the four linear parts of the membership function. They assume the following form:

$$A(x) = \begin{cases} 0 & \text{if } x < a \\ \dfrac{x-a}{m-a} & \text{if } x \in [a, m] \\ 1 & \text{if } x \in [m, n] \\ \dfrac{b-x}{b-n} & \text{if } x \in [n, b] \\ 0 & \text{if } x > b \end{cases} \quad (6.12)$$

Using an equivalent notation, we can rewrite A as follows: $A(x, a, m, n, b) = \max\{\min[(x-a)/(m-a), 1, (b-x)/(b-n)], 0\}$. Note that elements in $[m, n]$ are non-distinguishable as this region is described by a characteristic function.

S-Membership Functions

These functions are of the form

$$A(x) = \begin{cases} 0 & \text{if } x \le a \\ 2\left(\dfrac{x-a}{b-a}\right)^2 & \text{if } x \in [a, m] \\ 1-2\left(\dfrac{x-b}{b-a}\right)^2 & \text{if } x \in [m, b] \\ 1 & \text{if } x > b \end{cases} \quad (6.13)$$

The point $m = (a + b)/2$ is referred to as the crossover point.

Gaussian Membership Functions

These membership functions are described by the following relationship:

$$A(x, m, \sigma) = \exp\left(-\frac{(x-m)^2}{\sigma^2}\right)$$ (6.14)

Gaussian membership functions are described by two important parameters. The modal value (m) represents the typical element of A, while σ denotes the spread of A. Higher values of σ corresponds to larger spreads of the fuzzy sets.

One of the interesting properties of membership functions concerns sensitivity of fuzzy sets, which is typically expressed by computing the derivative of the membership function (assuming that the derivative exists) and taking its absolute value $|dA/dx|$. It expresses an extent to which fuzzy set changes its characteristics and over which region of the universe of discourse. From this perspective, we see that the piecewise linear membership function exhibits the same value of sensitivity to the left and the right from its modal value (which is equal to the absolute values of the slopes of the linear pieces of the membership function). This is not the case when considering the parabolic or Gaussian membership functions where the sensitivity depends upon the location in the universe of discourse X as well as the range of membership grades. If the requirement on the changes of membership is also imposed as a part of the estimation problem when building fuzzy sets, they can be taken into consideration in the formation of membership functions.

Support

Support of a fuzzy set A, denoted by Supp (A), is a set of all elements of X with non-zero membership degrees in A:

$$\text{Supp}(A) = \{x \in X | A(x) > 0\}$$ (6.15)

Core

The core of a fuzzy set A, Core(A), is a set of all elements of the universe that are typical for A, namely, they come with membership grades equal to 1:

$$\text{Core}(A) = \{x \in X | A(x) = 1\}$$ (6.16)

While core and support are somewhat extreme (in the sense that they identify the elements of A that exhibit the strongest and the weakest linkages with A), we may be also interested in characterizing sets of elements that come with some intermediate membership degrees. A notion of a so-called α-cut offers here an interesting insight into the nature of fuzzy sets.

α-Cut

The α-cut of a fuzzy set A, denoted by A_α, is a set consisting of the elements of the universe whose membership values are equal to or exceed a certain threshold level α where $\alpha \in [0,1]$. Formally speaking, we have $A_\alpha = \{x \in X | A(x) \geq \alpha\}$. A strong α-cut differs from the α-cut in the sense that it identifies all elements in X for which we have the following equality $A_\alpha = \{x \in X | A(x) > \alpha\}$. An illustration of the concept of the α-cut and strong α-cut is presented in Figure 6.5. Both support and core are limited cases of α-cuts and strong α-cuts. For $\alpha = 0$ and the strong α-cut, we arrive at the

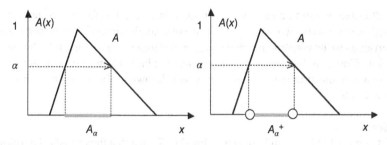

Figure 6.5 Examples of α-cut and strong α-cut.

concept of the support of A. The threshold $\alpha = 1$ means that the corresponding α-cut is the core of A.

We can characterize fuzzy sets by counting their elements and brining a single numeric quantity as a meaningful descriptor of this count. While in case of sets this sounds convincing, here we have to take into account different membership grades. In the simplest form, this counting comes under the name of cardinality.

Fuzzy sets offer an important conceptual and operational feature of information granules by endowing their formal models by gradual degrees of membership. We are interested in exploring relationships between fuzzy sets and sets. While sets come with the binary (yes–no) model of membership, it could be worth investigating whether they are indeed some special cases of fuzzy sets and if so, in which sense a set could be treated as a suitable approximation of some given fuzzy set. This could shed light on some related processing aspects. To gain a detailed insight into this matter, we recall here a concept of an α-cut and a family of α-cuts and show that they relate to fuzzy sets in an intuitive and transparent way. Let us revisit the semantics of α-cuts: an α-cut of A embraces all elements of the fuzzy set whose degrees of belongingness (membership) to this fuzzy set are at least equal to α. (Pedrycz *et al.*, 2009). In this sense, by selecting a sufficiently high value of α, we identify (tag) elements of A that belongs to it to a significant extent and thus could be sought as those substantially representative of the concept conveyed by A. Those elements of X exhibiting lower values of the membership grades are suppressed, so this allows us to selectively focus on the elements with the highest degrees of membership while dropping the others.

For α-cuts A_α the following properties hold:

a. $A_0 = X$

b. If $\alpha \leq \beta$ then $A_\alpha \supseteq A_\beta$ (6.17)

The first property tells us that if we allow for the zero value of α, then all elements of X are included in this α-cut (0-cut, to be more specific). The second property underlines the monotonic character of the construct: higher values of the threshold imply that more elements are accepted in the resulting α-cuts. In other words, we may say that the level sets (α-cuts) A_α form a nested family of sets indexed by some parameter (α). If we consider the limit value of α, that is, $\alpha = 1$, the corresponding α-cut is nonempty iff A is a normal fuzzy set.

It is also worth to remember that α-cuts, in contrast to fuzzy sets, are sets. We showed how for some given fuzzy set, its α-cut could be formed. An interesting question arises as to the construction that could be realized when moving into the opposite direction. Could we "reconstruct" a fuzzy set on a basis of an infinite family of sets? The answer to this problem is offered in what is known as the representation theorem for fuzzy sets:

Theorem
Let $\{A_\alpha\}$ $\alpha \in [0,1]$ be a family of sets defined in X such that they satisfy the following properties:

a. $A_0 = X$

b. If $\alpha \leq \beta$ then $A_\alpha \supseteq A_\beta$

c. For the sequence of threshold values $\alpha_1 \leq \alpha_2 \leq \cdots$ such that $\lim \alpha_n = \alpha$, we have $A_\alpha = \bigcap_{n=1}^\infty A_{\alpha_n}$.

Then there exists a unique fuzzy set B defined in X such that $B_\alpha = A_\alpha$ for each $\alpha \in [0,1]$.

6.4.2   Triangular Norms and Triangular Conorms as Models of Operations on Fuzzy Sets

Logic operations on fuzzy sets concern manipulation of their membership functions. Therefore they are domain dependent, and different contexts may require their different realizations. For instance, since operations provide ways to combine information, they can be performed differently in image processing, control, and diagnostic systems. When contemplating the realization of operations of intersection and union of fuzzy sets, we should require a satisfaction of the following collection of intuitively appealing properties: (a) commutativity, (b) associativity, (c) monotonicity, and (d) identity.

The last requirement of identity takes on a different form depending on the operation. In the case of intersection, we anticipate that an intersection of any fuzzy set with the universe of discourse X should return this fuzzy set. For the union operations, the identity implies that the union of any fuzzy set and an empty fuzzy set returns this fuzzy set.

Thus any binary operator $[0,1] \times [0,1] \rightarrow [0,1]$, which satisfies the collection of the requirements outlined previously, can be regarded as a potential candidate to realize the intersection or union of fuzzy sets. Note also that identity acts as boundary conditions, meaning that when confining to sets, the previously stated operations return the same results as encountered in set theory. In general, idempotency is not required; however the realizations of union and intersection could be idempotent as this happens for the operations of minimum and maximum where $\min(a, a) = a$ and $\max(a, a) = a$.

In the theory of fuzzy sets, triangular norms offer a general class of realizations of intersection and union. Originally they were introduced in the context of probabilistic metric spaces (Schweizer and Sklar, 1983). t-Norms arise a family of operators modeling intersection of fuzzy sets. Given a t-norm, a dual operator called a t-conorm (or s-norm) can be derived using the relationship $x\,s\,y = 1 - (1 - x)\,t\,(1 - y), \forall x, y \in [0,1]$, which is the De Morgan law. Triangular conorms provide generic models for the union of fuzzy sets. t-Conorms can also be described by an independent axiomatic system.

A triangular norm, t-norm for brief, is a binary operation t, $[0,1] \times [0,1] \to [0,1]$, that satisfies the following properties:

Commutativity:	$a\,t\,b = b\,t\,a$
Associativity:	$a\,t(b\,t\,c) = (a\,t\,b)\,t\,c$
Monotonicity:	if $b \le c$ then $a\,t\,b \le a\,t\,c$
Boundary conditions:	$a\,t\,1 = a$
	$a\,t\,0 = 0$

where a, b, $c \in [0,1]$.

Let us elaborate on the meaning of these requirements vis-à-vis the use of t-norms as models of operators of union and intersection of fuzzy sets. There is a one-to-one correspondence between the general requirements outlined previously and the properties of t-norms. The first three reflect the general character of set operations. Boundary conditions stress the fact all t-norms attain the same values at boundaries of the unit square $[0,1] \times [0,1]$. Thus, for sets, any t-norm produces the same result that coincides with the one we could have expected in set theory when dealing with intersection of sets, that is, $A \cap X = A$, $A \cap \emptyset = \emptyset$. In the existing plethora of t-norms and t-conorms, we consider several representative families of t-norms and t-conorms, including those triangular norms endowed with some parameters; see Tables 6.1 and 6.2. The t-norms reported here come with their duals:

Minimum:	$a\,t_m b = \min(a, b) = a \hat{\ } b$
Product:	$a\,t_p b = ab$
Lukasiewicz:	$a\,t_l b = \max(a + b - 1, 0)$
Drastic product:	$a t_d b = \begin{cases} a & \text{if } b = 1 \\ b & \text{if } a = 1 \\ 0 & \text{otherwise} \end{cases}$

TABLE 6.1 Selected examples of t-norms and t-conorms

Names	t-Norms	t-Conorms
Logical	$T_1(x_1, x_2) = \min(x_1, x_2)$	$S_1(x_1, x_2) = \max(x_1, x_2)$
Hamacher	$T_2(x_1, x_2) = \dfrac{x_1 x_2}{x_1 + x_2 - x_1 x_2}$	$S_2(x_1, x_2) = \dfrac{x_1 + x_2 - 2x_1 x_2}{1 - x_1 x_2}$
Algebraic	$T_3(x_1, x_2) = x_1 x_2$	$S_3(x_1, x_2) = x_1 + x_2 - x_1 x_2$
Einstein	$T_4(x_1, x_2) = \dfrac{x_1 x_2}{1 + (1 - x_1)(1 - x_2)}$	$S_4(x_1, x_2) = \dfrac{x_1 + x_2}{1 + x_1 x_2}$
Lukasiewicz	$T_5(x_1, x_2) = \max(x_1 + x_2 - 1, 0)$	$S_5(x_1, x_2) = \min(x_1 + x_2, 1)$
Drastic	$T_6(x_1, x_2) = \begin{cases} x_1 & x_2 = 1 \\ x_2 & x_1 = 1 \\ 0 & x_1, x_2 < 1 \end{cases}$	$S_6(x_1, x_2) = \begin{cases} x_1 & x_2 = 0 \\ x_2 & x_1 = 0 \\ 1 & x_1, x_2 > 0 \end{cases}$
Triangular 1	$T_7(x_1, x_2) = \dfrac{2}{\pi} \cot^{-1}\left[\cot\dfrac{\pi x_1}{2} + \cot\dfrac{\pi x_2}{2}\right]$	$S_7(x_1, x_2) = \dfrac{2}{\pi} \tan^{-1}\left[\tan\dfrac{\pi x_1}{2} + \tan\dfrac{\pi x_2}{2}\right]$
Triangular 2	$T_8(x_1, x_2) = \dfrac{2}{\pi} \arcsin\left(\sin\dfrac{\pi x_1}{2}\sin\dfrac{\pi x_2}{2}\right)$	$S_8(x_1, x_2) = \dfrac{2}{\pi} \arccos\left(\cos\dfrac{\pi x_1}{2}\cos\dfrac{\pi x_2}{2}\right)$

TABLE 6.2 Selected examples of parametric t-norms and t-conorms

Names	t-Norms	t-Conorms
Sugeno–Weber	$T_W(x_1,x_2)=\begin{cases} T_6(x_1,x_2) & \text{if } p=-1 \\ \max\left(\dfrac{x_1+x_2-1+px_1x_2}{1+p},\,0\right) & \text{if } p\in(-1,+\infty) \end{cases}$	$S_W(x_1,x_2)=\begin{cases} S_6(x_1,x_2) & \text{if } p=-1 \\ \min(x_1+x_2+px_1x_2,1) & \text{if } p\in(-1,+\infty) \end{cases}$
Schweizer–Sklar	$T_S(x_1,x_2)=\begin{cases} (x_1^p+x_2^p-1)^{1/p} & \text{if } p\in(-\infty,0) \\ T_3(x_1,x_2) & \text{if } p=0 \\ \max\left((x_1^p+x_2^p-1)^{1/p},\,0\right) & \text{if } p\in(0,+\infty) \end{cases}$	$S_S(x_1,x_2)=\begin{cases} 1-((1-x_1)^p+(1-x_2)^p-1)^{1/p} & \text{if } p\in(-\infty,0) \\ S_3(x_1,x_2) & \text{if } p=0 \\ 1-\max\left(((1-x_1)^p+(1-x_2)^p-1)^{1/p},\,0\right) & \text{if } p\in(0,+\infty) \end{cases}$
Yager	$T_Y(x_1,x_2)=\begin{cases} T_6(x_1,x_2) & \text{if } p=0 \\ \max\left(1-((1-x_1)^p+(1-x_2)^p)^{1/p},\,0\right) & \text{if } p\in(0,+\infty) \end{cases}$	$S_Y(x_1,x_2)=\begin{cases} S_6(x_1,x_2) & \text{if } p=0 \\ \max\left(1-(x_1^p+x_2^p)^{1/p},\,0\right) & \text{if } p\in(0,+\infty) \end{cases}$
Hamacher	$T_H(x_1,x_2)=\begin{cases} 0 & \text{if } p=x_1=x_2=0 \\ \dfrac{x_1x_2}{p+(1-p)(x_1+x_2-x_1x_2)} & \text{otherwise} \end{cases}$	$S_H(x_1,x_2)=\begin{cases} 1 & \text{if } p=0\, x_1=x_2=1 \\ \dfrac{x_1+x_2-x_1x_2-(1-p)x_1x_2}{1-(1-p)x_1x_2} & \text{otherwise} \end{cases}$
Frank	$T_F(x_1,x_2)=\begin{cases} T_1(x_1,x_2) & \text{if } p=0 \\ T_3(x_1,x_2) & \text{if } p=1 \\ \log_p\left(1+\dfrac{(p^{x_1}-1)(p^{x_2}-1)}{p-1}\right) & \text{otherwise} \end{cases}$	$S_F(x_1,x_2)=\begin{cases} S_1(x_1,x_2) & \text{if } p=0 \\ S_3(x_1,x_2) & \text{if } p=1 \\ 1-\log_p\left(1+\dfrac{(p^{1-x_1}-1)(p^{1-x_2}-1)}{p-1}\right) & \text{otherwise} \end{cases}$
Dombi	$T_D(x_1,x_2)=\begin{cases} T_6(x_1,x_2) & \text{if } p=0 \\ \dfrac{1}{1+\left(\left(\frac{1-x_1}{x_1}\right)^p+\left(\frac{1-x_2}{x_2}\right)^p\right)^{1/p}} & \text{otherwise} \end{cases}$	$S_D(x_1,x_2)=\begin{cases} S_6(x_1,x_2) & \text{if } p=0 \\ 1-\dfrac{1}{1+\left(\left(\frac{x_1}{1-x_1}\right)^p+\left(\frac{x_2}{1-x_2}\right)^p\right)^{1/p}} & \text{otherwise} \end{cases}$
Dubois–Prade	$T_P(x_1,x_2)=\dfrac{x_1x_2}{\max(x_1,x_2,p)}\qquad p\in[0,1]$	$S_P(x_1,x_2)=\dfrac{x_1+x_2-x_1x_2-\min(x_1,x_2,1-p)}{\max(1-x_1,1-x_2,p)}\qquad p\in[0,1]$
General Dombi	$T_G(x_1,x_2)=\begin{cases} T_6(x_1,x_2) & \text{if } p=0 \\ \dfrac{1}{1+\left(\frac{1}{a}\left(\left(1+a\left(\frac{1-x_1}{x_1}\right)^p\right)\left(1+a\left(\frac{1-x_2}{x_2}\right)^p\right)-1\right)\right)^{1/p}} & \text{otherwise} \end{cases}$ $a\in(0,+\infty)$	$S_G(x_1,x_2)=\begin{cases} S_6(x_1,x_2) & \text{if } p=0 \\ 1-\dfrac{1}{1+\left(\frac{1}{a}\left(\left(1+a\left(\frac{x_1}{1-x_1}\right)^p\right)\left(1+a\left(\frac{x_2}{1-x_2}\right)^p\right)-1\right)\right)^{1/p}} & \text{otherwise} \end{cases}$ $a\in(0,+\infty)$

In general, t-norms cannot be linearly ordered. One can demonstrate though that the min (t_m) t-norm is the largest t-norm, while the drastic product is the smallest one. They form the lower and upper bounds of the t-norms in the following sense:

$$a\,t_d b \leq a\,t\,b \leq a\,t_m b = \min(a, b) \tag{6.18}$$

Triangular conorms are functions s: $[0,1] \times [0,1] \to [0,1]$ that serve as generic realizations of the union operator of fuzzy sets. Similarly as triangular norms, conorms provide the highly desirable modeling flexibility needed to construct fuzzy models. Triangular conorms can be viewed as dual operators to the t-norms and, as such, are explicitly defined with the use of De Morgan laws. We may characterize them in a fully independent fashion by offering the following definition.

A triangular conorm (s-norm) is a binary operation s: $[0,1] \times [0,1] \to [0,1]$ that satisfies the following requirements:

Commutativity: $a\,s\,b = b\,s\,a$

Associativity: $a\,s\,(b\,s\,c) = (a\,s\,b)\,s\,c$

Monotonicity: if $b \leq c$ then $a\,s\,b \leq a\,s\,c$

Boundary conditions: $a\,s\,0 = a$
$a\,s\,1 = 1$

$a,b,c \in [0,1]$.

One can show that s: $[0,1] \times [0,1] \to [0,1]$ is a t-conorm iff there exists a t-norm (dual t-norm) such that for $\forall\, a,b \in [0,1]$, we have

$$a\,s\,b = 1 - (1-a)\,t\,(1-b) \tag{6.19}$$

For the corresponding dual t-norm, we have

$$a\,t\,b = 1 - (1-a)\,s\,(1-b) \tag{6.20}$$

The duality expressed by (6.19) and (6.20) can be helpful in delivering as an alternative definition of t-conorms. This duality allows us to deduce the properties of t-conorms on the basis of the analogous properties of t-norms. Notice that after rewriting (6.19) and (6.20), we obtain

$$(1-a)\,t\,(1-b) = 1 - a\,s\,b \tag{6.21}$$

$$(1-a)\,s\,(1-b) = 1 - a\,t\,b \tag{6.22}$$

These two relationships can be expressed symbolically as

$$\bar{A} \cap \bar{B} = \overline{A \cup B} \tag{6.23}$$

$$\bar{A} \cup \bar{B} = \overline{A \cap B} \tag{6.24}$$

that are nothing but the De Morgan laws well known in set theory. As shown, they are also satisfied for fuzzy sets.

The boundary conditions mean that all t-conorms behave similarly at the boundary of the unit square $[0,1] \times [0,1]$. Thus, for sets, any t-conorm returns the same result as encountered in set theory.

6.5 CHARACTERIZATION OF INFORMATION GRANULES: COVERAGE AND SPECIFICITY

The characterization of information granules is more challenging in comparison with a way in which a single numeric entity could be fully described. This is not surprising considering the abstract nature of the granule. In what follows, we introduce two measures that are of practical relevance and offer a useful insight into the nature and further usage of the granule in various constructs.

In a descriptive way, a level of abstraction captured by some information granules A is associated with the number of elements (data) embraced by the granule. For instance, these elements are the elements of the space or a collection of some experimental data. A certain measure of cardinality, which counts the number of elements involved in the information granule, forms a sound descriptor of information granularity. The higher the number of elements embraced by the information granule, the higher the abstraction of this granule and the lower its specificity becomes. Starting with a set theory formalism where A is a set, its cardinality is computed in the form of the following sum in the case of a finite universe of discourse $X = \{x_1, x_2, \ldots, x_n\}$:

$$\text{card}(A) = \sum_{i=1}^{n} A(x_i) \qquad (6.25)$$

or an integral (when the universe is infinite and the integral of the membership function itself does exists):

$$\text{card}(A) = \int_X A(x)dx \qquad (6.26)$$

where $A(x)$ is a formal description of the information granule (say, in the form of the characteristic function or membership function). For a fuzzy set, we count the number of its elements, but one has to bear in mind that each element may belong to a certain degree of membership, so the calculations carried out previously involve the degrees of membership. In this case, one commonly refers to (6.25)–(6.26) as a σ-count of A. For rough sets, one can proceed in a similar manner as aforementioned by expressing the granularity of the lower and upper bounds of the rough set (roughness). In case of probabilistic information granules, one can consider its standard deviation as a sound descriptor of information granularity. The higher the coverage (cardinality), the higher the level of abstraction being associated with the information granule.

The specificity is about the level of detail being captured by the information granule. As the name stipulates, the specificity indicates how detailed information granule is. In other words, the specificity can be determined by assessing the "size" of this fuzzy set. To explain the notion, we start with a case when information granule is an interval $[a, b]$; see Figure 6.6a. The specificity can be defined in the following way $1 - (b - a)/\text{range}$ where "range" is expressed as $y_{max} - y_{min}$ with y_{max} and y_{min} being the extreme values of assumed by the variable. This definition adheres to our perception of specificity: the broader the interval, the lower its specificity. In the boundary cases, when $a = b$, the specificity attains its maximal value equal to 1,

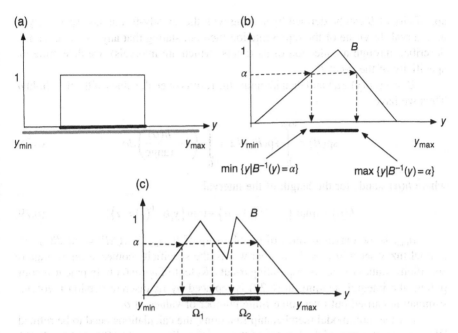

Figure 6.6   Determination of specificity of (a) interval (set), (b) unimodal fuzzy set, and (c) multimodal fuzzy set.

whereas when the interval spreads across the entire space (range), the specificity value is 0. The aforementioned definition is a special case. In a general way, specificity is defined as a measure adhering to the conditions we have already outlined previously:

$$\text{sp}: A \rightarrow [0, 1] \qquad (6.27)$$

1. Boundary conditions $\text{sp}(\{x\}) = 1$, $\text{sp}(X) = 0$. A single-element information granule is the most specific. The specificity of the entire space is the minimal one (here we can request that it is equal to 0, but this is not required in all situations).

2. Monotonicity if $A \subset B$, then $\text{sp}(A) \geq \text{sp}(B)$. This reflects our intuition that the more detailed information granule comes with the higher specificity. One has to note here that the requirement is general in the sense that the definition of inclusion and the details of computing the specificity are dependent upon the formalism of information granules. In the aforementioned example, we considered intervals; see Figure 6.6a. Here the inclusion of intervals is straightforward; apparently $A = [a, b]$ is included in $B = [c, d]$ if $a \geq c$ and $b \leq d$. The specificity defined previously is just an example; any decreasing function of the length of the interval could serve as a viable alternative. For instance, one can consider $\exp(-|b - a|)$ as the specificity of A.

The technical details of specificity have to be redefined when considering some other formalisms of information granules. Consider a given fuzzy set B. The specificity of B can be defined by starting with the already-formulated specificity of an interval. In virtue of the representation theorem stating that any fuzzy set can be described through a collection of its α-cuts (which are intervals), we determine the specificity of the α-cut B_α

$B_\alpha = \{y| B(y) \geq \alpha\}$ and then integrate the results over all values of the threshold α. Thus we have

$$\text{sp}(B) = \int\limits_0^{\alpha_{\max}} \text{sp}(B_\alpha) d\alpha = \int\limits_0^{\alpha_{\max}} \left(1 - \frac{h(\alpha)}{\text{range}} \right) d\alpha \qquad (6.28)$$

where $h(\alpha)$ stands for the length of the interval

$$h(\alpha) = \left| \max\{y|B^{-1}(y) = \alpha\} - \min\{y|B^{-1}(y) = \alpha\} \right| \qquad (6.29)$$

$\alpha_{\max}$ is the maximal value of membership of B, $\alpha_{\max} = hgt(B) = \sup_y B(y)$. For normal fuzzy set B, $\alpha_{\max} = 1$. In other words, the specificity comes as an average of specificity values of the corresponding α-cuts. Refer to Figure 6.6b. In practical computing, the integral standing in (6.28) is replaced by its discrete version involving summation carried out over some finite number of values of α.

For the multimodal membership functions, the calculations need to be refined. We consider the sum of the lengths of the α-cuts as illustrated in Figure 6.6c. In this way, one has

$$h(\alpha) = \text{length}(\Omega_1) + \text{length}(\Omega_2) + \cdots + \text{length}(\Omega_n) \qquad (6.30)$$

To visualize coverage and specificity as well as emphasize the relationships between these two characteristics, we assume that the data are governed by the Gaussian probability density function $p(x)$ with a zero mean and some standard deviation. As the interval is symmetric, we are interested in determining an optimal value of its upper bound b. The coverage provided by the interval $[0, b]$ is expressed as the integral of the probability function

$$\text{cov}([0, b]) = \int\limits_0^b p(x) dx \qquad (6.31)$$

while the specificity (assuming that the $x_{\max}$ is set as 3σ) is defined in the following form:

$$\text{sp}([0, b]) = 1 - \frac{b}{x_{\max}} \qquad (6.32)$$

The plots of the coverage and the specificity measures being regarded as a function of b are shown in Figure 6.7. One can note that the coverage is a nonlinear and monotonically increasing function of b, while the specificity (in the form specified previously) is a linearly decreasing function of b. As it becomes intuitively apparent, the increase in the coverage results in lower values of the specificity measure and vice versa. We will consider them as two essential criteria guiding the development of information granules.

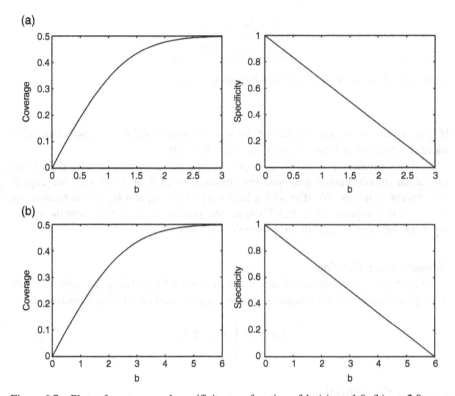

Figure 6.7   Plots of coverage and specificity as a function of b: (a) $\sigma = 1.0$, (b) $\sigma = 2.0$.

6.6   MATCHING INFORMATION GRANULES

The problem of quantifying how close (similar) two information granules A and B defined in the same space are becomes of paramount relevance. We start with A and B coming in the form of intervals $A = [a^-, a^+]$ and $B = [b^-, b^+]$. We introduce two operations, namely, join and meet, described as follows:

$$\text{Join } A \oplus B = [\min(a^-, b^-), \max(a^+, b^+)]$$
$$\text{Meet } A \otimes B = [\max(a^-, b^-), \min(a^+, b^+)]$$

(6.33)

For the meet operation, we consider that these two intervals are not disjoint; otherwise it is said that the meet is empty. The essence of these two operations is clarified in Figure 6.8.

To quantify a degree to which these intervals A and B are similar (match), we propose the following definition of a degree of matching, $\xi(A, B)$, coming in the form of the following ratio:

$$\xi(A, B) = \frac{|A \otimes B|}{|A \oplus B|}$$

(6.34)

The crux of this definition is to regard a length of the meet, $|A \otimes B|$, as a measure of overlap of the intervals and calibrate this measure by taking the length of the join,

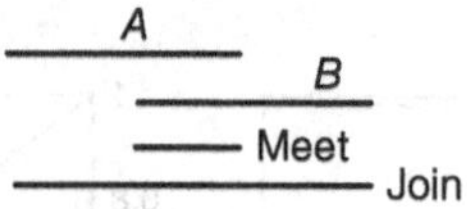

Figure 6.8   Join and meet of interval information granules.

$|A \oplus B|$. Interestingly, the essence of the aforementioned expression associates with Jaccard's coefficient being used to quantify similarity.

The definition, originally developed for intervals, can be generalized for other formalism of information granules. For instance, in case of fuzzy sets, we engage the concept of α-cuts. We start with a family of α-cuts, A_α and B_α, and determine the corresponding sequence $\xi(A_\alpha, B_\alpha)$. There are two possible ways of describing the matching as an information granule or a numeric description.

Numeric Quantification
We aggregate the elements of the above sequence by forming a single numeric descriptor. The use of the integral is a sound aggregation alternative considered here:

$$\xi(A, B) = \int_0^1 \xi(A_\alpha, B_\alpha) d\alpha \qquad (6.35)$$

Granular Quantification
In contrast to the previous matching approach, the result of matching is a fuzzy set defined over the unit interval. This characterization is more comprehensive as we obtain a non-numeric result. There is a certain disadvantage, though. As the results are information granules, one has to invoke some techniques of ranking information granules.

In case two information granules are described in a highly dimensional space, the aforementioned constructs can be involved here by determining the matching for each variable and then computing the average of the partial results obtained in this way.

6.7   CONCLUSIONS

The chapter serves as an introduction to the concepts of information granules and granular computing. Information granules form a tangible way of realizing abstraction and delivering a synthetic view at the real-world phenomena. The plethora of formal frameworks of information granules is visible: there are numerous alternatives starting from sets (and interval calculus) and embracing fuzzy sets, rough sets, probabilities, and shadowed sets, to name a few commonly considered alternatives. The calculus of information granules is presented, and the characterization of granules with the aid of the fundamental notions of coverage and specificity is provided.

REFERENCES

G. Alefeld and J. Herzberger, *Introduction to Interval Computations*, New York, Academic Press, 1983.

A. Bargiela and W. Pedrycz, *Granular Computing: An Introduction*, Dordrecht, Kluwer Academic Publishers, 2003.

A. Bargiela and W. Pedrycz, Granular mappings, *IEEE Transactions on Systems, Man, and Cybernetics Part A* 35(2), 2005, 292–297.

A. Bargiela and W. Pedrycz, Toward a theory of granular computing for human-centered information processing, *IEEE Transactions on Fuzzy Systems* 16(2), 2008, 320–330.

D. Dubois and H. Prade, Outline of fuzzy set theory: An introduction, In *Advances in Fuzzy Set Theory and Applications*, M. M. Gupta, R. K. Ragade, and R. R. Yager (eds.), Amsterdam, North-Holland, 1979, 27–39.

D. Dubois and H. Prade, The three semantics of fuzzy sets, *Fuzzy Sets and Systems* 90, 1997, 141–150.

D. Dubois and H. Prade, An introduction to fuzzy sets, *Clinica Chimica Acta* 70, 1998, 3–29.

K. Hirota, Concepts of probabilistic sets, *Fuzzy Sets and Systems* 5(1), 1981, 31–46.

P. Klement, R. Mesiar, and E. Pap, *Triangular Norms*, Dordrecht, Kluwer Academic Publishers, 2000.

G. Klir and B. Yuan, *Fuzzy Sets and Fuzzy Logic: Theory and Applications*, Upper Saddle River, Prentice-Hall, 1995.

X. Liu and W. Pedrycz, *Axiomatic Fuzzy Set Theory and Its Applications*, Berlin, Springer-Verlag, 2009.

R. Moore, *Interval Analysis*, Englewood Cliffs, Prentice Hall, 1966.

R. Moore, R. B. Kearfott, and M. J. Cloud, *Introduction to Interval Analysis*, Philadelphia, SIAM, 2009.

H. Nguyen and E. Walker, *A First Course in Fuzzy Logic*, Boca Raton, Chapman Hall, CRC Press, 1999.

Z. Pawlak, Rough sets, *International Journal of Information and Computer Science* 11(15), 1982, 341–356.

Z. Pawlak, Rough sets and fuzzy sets, *Fuzzy Sets and Systems* 17(1), 1985, 99–102.

Z. Pawlak, *Rough Sets. Theoretical Aspects of Reasoning About Data*, Dordrecht, Kluwer Academic Publishers, 1991.

Z. Pawlak and A. Skowron, Rough sets and boolean reasoning, *Information Sciences* 177(1), 2007a, 41–73.

Z. Pawlak and A. Skowron, Rudiments of rough sets, *Information Sciences* 177(1), 2007b, 3–27.

W. Pedrycz, Shadowed sets: Representing and processing fuzzy sets, *IEEE Transactions on Systems, Man, and Cybernetics Part B* 28, 1998, 103–109.

W. Pedrycz, Interpretation of clusters in the framework of shadowed sets, *Pattern Recognition Letters* 26(15), 2005, 2439–2449.

W. Pedrycz and A. Bargiela, Granular clustering: A granular signature of data, *IEEE Transactions on Systems, Man, and Cybernetics* 32, 2002, 212–224.

A. Pedrycz, F. Dong, and K. Hirota, Finite α cut-based approximation of fuzzy sets and its evolutionary optimization, *Fuzzy Sets and Systems* 160, 2009, 3550–3564.

W. Pedrycz and A. Gacek, Temporal granulation and its application to signal analysis, *Information Sciences* 143(1–4), 2002, 47–71.

W. Pedrycz and F. Gomide, *Fuzzy Systems Engineering: Toward Human-Centric Computing*, Hoboken, NJ, John Wiley & Sons, Inc., 2007.

B. Schweizer and A. Sklar, *Probabilistic Metric Spaces*, New York, North-Holland, 1983.

L. A. Zadeh, Fuzzy sets, *Information and Control* 8, 1965, 33–353.

L. A. Zadeh, The concept of linguistic variables and its application to approximate reasoning I, II, III, *Information Sciences* 8, 1975, 199–249, 301–357, 43–80.

L. A. Zadeh, Fuzzy sets as a basis for a theory of possibility, *Fuzzy Sets and Systems* 1, 1978, 3–28.

L. A. Zadeh, Towards a theory of fuzzy information granulation and its centrality in human reasoning and fuzzy logic, *Fuzzy Sets and Systems* 90, 1997, 111–117.

L. A. Zadeh, From computing with numbers to computing with words-from manipulation of measurements to manipulation of perceptions, *IEEE Transactions on Circuits and Systems* 45, 1999, 105–119.

L. A. Zadeh, Toward a generalized theory of uncertainty (GTU)—an outline, *Information Sciences* 172, 2005, 1–40.

CHAPTER 7

INFORMATION GRANULES: FUNDAMENTAL CONSTRUCTS

We focus on the fundamental constructs of granular computing (Pedrycz and Bargiela, 2002; Bargiela and Pedrycz, 2003; Pedrycz, 2013), which are of importance when processing information granules and formulating some general ways in which the results are developed and interpreted. The concepts and algorithms presented here are of direct usage in various problems of pattern recognition and data quality. There are two main categories of the algorithms. In the first one, we introduce a principle of justifiable granularity along with its generalizations to deliver a way of forming information granules on a basis of available experimental evidence. It is emphasized how information granules are developed on the basis of some well-defined and intuitively supported objectives. It is also underlined in a general way in which information granules of higher type arise. In the second category of approaches, we show that information granularity can be regarded as a useful design asset augmenting the existing numeric constructs by forming the so-called granular mappings considered and studied in system modeling and pattern recognition. Subsequently, we discuss a direct use of information granules in classification schemes.

7.1 THE PRINCIPLE OF JUSTIFIABLE GRANULARITY

The principle of justifiable granularity (Pedrycz, 2013; Pedrycz and Homenda, 2013; Pedrycz and Wang, 2016; Pedrycz et al., 2016) delivers a comprehensive conceptual and algorithmic setting to develop information granules. The principle is general in the sense it shows a way of forming information granules without being restricted to certain formalism in which granules are formalized. Information granules are built by considering available experimental evidence.

Let us start with a simple scenario using which we illustrate the key components of the principle and its underlying motivation.

Pattern Recognition: A Quality of Data Perspective, First Edition. Władysław Homenda and Witold Pedrycz.
© 2018 John Wiley & Sons, Inc. Published 2018 by John Wiley & Sons, Inc.

Consider a collection of one-dimensional numeric real-number data of interest (for which an information granule is to be formed): $X = \{x_1, x_2, ..., x_N\}$. Denote the largest and the smallest element in X by x_{min} and x_{max}, respectively. On a basis of this experimental evidence X, we form an interval information granule A so that it satisfies the requirements of coverage and specificity. The first requirement implies that the information granule is justifiable, namely, it embraces (covers) as many elements of X as possible and can be sought as a sound representative. The quest for meeting the requirement of the well-defined semantics is quantified in terms of high specificity of A. In other words, for a given X, the interval A has to satisfy the requirement of high coverage and specificity; these two concepts have already been discussed in the previous chapter. In other words, the construction of $A = [a, b]$ leads to the optimization of its bounds a and b so that at the same time we maximize the coverage and specificity. It is known that these requirements are in conflict: the increase in the coverage values leads to lower values of specificity. To transform the two-objective optimization problem in a scalar version of the optimization, we consider the performance index built as a product of the coverage and specificity,

$$V(a, b) = \mathrm{cov}(A) * \mathrm{sp}(A) \tag{7.1}$$

and determine the solution (a_{opt}, b_{opt}) such that $V(a, b)$ becomes maximized.

The ensuing approach can be established as a two-phase algorithm. In phase one, we proceed with a formation of a numeric representative of X, say, a mean, median, or modal value (denoted here by r) that can be regarded as a rough initial representative of X. In the second phase, we separately determine the lower bound (a) and the upper bound (b) of the interval by maximizing the product of the coverage and specificity as specified by the optimization criterion. This simplifies the process of building the granule as we encounter two separate optimization tasks:

$$a_{opt} = \arg\mathrm{Max}_a V(a) \quad V(a) = \mathrm{cov}([a, r]) * \mathrm{sp}([a, r])$$
$$b_{opt} = = \arg\mathrm{Max}_b V(b) \quad V(b) = \mathrm{cov}([r, b]) * \mathrm{sp}([r, b]) \tag{7.2}$$

We calculate $\mathrm{cov}([r, b]) = \mathrm{card}\ \{x_k | x_k \in [r, b]\}/N$. The specificity model has to be provided in advance. Its simplest version is expressed as $\mathrm{sp}([r, b]) = 1 - |b - r|/(x_{max} - r)$. By sweeping through possible values of b positioned within the range $[r, x_{max}]$, we observe that the coverage is a stair-wise increasing function, whereas the specificity decreases linearly; see Figure 7.1. The maximum of the product can be easily determined.

The determination of the optimal value of the lower bound of the interval a is completed in the same way as earlier. We determine the coverage by counting the data located to the left from the numeric representative r, namely, $\mathrm{cov}([a, r]) = \mathrm{card}\ \{x_k | x_k \in [a, r]\}/N$, and computing the specificity as $\mathrm{sp}([a, r]) = 1 - |a - r|/(r - x_{min})$.

Some additional flexibility can be added to the optimized performance index by adjusting the impact of the specificity in the construction of the information granule. This is done by bringing a weight factor ξ as follows:

$$V(a, b) = \mathrm{cov}(A) * \mathrm{sp}(A)^{\xi} \tag{7.3}$$

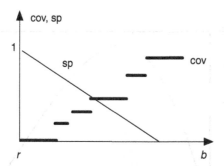

Figure 7.1 Example plots of coverage and specificity (linear model) regarded as a function of b.

Note that the values of ξ lower than 1 discount the influence of the specificity; in the limit case, this impact is eliminated when $\xi = 0$. The value of ξ set to 1 returns the original performance index, whereas the values of ξ greater than 1 stress the importance of specificity by producing results that are more specific.

In case the data are governed by some given probability function $p(x)$, the coverage is computed as an integral $\mathrm{cov}([r,\ b]) = \int_{r}^{b} p(x)dx$ and the specificity expressed in the form

$$\mathrm{sp}([r,b]) = 1 - |b-r|/(r - x_{\max})\frac{b}{x_{\max}}$$

As an illustrative example, consider the data governed by the Gaussian probability density function $p(x)$ with a zero mean and some standard deviation σ; $x_{\max}$ is set as 3σ. The corresponding plots of the product of coverage and specificity are displayed in Figure 7.2. The individual plots of coverage and specificity are presented in detail in Chapter 6. The obtained functions are smooth and exhibit clearly visible maximal values. For $\sigma = 1$, b_{opt} is equal to 1.16, while for $\sigma = 2$, the optimal location of the upper bound is 2.31, $b_{\mathrm{opt}} = 2.31$. The values of b are moved toward higher values to reflect higher dispersion of the available data.

In case of n-dimensional multivariable data, $X = \{x_1, x_2, ..., x_N\}$, the principle is realized in a similar manner. For convenience, we assume that the data are normalized to $[0,1]$, meaning that each coordinate of the normalized x_k assumes values positioned in $[0,1]$. The numeric representative r is determined first and then the information granule is built around it. The coverage is expressed in the form of the following count:

$$\mathrm{cov}(A) = \mathrm{card}\left\{x_k \,\middle|\, \|x_k - r\|^2 \leq n\rho^2\right\} \tag{7.4}$$

Note that the geometry of the resulting information granule is implied by the form of the distance function $\|.\|$ used in (7.4). For the Euclidean distance, the granule is a circle. For the Chebyshev one, we end up with hyperrectangular shapes. The specificity is expressed as $\mathrm{sp}(A) = 1 - \rho$. For these two distance functions, the corresponding plots for a two-dimensional case are illustrated in Figure 7.3.

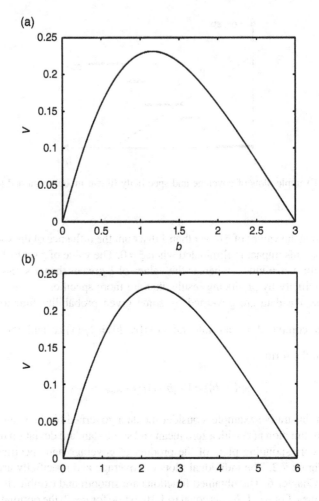

Figure 7.2 $V(b)$ as a function b: (a) $\sigma = 1.0$ and (b) $\sigma = 2.0$.

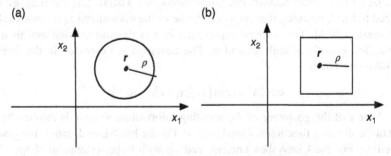

Figure 7.3 Development of information granules in the two-dimensional case when using two distance functions: (a) Euclidean distance and (b) Chebyshev distance.

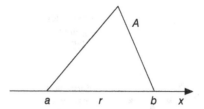

Figure 7.4 Triangular membership function with adjustable (optimized) bounds a and b.

So far we have presented the development of the principle of justifiable granularity when building an interval information granule.

When constructing an information granule in the form of a fuzzy set, the implementation of the principle has to be modified. Considering some predetermined form of the membership function, say, a triangular one, the parameters of this fuzzy set (lower and upper bounds, a and b) are optimized. See Figure 7.4.

The coverage is replaced by a σ-count by summing up the membership grades of the data in A (in what follows we are concerned with the determination of the upper bound of the membership function, viz., b):

$$\mathrm{cov}(A) = \sum_{k:x_k > r} A(x_k) \tag{7.5}$$

The coverage computed to determine the lower bound (a) is expressed in the form

$$\mathrm{cov}(A) = \sum_{k:x_k < r} A(x_k) \tag{7.6}$$

The specificity of the fuzzy set is discussed in Chapter 6; recall that it is computed as an integral of the specificity values of the α-cuts of A.

7.1.1 General Observation

A collection of experimental data, numerical data, namely, type-0 information granules, leads to a single information granule of elevated type in comparison to the available experimental evidence we have started with. This comes as a general regularity of elevation of type of information granularity (Pedrycz, 2013): data of type-1 transform into type-2, data of type-2 into type-3, and so on. Hence we talk about type-2 fuzzy sets, granular intervals, and imprecise probabilities. Let us recall that type-2 information granule is a granule whose parameters are information granules rather than numeric entities. Figure 7.5 illustrates a hierarchy of information granules built successively by using the principle of justifiable granularity.

The principle of justifiable granularity applies to various formalisms of information granules, making this approach substantially general.

Several important variants of the principle are discussed later where its generic version becomes augmented by available domain knowledge.

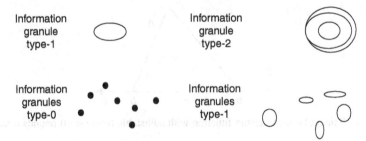

Figure 7.5 Aggregation of experimental evidence through the principle of justifiable granularity: an elevation of type of information granularity.

7.1.2 Weighted Data

The data can come in a weighted format, meaning that each data point x_k is associated with a weight w_k, assuming values in $[0,1]$ and quantifying the relevance (importance) of the data. The higher the value of the weight w_k is, the higher the importance of x_k becomes. Apparently, this situation generalized the previously discussed in which all w_ks can be treated as equal to 1. Computing the coverage is modified to accommodate the varying values of the weight. When forming an interval information granule, we consider the sum of the weights leading to the coverage expressed in the following form (we are concerned with the optimization of the upper bound of the interval $[a,b]$)

$$\operatorname{cov}([r,b]) = \sum_{k:x_k > r} w_k \qquad (7.7)$$

When building a fuzzy set, we additionally accommodate the values of the corresponding membership grades, thus computing the coverage in the form (again the computations are concerned with the optimization of the upper bound of the support of A):

$$\operatorname{cov}(A) = \sum_{k:x_k > r} \min(A(x_k), w_k) \qquad (7.8)$$

Note that the minimum operation leads to the conservative way of determining the contribution of x_k in the computing the coverage.

The definition of specificity and its computing is kept unchanged.

The approach presented here can be referred to a filter-based (or context-based) principle of justifiable granularity. The weights associated with the data play a role of filter delivering some auxiliary information about the data for which an information granule is being constructed.

7.1.3 Inhibitory Data

In a number of problems, especially in classification tasks, where we usually encounter data (patterns) belonging to several classes, an information granule is built for the data belonging to a given class. In terms of coverage, the objective is to embrace (cover) as much experimental evidence behind the given class, but at the same time an inclusion of data of inhibitory character (those coming from other classes) has to be penalized. This leads us to the modification of the coverage to accommodate the data of the inhibitory character. Consider the interval information granule and focus on the optimization of the upper bound. As usual, the numeric representative is determined by taking a weighted average of the excitatory data (r). Along with the excitatory data to be represented (x_k, w_k), the inhibitory data come in the form of the pairs (z_k, v_k). The weights w_k and v_k assume the values in the unit interval. The computation of the coverage has to take into consideration the discounting nature of the inhibitory data, namely,

$$\text{cov}([r,b]) = \max\left(0, \sum_{k:x_k \geq r} w_k - \gamma \sum_{k:z_k \in [r,b]} v_k\right) \tag{7.9}$$

The alternative version of the coverage can be expressed as follows:

$$\text{cov}([r,b]) = \sum_{\substack{k:x_k \geq r, \\ z_k \in [r,b]}} [\max(0, w_k - \gamma v_k)] \tag{7.10}$$

As seen earlier, the inhibitory data reduce to the reduction of the coverage; the nonnegative parameter γ is used to control an impact coming from the inhibitory data. The specificity of the information granule is computed in the same way as done previously.

We consider data governed by the normal probability function $N(0,1)$ ($p_1(x)$). The inhibitory data are also governed by the normal distribution $N(2,2)$ ($p_2(x)$). The plots of these pdfs are shown in Figure 7.6. In virtue of the given pdfs, the coverage is easily

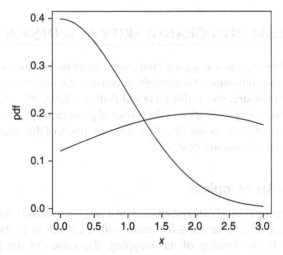

Figure 7.6   Plots of pdfs of the data for which the principle of justifiable granularity is applied. Shown are also inhibitory data (governed by p_2).

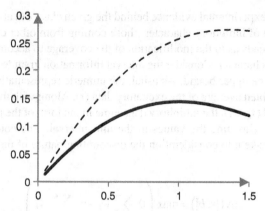

Figure 7.7   Plots of $V(b)$ for $\gamma = 1$ (solid line) and $\gamma = 0$ (dotted line).

computed as follows: (here $\gamma = 1$), $\mathrm{cov} = \max(0, \int\limits_0^b p_1(x)dx - \int\limits_0^b p_2(x)dx)$. We are interested in forming an optimal interval $[0,b]$ with the upper bound being optimized.

The optimal value of b is determined by maximizing the product of the aforementioned coverage and specificity (which in this case is taken as $1 - b/4$): $V = b_{\mathrm{opt}} = \arg \mathrm{Max}_b V(b)$.

The plots of the maximized performance index V versus values of b are displayed in Figure 7.7. The maximum is clearly visible as it is achieved for $b = 1.05$. For comparison, when $\gamma = 0$ (so no inhibitory data are taken into consideration), the optimal value of b becomes higher and equals to 1.35 (which is not surprising as we are not penalizing by the inhibitory data). In this case, the corresponding plot of V is also shown in Figure 7.7 (dotted curve).

7.2   INFORMATION GRANULARITY AS A DESIGN ASSET

The concept of the granular mapping (and model described by the mapping, in general) delivers a generalization of commonly encountered numeric mappings (models) no matter what their structure is (Bargiela and Pedrycz, 2003, 2005; Lu *et al.*, 2014). In this sense, this conceptualization in the format of granular mappings offers an interesting and practically convincing direction. The constructs of this nature are valid for any formalism of information granules.

7.2.1   Granular Mappings

A numeric mapping (model) M_0 constructed on a basis of a collection of training data (x_k, target_k), $x_k \in R^n$ and $\mathrm{target}_k \in R$, comes with a collection of its parameters a, where $a \in R^p$. In the buildup of the mapping, the values of the parameters are optimized yielding a vector a_{opt}. The estimation of the parameters is realized by minimizing a certain performance index Q (say, a sum of squared errors between target_k

and $M_0(x_k)$), namely, $a_{opt} = \arg \mathrm{Min}_a\, Q(a)$. To compensate for inevitable errors of the model (as the values of the index Q are never equal identically to zero), we make the parameters of the model information granules, resulting in a vector of information granules $A = [A_1\, A_2\, ...\, A_p]$. These granules are built around original numeric values of a. In other words, we say that the model is embedded in the *granular* parameter space. The elements of the vector a are generalized, the model becomes granular, and subsequently the results produced by the model are information granules as well. Formally speaking, we have

- Granulation of parameters of the model $A = G(a)$, where G stands for the mechanisms of forming information granules, namely, building an information granule around the numeric parameter
- Result of the granular model for any x producing the corresponding information granule Y, $Y = M(x, A) = G(M(x)) = M(x, G(a))$

Figure 7.8 illustrates the underlying idea. The nonlinear mapping M endowed with some parameters approximates the data. We make the parameters of M in the form of intervals, thus building the output of the mapping in the form of some interval. Note that the envelope of $y = M(x, a)$ denoted as Y (with the bounds y^- and y^+) covers most the data (leaving some out, which are of a clear outlier nature).

As a concise illustration, consider a linear mapping in the following form: $y - y_0 = a(x - x_0)$, where $x_0 = 2.2$, $y_0 = 2.5$, and $a = 2.0$. If we admit an interval generalization of the numeric parameter by forming an interval A distributed around 2, namely, $A = [a^-, a^+] = [1.7, 2.6]$, we obtain the interval $Y = [y^-, y^+]$ whose bounds are computed in the following way (Moore, 1966; Moore *et al.*, 2009):

$$y^- = y_0 + \min(a^-(x - x_0), a^+(x - x_0))$$
$$y^+ = y_0 + \max(a^-(x - x_0), a^+(x - x_0))$$

See Figure 7.9.

As another example, let us consider a nonlinear function $y - y_0 = a*\sin(b*(x - x_0))$ in the range of arguments in $[x_0, x_0 + \pi/4]$ where $x_0 = 0.9$ and $y_0 = -0.6$. The numeric

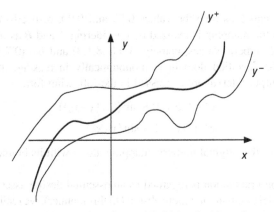

Figure 7.8   Original numeric mapping along with the interval bounds.

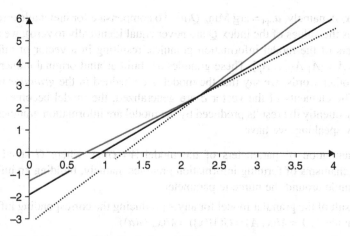

Figure 7.9 Linear mapping (dark line) along with its interval (granular) generalization; gray and dotted lines show the upper and lower bounds, respectively.

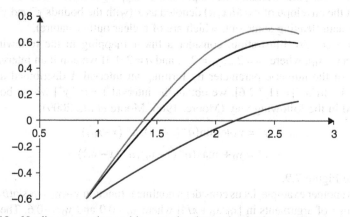

Figure 7.10 Nonlinear mapping and its granular (interval-valued) generalization.

parameters of a and b assume the values 0.95 and 0.90, respectively. The granular augmentation of the mapping is realized by considering A and B as intervals spanned over the original numeric values, namely, $A = [0.8,1.3]$ and $B = [0.7,1]$. The function considered in the specified domain is monotonically increasing, meaning that the bounds of the output interval are expressed in the following form:

$$y^- + 0.6 = 0.8 \sin(0.7(x-x_0))$$
$$y^+ + 0.6 = 1.3 \sin(x-x_0)$$

The plots of the original numeric mapping along with its bounds are displayed in Figure 7.10.

Information granulation is regarded as an essential design asset. By making the results of the model granular (and more abstract in this manner), we realize a better alignment of $G(M_0)$ with the data. Intuitively, we envision that the output of the granular

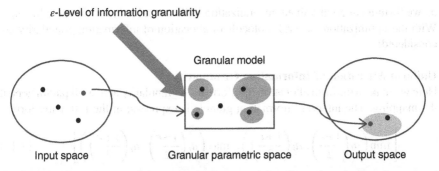

Figure 7.11   From fuzzy models to granular fuzzy models: a formation of a granular space of parameters.

model *covers* the corresponding target. Formally, let cov(target, Y) denote a certain coverage predicate (either being Boolean or multivalued) and quantify an extent to which target is included (covered) in Y. The definition of coverage was introduced in Chapter 6.

The design asset is supplied in the form of a certain allowable level of information granularity ε, which is a certain nonnegative parameter being provided in advance. We allocate (distribute) the design asset by forming intervals around the parameters of the model. This is done in a way so that both the coverage measure and the specificity are maximized. The overall level of information granularity ε serves here as a constraint to be satisfied when allocating information granularity across the parameters of the model, namely, $\sum_{i=1}^{p}\varepsilon_i = p\varepsilon$. The maximization of both the objectives is translated into the product of the coverage and specificity. In the sequel, the constraint-based optimization problem reads as follows:

$$V(\varepsilon_1,\varepsilon_2,\ldots,\varepsilon_n) = \left(\frac{1}{N}\sum_{k=1}^{N}\text{cov}(\text{target}_k, Y_k)\right)\left(\frac{1}{N}\sum_{k=1}^{N}\text{sp}(Y_k)\right)$$

$$\max_{\varepsilon_1,\varepsilon_2,\ldots,\varepsilon_n} V(\varepsilon_1,\varepsilon_2,\ldots,\varepsilon_n)$$

subject to

$$\sum_{i=1}^{p}\varepsilon_i = p\varepsilon \quad \text{and} \quad \varepsilon_i \geq 0 \qquad (7.11)$$

The monotonicity property of the coverage measure is obvious: the higher the values of ε, the higher the resulting coverage and the lower the specificity of the result. As the two contributing criteria are evidently in conflict, one may anticipate that there is a sound compromise that might be established.

The underlying concept of the allocation of information granularity is succinctly displayed in Figure 7.11.

7.2.2   Protocols of Allocation of Information Granularity

The numeric parameters of the mapping are elevated to their granular counterparts with intent of maximizing the performance index. Considering the i-th parameter

a_i, we form its interval-valued generalization built around its numeric value $[a_i^-, a_i^+]$. With the optimization, several protocols of allocation of information granularity are considered:

Uniform Allocation of Information Granularity
Here we make a uniform allocation of information granularity across all parameters of the mapping. The interval information granule is expressed in the following form:

$$\left[\min\left(a_i\left(\frac{1-\varepsilon}{2}\right), a_i\left(\frac{1+\varepsilon}{2}\right) \right), \ \max\left(a_i\left(\frac{1-\varepsilon}{2}\right), a_i\left(\frac{1+\varepsilon}{2}\right) \right) \right] \qquad (7.12)$$

In case $a_i = 0$, the interval is built around zero as $[-\varepsilon/2, \varepsilon/2]$. Note that the balance of information granularity as expressed in (7.12) is obviously satisfied. There is no optimization here, and this protocol can serve as a reference scenario that helps quantify the performance when compared with realizing some mechanisms of allocation of information granularity.

Symmetric Allocation of Information Granularity
Here information granularity is allocated in the following form

$$[\min(a_i(1-\varepsilon_i/2), a_i(1+\varepsilon_i/2)), \ \max(a_i(1-\varepsilon_i/2), a_i(1+\varepsilon_i/2))] \qquad (7.13)$$

In this way an information granule is created symmetrically around a_i.

Asymmetric Allocation of Information Granularity
In this scenario, we admit a formation of information granules in a more flexible manner as follows:

$$\left[\min\left(a_i\left(1-\varepsilon_i^-\right), a_i\left(1+\varepsilon_i^-\right)\right), \ \max\left(a_i\left(1-\varepsilon_i^+\right), a_i\left(1+\varepsilon_i^+\right)\right) \right] \qquad (7.14)$$

The number of parameters in this situation is larger than in the previous one as information granules are positioned asymmetrically. In this case we also require the satisfaction of the total balance of information granularity.

7.2.3 Granular Aggregation: An Enhancement of Aggregation Operations through Allocation of Information Granularity

In what follows, we look at a special case of the allocation of information granularity by looking at a concept of granular aggregation being a generalization of the well-known schemes of data (evidence) aggregation. The proposed approach works in the presence of a collection of the pairs of the data $(x(1), target(1)), (x(2), target(2)), \ldots, (x(N), target(N))$ where $x(k)$ is an n-dimensional vector in the unit hypercube $[0,1]^n$ and $target(k)$ is in the $[0,1]$ is regarded as an experimentally available result of aggregation of the components of $x(k)$. No matter what formal model of the aggregation of the elements of $x(k)$s is envisioned, it is very unlikely that the model returns the result that coincides with the value of the $target(k)$. Denote the aggregation formula

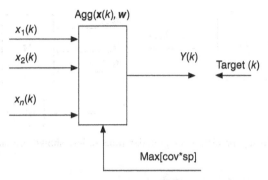

Figure 7.12   Allocation of information granularity in aggregation problem.

as $Agg(w, x(k))$ where w is a weight vector of the aggregation operation. The values of w are not available and need to be optimized so that the aggregation mechanism returns the results as close as possible to the required target(k). For instance, an initial optimization scheme can optimize the vector of weights w so that the following distance becomes minimized:

$$Q = \sum_{k=1}^{N} (\text{target}(k) - \text{Agg}(w, x(k)))^2 \qquad (7.15)$$

namely, $w_{opt} = \arg \text{Min}_w Q$. Evidently, it is unlikely to derive w such that Q attains the zero value. Nevertheless we can regard w_{opt} as an initial numeric estimate of the weight vector and further optimize it by making the weights granular (say interval valued) by invoking the criterion guiding the optimal allocation of information granularity. In other words, w_{opt} is made granular, leading to interval-valued w so that the product of the coverage and specificity attains the maximal value. The architecture of the overall system is illustrated in Figure 7.12. The optimal allocation of information granularity is realized by following one of the protocols discussed earlier.

7.3   SINGLE-STEP AND MULTISTEP PREDICTION OF TEMPORAL DATA IN TIME SERIES MODELS

We construct a single-step nonlinear prediction model M on a basis of a finite horizon of past data (third-order prediction model):

$$x_{k+1} = M(x_k, x_{k-1}, x_{k-2}, a) \qquad (7.16)$$

where a is a vector of parameters of the model. Owing to the performance of the model (nonzero values of the performance index), the result of prediction is inherently an information granule (the performance of the model can be assessed and the model is made granular by constructing a granular parameter space or granular output space). Consider the granular parameter space assuming that information granules A have

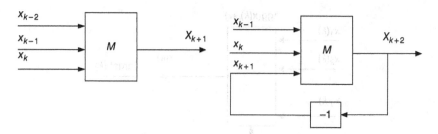

Figure 7.13 Multistep prediction in granular time series; shown are successive steps of prediction.

been formed. The one-step prediction (one-step ahead prediction) is carried out as follows:

$$X_{k+1} = M(x_k, x_{k-1}, x_{k-2}, A) \tag{7.17}$$

By virtue of the granular parameters of the model, the prediction result X_{k+1} becomes an information granule. Then to predict X_{k+2}, we proceed to an iterative fashion; refer to Figure 7.13 for more details.

In other words, we have

$$X_{k+2} = M(X_{k+1}, x_k, x_{k-1}, A) \tag{7.18}$$

In the next iteration one has

$$X_{k+3} = M(X_{k+2}, X_{k+1}, x_k, A) \tag{7.19}$$

In the consecutive prediction step, we rely on the results of prediction produced by the model and thus formed as information granules.

There is an accumulation of information granularity, which manifests vividly in situations when one proceeds with prediction completed in longer prediction horizon. This also helps us assess a maximal length of possible prediction horizon: once the specificity of the prediction is lower than some predetermined threshold, one cannot exceed a certain length of the prediction interval (long-term prediction).

7.4 DEVELOPMENT OF GRANULAR MODELS OF HIGHER TYPE

In system modeling, granular models arise in a hierarchical way by raising a level of information granularity at successive layers of the architecture. The essence of the formation of such models can be concisely elaborated as follows; refer also to Figure 7.14. The data available for the design of the model come as a collection of pairs (x_k, target_k), $k = 1, 2, ..., N$.

An initial model M is numeric, namely, in terms of the terminology used, here is a granular model of type-0. Through the process of an optimal allocation of information granularity, we form a granular model of type-1 by making the numeric parameters of the previous model granular. The level of information granularity used

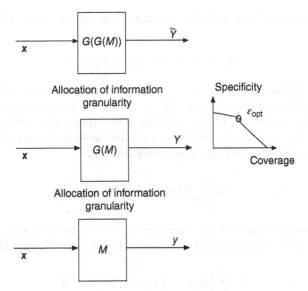

Figure 7.14   Forming granular models of higher type: a concept.

when forming the granular model $G(M)$ is selected on a basis of the coverage–specificity characteristics of the model. As shown in Figure 7.14, the optimal value of the level of information granularity is the one for which there is a substantial increase in the coverage value while still retaining the specificity value or eventually accepting its quite limited drop.

When assessing the performance of the granular model in terms of the coverage, one may still have a number of data that are not "covered" by the information granules of the type-1 model. More formally, the set of outlier O_1 comes in the form

$$O_1 = \{(x_k, \text{target}_k) | \text{target}_k \notin GM(x_k)\} \qquad (7.20)$$

Those data can be regarded as type-1 granular outliers. To increase the coverage of the granular model, one elevates the level of type of information granules by making them type-2 information granules, which yield type-2 granular models denoted as $G(G(M))$. For instance, when the parameters are intervals, type-2 information granules become granular intervals (viz., intervals whose bounds are information granules rather than single numeric values). In case of parameters in terms of fuzzy sets, the parameters are type-2 fuzzy sets. Again by invoking an optimal allocation of information granularity with the criterion of coverage determined for O_1, some outliers are covered by the results produced by $G(G(M))$. The remaining ones form O_2:

$$O_2 = \{(x_k, \text{target}_k) \in O_1 | \text{target}_k \notin G(G(M))(x_k)\} \qquad (7.21)$$

Those data can be thought as type-2 granular outliers.

The formation of granular models of higher type offers a general way of moving toward granular parameters of higher type. In what follows, we look into more details by focusing on rule-based models. Consider a collection of fuzzy rules described in the form

$$-\text{if } x \text{ is } B_i \text{ then } y = a_{i0} + a_i^T (x - v_i) \qquad (7.22)$$

where v_i is a modal value of the fuzzy set B_i built in the input space. The parameters of the conclusion part describe a local linear model that is centered around v_i. Note that the linear function in (7.22) describes a hyperplane that rotates around the prototype v_i. For any input x, the rules are invoked at some degrees of membership $B_i(x)$, and the corresponding output is a weighted sum:

$$\hat{y} = \sum_{i=1}^{c} B_i(x) \left[a_{i0} + a_i^T (x - v_i) \right] \qquad (7.23)$$

We construct a granular parameter space of the model by making and optimizing the parameters of the local models. The granular rules come in the following form:

$$\hat{Y} = \sum_{i=1}^{c} B_i(x) \otimes \left[A_{i0} \oplus A_i^T \otimes (x - v_i) \right] \qquad (7.24)$$

where A_{i0} and A_i are granular parameters spanned around the original numeric values of these parameters.

An illustrative example offers a detailed insight into the nature of the construct; here we confine ourselves to a one-dimensional input space where $X = [-3,4]$. There are four rules of the model ($c = 4$). As the fuzzy sets B_i are constructed with the use of the FCM algorithm, their membership functions come in the well-known form

$$B_i(x) = \cfrac{1}{\sum_{j=1}^{4} \left(\|x - v_i\| / \|x - v_j\| \right)^2} \qquad (7.25)$$

The prototypes of the fuzzy sets in the condition parts of the rules assume the following values: $v_1 = -1.5$, $v_2 = -0.3$, $v_3 = 1.1$, and $v_4 = -2.3$. The calculations of the bounds of the interval-valued output are carried out as follows:

Lower bound:

$$y^- = \sum_{i=1}^{4} B_i(x) \left[a_{i0}^- + \min\left(a_i^- z_i, a_i^+ z_i \right) \right] \qquad (7.26)$$

Upper bound:

$$y^+ = \sum_{i=1}^{4} B_i(x) \left[a_{i0}^+ + \max\left(a_i^- z_i, a_i^+ z_i \right) \right] \qquad (7.27)$$

where $z_i = x - v_i$. The numeric values of the parameters of the local model are the following:

$a_{10} = 3.0$, $a_{20} = -1.3$, $a_{30} = 0.5$, $a_{40} = 1.9$, $a_1 = -1.5$, $a_1 = 0.8$, $a_1 = 2.1$, $a_1 = -0.9$.

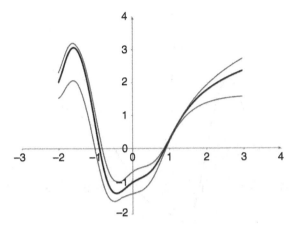

Figure 7.15 Interval-valued output of the rule-based fuzzy model; gray lines—the bounds of the interval.

Given are the interval-valued parameters of the local models:

$$A_{10} = [2.0, 3.1], A_{20} = [-1.5, -1.0], A_{30} = [0.45, 0.5], A_{40} = [1.7, 2.1]$$
$$A_1 = [-1.7, -1.3], A_2 = [0.4, 0.9], A_3 = [2.0, 2.1], A_4 = [-1.2, -0.6]$$

The plot of the interval-valued output of the model is given in Figure 7.15. All the data that are positioned beyond the lower and upper bound are regarded as type-2 outliers.

If the values of the constants in the linear models are kept intact as before (no intervals) and the obtained interval values of the parameters are given in the form

$$A_1 = [-2.0, -1.0], A_2 = [0.2, 1.5], A_3 = [1.0, 3.0], A_4 = [-3.0, 0.5]$$

then the characteristics of the rule-based model are named in Figure 7.16.

It is noticeable that for the input values coincide at the prototypes, the output assumes a single numeric value. This aligns well with an intuition as for such input values, only a single local model is activated and the constant is numeric. This effect does not occur in the previous case because of the interval-valued constants of the local models.

Consider that when applying the same process of expanding the model to cope with the granular outliers of type-1, the parameters of the coefficients of the intervals are elevated to granular intervals whose bounds are intervals themselves. For instance, the type-1 granular parameter $A_i = [a_i^- \, a_i^+]$ becomes elevated to type-2 information granule $A_i^\sim$ (granular interval) with granular bounds that is expressed as

$$A_i^\sim = \left[\underbrace{\left[a_i^{--}, a_i^{-+} \right]}_{\substack{\text{Granular} \\ \text{lower} \\ \text{bound}}} \underbrace{\left[a_i^{+-}, a_i^{++} \right]}_{\substack{\text{Granular} \\ \text{upper} \\ \text{bound}}} \right] \qquad (7.28)$$

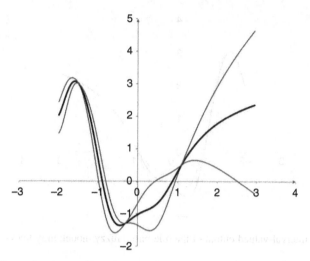

Figure 7.16   Interval-valued output of the rule-based fuzzy model: the bounds y^- and y^+ of the interval are shown in gray.

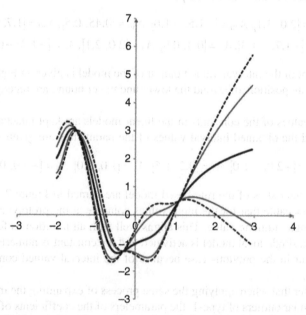

Figure 7.17   Characteristics of type-2 granular rule-based models; dotted lines show the bounds produced by the type-2 (granular) intervals.

Let us consider that the granular intervals were optimized and come in the following form:

$$A_1^\sim = \big[[-2.5, -1.5], [-1.2, 0.0]\big], A_2^\sim = [0.1, 0.5], [0.9, 1.7], A_3^\sim = [0.7, 1.1], [2.7, 3.1]\big],$$
$$A_4^\sim = \big[[-3.1, -1.7], [0.5, 1.4]\big]$$

The calculations of the bounds involve all the combinations of the bounds and yield the following expressions:

$$y^- = \sum_{i=1}^{4} B_i(x) \left[a_{i0}^- + \min\left(a_i^{--} z_i, a_i^{-+} z_i, a_i^{+-} z_i, a_i^{++} z_i \right) \right]$$

$$y^- = \sum_{i=1}^{4} B_i(x) \left[a_{i0}^+ + \max\left(a_i^{--} z_i, a_i^{-+} z_i, a_i^{+-} z_i, a_i^{++} z_i \right) \right] \qquad (7.29)$$

The plots of the bounds are shown in Figure 7.17. The data localized outside the bounds (dotted lines) are outliers of type-2.

7.5 CLASSIFICATION WITH GRANULAR PATTERNS

In this chapter, we consider an extension of existing mappings and pattern classifiers to cope with granular data. In general, in this situation, patterns are described by information granules such as intervals, fuzzy sets, rough sets, or probability functions. The ultimate question is how to use such data in the construction of classifiers and how to interpret classification results. There are no granular data in a real world so they have to be formed (abstracted) on the basis of the existing numeric data. With this regard, it is essential to identify and quantify the advantages of the use of granular data in the design of classifiers.

7.5.1 Formulation of the Classification Problem

Let us recall that a c-classification problem is commonly formulated as follows (Duda et al., 2001):

Given a labeled set of patterns $x_1, x_2, \ldots, x_N$ positioned in an n-dimensional feature space $X \subset R^n$, construct a mapping (classifier) F from the feature space to the space of labels $\omega_1, \omega_2, \ldots, \omega_c$, namely, $\{0,1\}^c$ or $[0,1]^c$:

$$F : X \rightarrow \{0, 1\}^c \qquad (7.30)$$

(binary classification with binary class assignment)
or

$$F : X \rightarrow [0, 1]^c \qquad (7.31)$$

(fuzzy classification with membership grades).

The essence of the design of the classifier is to develop (optimize) a mapping minimizing a classification error.

It is worth noting that in spite of the existing diversity of linear or nonlinear classifiers, all of them are *numeric* in the sense we encounter numeric values of features $X \subset R^n$.

Granular classifiers are generalizations of (numeric) classifiers where a granular feature space is considered. The mapping GF is now realized as follows:

$$GF: G(X) \rightarrow [0, 1]^c \qquad (7.32)$$

where GF stands for the granular classifier and $G(.)$ denotes a granular feature space. Patterns in this space are information granules.

There are some conceptually and computationally appealing motivating factors behind the emergence of granular data. First, individual numeric data are impacted by noise. Second, a rate at which they are available (say, in the form of data streams) is high, and it becomes advantageous to consider some way of reduction (compression). If such data were described at the higher level of abstraction forming a collection of granular data, they could be more manageable and meaningful. The number of patterns to be used in classification gets substantially lower, facilitating a design of a classifier. Third, granular data are reflective of the quality of the original data. Here two aspects are worth highlighting: (i) information granularity comes as a result of data imputation and in this way it helps distinguish the complete data and those that have been imputed, and (ii) granular patterns arise as a result of balancing imbalanced data.

7.5.2 From Numeric Data to Granular Data

Information granules are constructed on a basis of numeric representatives (prototypes) produced through clustering or fuzzy clustering (Kaufmann and Rousseeuw, 1990; Gacek and Pedrycz, 2015). They can be also selected randomly. Denote them by $v_1, v_2, ..., v_c$.

Depending on the nature of the available data, two general design scenarios are considered:

Unlabeled Data
The patterns are located in the n-dimensional space of real numbers. Some preliminary processing is completed by selecting a subset of data (say, through clustering or some random mechanism) and building information granules around them. Such information granules could be intervals or set-based constructs or probabilities (estimates of probability density functions). The principle of justifiable granularity arises here as a viable alternative to be considered in this setting.

Labeled Data
In this scenario, as before we involve the principle of justifiable granularity and in the construction of the granule, a mechanism of supervision is invoked. The obtained granule is endowed with its content described by the number of patterns belonging to different classes.

Proceeding with the formation of information granules carried out in unsupervised mode, using the principle of justifiable granularity, we construct information granules $V_1, V_2, ..., V_M$, which are further regarded as granular data. As usual, the two characteristics of granules are considered here.

We determine the coverage

$$\text{cov}(V_i) = \{x_k \mid \|x_k - v_i\| \le \rho_i\} \tag{7.33}$$

where $\rho_i \in [0, 1]$ is the size (diameter) of the information granule. The specificity of the granule is expressed in the form

$$\text{Sp}(V_i) = 1 - \rho_i \tag{7.34}$$

Proceeding with details, the coverage involves the distance ‖.‖, which can be specified in various ways. Here we recall the two commonly encountered examples such as the Euclidean and Chebyshev distances. This leads to the detailed formulas of the coverage

$$\text{cov}(V_i) = \left\{ x_k \,\middle|\, \frac{1}{n}\sum_{j=1}^{n} \frac{(x_{kj} - v_{ij})^2}{\sigma_j^2} \le n\rho_i^2 \right\} \tag{7.35}$$

$$\text{cov}(V_i) = \left\{ x_k \,\middle|\, \max_{j=1,2,\dots,n} |x_{kj} - v_{ij}| \le \rho_i \right\} \tag{7.36}$$

The size of the information granule ρ_i is optimized by maximizing the product of the coverage and specificity producing an optimal size of the granule

$$\rho_i = \arg\text{Max}_{\rho \in [0,\,1]}\left[\text{Cov}(V_i) * \text{Sp}(V_i)\right] \tag{7.37}$$

The impact of the specificity criterion can be weighted by bringing a nonzero weight coefficient β:

$$\rho_i = \arg\text{Max}_{\rho \in [0,\,1]}\left[\text{Cov}(V_i) * \text{Sp}^{\beta}(V_i)\right] \tag{7.38}$$

The higher the value of β, the more visible the impact of the specificity on the constructed information granule.

All in all, as a result we form M information granules characterized by the prototypes and the corresponding sizes, namely, $V_1 = (v_1, \rho_1)$, $V_2 = (v_2, \rho_2)$, ..., $V_M = (v_M, \rho_M)$. Note that the granular data formed in this way can be further clustered as discussed in Chapter 8.

In the supervised mode of the development of information granules, one has to take into consideration class information. It is very likely that patterns falling within the realm of information granules may belong to different classes. The class content of information granule can be regarded as a probability vector p with the entries

$$[p_1 p_2 \dots p_c] = \left[\frac{n_1}{n_1 + n_2 + \cdots + n_c} \; \frac{n_2}{n_1 + n_2 + \cdots + n_c} \cdots \frac{n_c}{n_1 + n_2 + \cdots + n_c} \right].$$

Here $n_1, n_2, \dots, n_c$ are the number of patterns embraced by the information granule and belonging to the corresponding class of patterns. Alternatively, one can consider a class membership vector being a normalized version of p in which all coordinates are divided by the highest entry of this vector yielding $[p_1/\max(p_1, p_2, \dots, p_c)\ p_2/\max(p_1, p_2, \dots, p_c) \cdots p_c/\max(p_1, p_2, \dots, p_c)]$.

The class content plays an integral component in the description of information granule. With this regard, an overall scalar characterization of the heterogeneity of information granule comes in the form of entropy (Duda *et al.*, 2001):

$$H(\boldsymbol{p}) = -\frac{1}{c}\sum_{i=1}^{c} p_i \log_2 p_i \qquad (7.39)$$

When constructing information granule, the optimization criterion is augmented by the entropy component, which is incorporated in the already existing product of coverage and specificity, namely,

$$\rho_i = \arg \text{Max}_{\rho \in [0,\,1]} [\text{cov}(\boldsymbol{V}_i) * \text{Sp}(\boldsymbol{V}_i) * H(\boldsymbol{p})] \qquad (7.40)$$

Note that if the formation of information granule involves numeric patterns belonging to the same class, then the entropy component is equal to 1 and does not impact the overall performance index.

As a result of the use of the principle of justifiable granularity modified as shown earlier, we obtain M granular data in the following triple $(\boldsymbol{v}_i, \rho_i, \boldsymbol{p}_i)$, $i = 1,2,\ldots, M$ described in terms of the numeric prototypes, the size ρ_i, and the class content $\boldsymbol{p}_i$. The Boolean manifestation of the same information granule is provided as $(\boldsymbol{v}_i, \rho_i, \boldsymbol{I}_i)$ where $\boldsymbol{I}_i$ is a Boolean vector $\boldsymbol{I}_i = [0\,0\ldots0\,1\,0\ldots.0]$ with a single nonzero entry whose coordinate (class index) corresponds to an index of the class for which the entry of $\boldsymbol{p}_i$ is the highest.

7.5.3   Granular Classifiers: Augmentation Issues

The classifiers used to work with numeric data that need to be augmented to cope with the granular nature of the pattern. The essence is to construct a new feature space, which accommodates parameters of information granules constructed so far. The consecutive step is to realize a classifier mapping the granular patterns; see Figure 7.18.

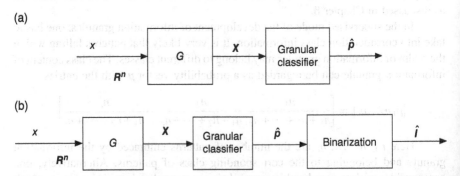

Figure 7.18   An overview of the design of granular classifiers; note a functionality of the preprocessing module forming a granular feature space. Two modes of performance evaluation: (a) considering class membership and (b) considering binary classification.

By virtue of the multi-label character of the granular patterns, the output of the classifier is c-dimensional by returning a vector of class membership $\boldsymbol{p}$. The classification error is expressed and subsequently optimized in two different ways:

1. By considering the probabilistic information about the granular patterns. The distance between $\boldsymbol{p}_k$ and $\hat{\boldsymbol{p}}_k$ being produced by the classifier for patterns $x_1, x_2, \ldots, x'_M$, belonging to the training set, is minimized:

$$Q = \frac{1}{M'} \sum_{k=1}^{M'} \|\boldsymbol{p}_k - \hat{\boldsymbol{p}}_k\|^2 \qquad (7.41)$$

2. By taking into consideration the Boolean manifestation of the granular patterns, namely, by considering a count of misclassifications expressed by differences between the vectors $\boldsymbol{I}_k$ and $\hat{\boldsymbol{I}}_k$,

$$Q = \frac{1}{M'} \sum_{k=1}^{M'} \sum_{j=1}^{n} |I_{kj} - \hat{I}_{kj}|^2 \qquad (7.42)$$

The performance of the classifier is assessed for the training and testing datasets. No overfitting is encountered, as the set of patterns is far smaller than in the case of numeric data so no specific precautions are taken with this regard.

7.6   CONCLUSIONS

In this chapter, we focused on the conceptual and algorithmic developments of information granules. The principle of justifiable granularity delivers a general conceptual and algorithmic setting of building information granules. It is important to stress two key features that exhibit far-reaching implications. First, the principle motivates the construction of information granules on the basis of the essentials of these entities, namely, experimental justification and semantic content. Second, it applies across various formalisms of information granules. The discussion presented here casts the commonly encountered fuzzy sets of type-2 (and interval-valued fuzzy sets, in particular) in an original setting in terms of their origin and the development by stressing their emergence in association with all practical situations where emerge various sources of data (knowledge) require a certain type of aggregation. As a result, information granules of higher type emerge as a genuine and fully legitimized necessity to capture and quantify the diversity of information granules involved in the aggregation process.

The concepts of granular mappings are the cornerstone of granular calculations by expanding upon numerical mappings (being at the center of constructs of classifiers and predictors) and augmenting originally available numeric results. Again, as in case of the principle of justifiable granularity, the quality of the information granules is evaluated and optimized vis-à-vis the coverage of the experimental data and their specificity.

REFERENCES

A. Bargiela and W. Pedrycz, *Granular Computing: An Introduction*, Dordrecht, Kluwer Academic Publishers, 2003.

A. Bargiela and W. Pedrycz, Granular mappings, *IEEE Transactions on Systems, Man, and Cybernetics-Part A*, 35(2), 2005, 292–297.

R. O. Duda, P. E. Hart, and D. G. Stork, *Pattern Classification*, 2nd ed., New York, John Wiley & Sons, Inc., 2001.

A. Gacek and W. Pedrycz, Clustering granular data and their characterization with information granules of higher type, *IEEE Transactions on Fuzzy Systems*, 23, 4, 2015, 850–860.

L. Kaufmann and P. J. Rousseeuw, *Finding Groups in Data: An Introduction to Cluster Analysis*, New York, John Wiley & Sons, Inc., 1990.

W. Lu, W. Pedrycz, X. Liu, J. Yang, and P. Li, The modeling of time series based on fuzzy information granules, *Expert Systems with Applications*, 41(8), 2014, 3799–3808.

R. Moore, *Interval Analysis*, Englewood Cliffs, NJ, Prentice Hall, 1966.

R. Moore, R. B. Kearfott, and M. J. Cloud, *Introduction to Interval Analysis*, Philadelphia, PA, SIAM, 2009

W. Pedrycz, *Granular Computing*, Boca Raton, FL, CRC Press, 2013.

W. Pedrycz and A. Bargiela, Granular clustering: A granular signature of data, *IEEE Transactions on Systems, Man and Cybernetics*, 32, 2002, 212–224.

W. Pedrycz and W. Homenda, Building the fundamentals of granular computing: A principle of justifiable granularity, *Applied Soft Computing*, 13, 2013, 4209–4218.

W. Pedrycz, A. Jastrzebska, and W. Homenda, Design of fuzzy cognitive maps for modeling time series, *IEEE Transactions on Fuzzy Systems*, 24(1), 2016, 120–130.

W. Pedrycz and X. Wang, Designing fuzzy sets with the use of the parametric principle of justifiable granularity, *IEEE Transactions on Fuzzy Systems*, 24(2), 2016, 489–496.

CHAPTER 8

CLUSTERING

Clustering (Jain and Dubes, 1988; Duda *et al.*, 2001; Jain, 2010) is about discovering and describing a structure in a collection of patterns (data). It is realized in an unsupervised manner and is often regarded as a synonym of learning without a teacher (unsupervised learning). In the setting of pattern recognition, clustering is one of the dominant and well-delineated domains in this area having its own agenda and contributing results to the successive processing scheme, especially classifiers and predictors.

There are a large number of diverse clustering methods being motivated by different objectives and producing results in quite distinct formats. In this chapter, we focus on two representative examples present among numerous approaches being representative of objective function-based clustering and hierarchical clustering. We elaborate on the essence of the discussed algorithms along with computing details. Some selected augmentations of the clustering techniques are discussed. The quality of clustering results is quantified in various ways; here we offer taxonomy of performance indicators by identifying so-called external and internal ones.

Clustering and clusters are closely related with information granules. In the sequel we identify a role of information granules and underline that clustering serves as a vehicle to construct granules. Furthermore it is shown that granular clustering concerns also ways of building clusters on a basis of data coming in the form of information granules. We also demonstrate how the principle of justifiable information granularity contributes to the concise description of information granules built on a basis of results of clustering.

8.1 FUZZY C-MEANS CLUSTERING METHOD

Let us briefly review the formulation of the fuzzy c-means (FCM) algorithm (Dunn, 1974; Bezdek, 1981), develop the algorithm, and highlight the main properties of the fuzzy clusters. Given a collection of n-dimensional dataset located in R^n, $X = \{x_k\}$, $k = 1, 2, \ldots, N$, the task of determining its structure—a collection of c

---

Pattern Recognition: A Quality of Data Perspective, First Edition. Władysław Homenda and Witold Pedrycz.
© 2018 John Wiley & Sons, Inc. Published 2018 by John Wiley & Sons, Inc.

clusters—is expressed as a minimization of the following objective function (performance index), Q being regarded as a sum of the squared distances:

$$Q = \sum_{i=1}^{c} \sum_{k=1}^{N} u_{ik}^{m} \|x_k - v_i\|^2 \tag{8.1}$$

where $v_1, v_2, \ldots, v_c$ are n-dimensional prototypes of the clusters and $U = [u_{ik}]$ stands for a partition matrix expressing a way of allocation of the data to the corresponding clusters; u_{ik} is the membership degree of data x_k in the ith cluster. The values of u_{ik} are arranged in a partition matrix U. The distance between the data x_k and prototype v_i is denoted by $\|.\|$. The fuzzification coefficient m assuming values greater than 1 expresses the impact of the membership grades on the individual clusters and produces certain geometry of the information granules.

A partition matrix satisfies two important and intuitively motivated properties:

a. $0 < \sum_{k=1}^{N} u_{ik} < N, \quad i = 1, 2, \ldots, c$ \hfill (8.2)

b. $\sum_{i=1}^{c} u_{ik} = 1, \quad k = 1, 2, \ldots, N$ \hfill (8.3)

The first requirement states that each cluster has to be nonempty and different from the entire set. The second requirement states that the sum of the membership grades should be confined to 1. Let us denote by U a family of partition matrices satisfying these two requirements (a)–(b).

The minimization of Q completed with respect to $U \in U$ and the prototypes v_i of $V = \{v_1, v_2, \ldots, v_c\}$ of the clusters. More explicitly, we write the optimization problem as follows:

$$\min Q \text{ with respect to } U \in U, v_1, v_2, \ldots, v_c \in R^n \tag{8.4}$$

The minimized objective function being a double sum of weighted distances taken over the space of data ($k = 1, 2, \ldots, N$) and clusters ($i = 1, 2, \ldots, c$) comes with a simple interpretation. The structure revealed by the clustering method and characterized by the partition matrix and the prototypes is the one for which the sum of distances is the lowest one. Say, if x_k associates with the ith cluster, the corresponding distance $\|x_k - v_i\|$ is small and the membership degree u_{ik} attains values close to 1. On the other hand, if x_k is located quite far from the prototype and the distance is quite high, this high value is discounted in the entire sum by making the corresponding membership value to the ith cluster close to 0, $u_{ik} = 0$, thus substantially alleviating its contribution to the overall sum.

From the optimization standpoint, there are two individual optimization tasks to be carried out separately for the partition matrix and the prototypes. The first one concerns the minimization with the constraints given by the requirement of the form (8.2)–(8.3), which holds for each data point x_k. The use of Lagrange multipliers transforms the problem into its constraint-free optimization version. The augmented objective function V formulated for each data point, $k = 1, 2, \ldots, N$, reads as

$$V = \sum_{i=1}^{c} u_{ik}^{m} d_{ik}^{2} + \lambda \left(\sum_{i=1}^{c} u_{ik} - 1 \right) \tag{8.5}$$

where we use an abbreviated form $d_{ik}^2 = ||x_k - v_i||^2$. Proceeding with formulation of the necessary conditions for the minimum of V for $k = 1, 2, \ldots, N$, one has

$$\frac{\partial V}{\partial u_{st}} = 0 \quad \frac{\partial V}{\partial \lambda} = 0 \tag{8.6}$$

$s = 1, 2, \ldots, c, t = 1, 2, \ldots, N$. Now we calculate the derivative of V with respect to the elements of the partition matrix in the following way:

$$\frac{\partial V}{\partial u_{st}} = m u_{st}^{m-1} d_{st}^2 + \lambda \tag{8.7}$$

From (8.7) set to zero and the use of the normalization condition (8.3), we calculate the membership grade u_{st} to be equal to

$$u_{st} = -\left(\frac{\lambda}{m}\right)^{\frac{1}{m-1}} d_{st}^{\frac{2}{m-1}} \tag{8.8}$$

We complete some rearrangements of the earlier expression by isolating the term including the Lagrange multiplier:

$$-\left(\frac{\lambda}{m}\right)^{\frac{1}{m-1}} = \frac{1}{\sum_{j=1}^{c} d_{jt}^{\frac{2}{m-1}}} \tag{8.9}$$

Inserting this expression into (8.8), we obtain the successive entries of the partition matrix

$$u_{st} = \frac{1}{\sum_{j=1}^{c} \left(\frac{d_{st}^2}{d_{jt}^2}\right)^{\frac{1}{m-1}}} \tag{8.10}$$

The optimization of the prototypes v_i is carried out assuming the Euclidean distance between the data and the prototypes that is $||x_k - v_i||^2 = \sum_{j=1}^{n} (x_{kj} - v_{ij})^2$. The objective function reads now as follows $Q = \sum_{i=1}^{c} \sum_{k=1}^{N} u_{ik}^m \sum_{j=1}^{n} (x_{kj} - v_{ij})^2$, and its gradient with respect to v_i, $\nabla_{v_i} Q$, made equal to 0 yields the system of linear equations:

$$\sum_{k=1}^{N} u_{ik}^m (x_{kt} - v_{st}) = 0, \quad s = 1, 2, \ldots, c; \quad t = 1, 2, \ldots, n \tag{8.11}$$

Thus

$$v_{st} = \frac{\sum_{k=1}^{N} u_{ik}^m x_{kt}}{\sum_{k=1}^{N} u_{ik}^m} \tag{8.12}$$

One should emphasize that the use of some other distance functions different from the Euclidean one and the family of Euclidean distances brings some computational complexity and the formula for the prototype cannot be presented in the concise manner as given earlier.

Quite commonly, as the data are multivariable and the individual coordinates (variables) exhibit different ranges, the variables are normalized or the weighted Euclidean distance (including the variances of the individual variables) is considered:

$$||\mathbf{x}_k - \mathbf{v}_i||^2 = \sum_{j=1}^{n} \frac{\left(x_{kj} - v_{ij}\right)^2}{\sigma_j^2} \tag{8.13}$$

One should comment that the distance function used in the objective function (8.1) directly entails the geometry of the produced clusters. The two other commonly encountered examples of distances such as the Hamming and the Chebyshev one are considered.

Hamming (City Block Distance)

$$||\mathbf{x}_k - \mathbf{v}_i|| = \sum_{j=1}^{n} |x_{kj} - v_{ij}| \tag{8.14}$$

Chebyshev

$$||\mathbf{x}_k - \mathbf{v}_i|| = \max_{j=1, 2, \ldots, n} |x_{kj} - v_{ij}| \tag{8.15}$$

They yield clusters whose geometry resembles diamond-like and hyperbox-like shapes (see Figure 8.1). To demonstrate this, we look at the distribution of points $\mathbf{x}$ whose distance from the origin is set to some constant ρ, namely, $\{\mathbf{x} \mid ||\mathbf{x} - 0|| \leq \rho\}$, and characterize the location of such points. In the two-dimensional case one has the detailed expressions, which entail the following:

- Hamming (city block) distance $\{(x_1, x_2) \mid |x_1| + |x_2| \leq \rho\}$
- Euclidean distance $\{(x_1, x_2) \mid x_1^2 + x_2^2 \leq \rho^2\}$
- Chebyshev distance $\{(x_1, x_2) \mid \max(|x_1|, |x_2|) \leq \rho\}$

As in the case of the Euclidean distance, weighted versions of the Hamming and Chebyshev distance can be considered. It is worth stressing that the type of distance

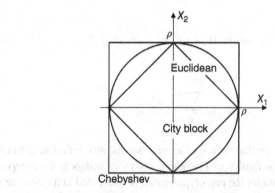

Figure 8.1 Geometry of data distributed from the origin at some constant distance ρ.

directly implies the geometry of the clusters generated by the method. In other words, by setting up a specific distance, we directly favor geometry of the clusters to be revealed by the FCM. Along with such geometric underpinnings, any distance function entails some optimization scenario requirements. With this regard, we note that the computing of the prototypes depends directly on the selection of the distance. Let us emphasize again that the closed-form formula (8.12) is valid only in case of the Euclidean distance; however the calculations for other distances are more complicated.

Overall, the FCM clustering is completed through a sequence of iterations where we start from some random allocation of data (a certain randomly initialized partition matrix) and carry out the following updates by successively adjusting the values of the partition matrix and the prototypes. The iterative process is repeated until a certain termination criterion has been satisfied. Typically, the termination condition is quantified by looking at the changes in the membership values of the successive partition matrices. Denote by U(iter) and U(iter + 1) the two partition matrices produced in two consecutive iterations of the algorithm. If the distance $\|U$(iter + 1) − U(iter)$\|$ is less than a small predefined threshold ε, then we terminate the algorithm. Typically, one considers the Chebyshev distance between the partition matrices, meaning that the termination criterion reads as follows:

$$\max_{i,k} |u_{ik}(\text{iter} + 1) - u_{ik}(\text{iter})| \le \varepsilon \tag{8.16}$$

The fuzzification coefficient exhibits a direct impact on the geometry of fuzzy sets generated by the algorithm. Typically, the value of m is assumed to be equal to 2.0. Lower values of m (that are closer to 1) yield membership functions that start resembling characteristic functions of sets; most of the membership values become localized around 1 or 0. The increase of the fuzzification coefficient ($m =$ 3, 4, etc.) produces "spiky" membership functions with the membership grades equal to 1 at the prototypes and a fast decline of the values when moving away from the prototypes.

In case m tends to 1, the partition matrix comes with the entries equal to 0 or 1 so the results come in the form of the Boolean partition matrix. The algorithm becomes then the well-known k-means clustering.

From the perspective of information granules built through clustering and fuzzy clustering, let us note that the partition matrix U can be sought as a collection of membership functions of the information granules occupying successive rows of the matrix. In other words, fuzzy sets A_1, A_2, ..., A_c with their membership values assumed for the data x_1, x_2, ..., x_N form the corresponding rows of U:

$$U = \begin{bmatrix} A_1 \\ A_2 \\ \vdots \\ A_c \end{bmatrix} = \begin{bmatrix} u_1 \\ u_2 \\ \vdots \\ u_c \end{bmatrix}$$

where $A_i(x_k) = u_{ik}$ and u_i stands for a vector of membership degrees of data localized in the ith cluster.

The constructed partition matrix offers also another insight into the revealed structure at the level of pairs of data, yielding a so-called proximity matrix and a linkage matrix of clusters.

Proximity Matrix

The proximity matrix $P = [p_{kl}]$, k, $l = 1, 2, \ldots, N$ implied by the partition matrix comes with the entries defined in the form

$$p_{kj} = \sum_{i=1}^{c} \min\left(p_{ik}, p_{ij}\right) \qquad (8.17)$$

where p_{kj} characterizes a level of closeness between pairs of individual data. In light of the earlier formula, proximity of some pairs of data (k, j) is equal to 1 if and only if these two points are identical. If the membership grades of such points are closer to each other, the corresponding entry of the proximity matrix attains higher values (Loia et al., 2007). Note that p_{kj} can be sought as an approximation of a value of some kernel function K at the pairs of the corresponding pair of points, $p_{kj} \approx K\left(x_k, x_j\right)$.

Linkage Matrix

The $c*c$-dimensional linkage matrix $L = [l_{ij}]$ offers a global characterization of association between two clusters i and j determined (averaged) over all data:

$$l_{ij} = 1 - \frac{1}{N} \sum_{k=1}^{N} \left|u_{ik} \equiv u_{jk}\right| \qquad (8.18)$$

where the earlier expression uses the Hamming distance between the corresponding rows of the partition matrix. Having this in mind, the formula reads as

$$l_{ij} = 1 - \frac{1}{N} \left\|u_i - u_j\right\| \qquad (8.19)$$

If the two corresponding rows (i and j) of the partition matrix are getting close to each other, then the clusters (information granules) become strongly linked (associated).

8.2 *k*-MEANS CLUSTERING ALGORITHM

The k-means clustering algorithm is one of the commonly considered clustering methods (Jain and Dubes, 1988). While we noted that it comes as a special case of the FCM for the limit case of the fuzzification coefficient, one can derive it by starting from the minimization of the objective function. It assumes the same form as given by (8.1) with the partition matrix taking on binary values coming from the two-element set {0, 1}. As the optimization of the partition matrix is realized independently for each data, Q is a linear function of u_{ik}. This immediately implies that the minimum of the expression $\sum_{i=1}^{c} u_{ik} \| x_k - v_i \|^2$ is achieved for u_{ik} expressed in the form

$$u_{ik} = \begin{cases} 1, & \text{if } i = \arg \min_{l=1,2,\ldots,c} \| x_k - v_l \| \\ 0, & \text{otherwise} \end{cases} \tag{8.20}$$

The prototypes are computed in the same way as in the case of the FCM, namely,

$$v_i = \sum_{k=1}^{N} u_{ik} x_k \Big/ \sum_{k=1}^{N} u_{ik} \tag{8.21}$$

Obviously, here the fuzzification coefficient m does not play any role as u_{ik} is either 0 or 1.

8.3 AUGMENTED FUZZY CLUSTERING WITH CLUSTERS AND VARIABLES WEIGHTING

Bringing some parameters to the optimized objective function could enhance the flexibility of the generic clustering algorithm. The augmentations are provided in two ways: (i) by weighting individual clusters and (ii) by weighting variables. The way of weighting the clusters, namely, expressing relevance of individual groups, is quantified by means of the c-dimensional weight vector w whose coordinates are values in [0,1]. The lower the weight value, the less influential is the corresponding cluster. The impact of the individual features (variables) is quantified in terms of the n-dimensional feature weight vector f where higher values of the coordinates of f imply higher relevance of the corresponding features (variables). In this way, some features are made more essential, and the features space can be reduced by identifying (and dropping) features coming with very low values of the weights.

Proceeding with the detailed objective function, we have

$$Q = \sum_{i=1}^{c} \sum_{k=1}^{N} u_{ik}^m w_i^2 \| x_k - v_i \|_f^2 \tag{8.22}$$

The weight vector f is included in the computing of the distance. In other words,

$$\| x_k - v_i \|_f^2 = \sum_{j=1}^{n} \left(x_{kj} - v_{ij} \right)^2 f_{j^2} \tag{8.23}$$

In the case of the weighted Euclidean distance, we have

$$\| x_k - v_i \|_f^2 = \sum_{j=1}^{n} \frac{\left(x_{kj} - v_{ij} \right)^2}{\sigma_j^2} f_{j^2} \tag{8.24}$$

with σ_j^2 standing for the standard deviation of the jth feature.

The weight vectors w and f satisfy the following constraints:

$$w_i \in [0, 1]$$
$$\sum_{i=1}^{c} w_i = 1$$
$$f_j \in [0, 1] \tag{8.25}$$
$$\sum_{j=1}^{n} f_j = 1$$

Formally speaking, the optimization (minimization) of Q is realized as follows:

$$\min_{U, V, w, f} Q(U, v_1, v_2, \dots, v_c, w, f) \tag{8.26}$$

where $V = \{v_1, v_2, \dots, v_c\}$. Given the constraints, the detailed optimization is structured as problems of constrained optimization, which engages Lagrange multipliers.

8.4 KNOWLEDGE-BASED CLUSTERING

In contrast to supervised learning, clustering is a typical mechanism of learning in an unsupervised environment. There are a lot of scenarios where there are some sources of additional information, which are taken into consideration in the learning process. Those techniques can be referred to as knowledge-based clustering (Pedrycz, 2005). For instance, one may encounter a limited number of data points, which were labeled, and their available class membership grades are organized in a certain partition matrix (either Boolean or fuzzy). This scenario can be referred to as clustering with partial supervision (Pedrycz and Waletzky, 1997; Pedrycz, 2005). Consider that a certain subset of data of X of size M, call it X_0, consists of labeled data with the partition matrix F. The original objective function is expanded by including an additional component expressing closeness of structure conveyed by the partition matrix U with the structural information conveyed by F:

$$Q = \sum_{i=1}^{c} \sum_{k=1}^{N} u_{ik}^2 \|x_k - v_i\|^2 + \kappa \sum_{i=1}^{c} \sum_{\substack{k=1 \\ x_k \in X_0}}^{N} (u_{ik} - f_{ik})^2 \|x_k - v_i\|^2 \tag{8.27}$$

The second term in (8.27) quantifies a level of agreement of U and F obtained in the labeled data. The scaling coefficient κ helps strike a balance between the structural findings coming as a result of clustering completed on X and the structural information available through the entries of F. If $\kappa = 0$, then the method reduces to the standard FCM algorithm. Another alternative where knowledge is acquired from different sources comes in the form of a collaborative clustering (Pedrycz and Rai, 2009).

8.5 QUALITY OF CLUSTERING RESULTS

In evaluating the quality of the clustering results, there are two main categories of quality indicators that might be classified as internal and external ones. The internal ones, which include a large number of alternatives, are referred to as cluster validity indices. In one way or another, they tend to reflect a general nature of constructed clusters involving their compactness and separation abilities. The external measures of quality are used to assess the obtained clusters vis-à-vis the abilities of the clusters to cope with the inherent features of the data. The resulting clusters are evaluated with respect to the abilities to represent the data themselves or the potential of the clusters in the construction of predictors or classifiers.

Cluster Validity Indices

There is a significant collection of indices used to quantify the performance of the obtained clusters (so-called cluster validity indices) (Milligan and Cooper, 1985; Rousseeuw, 1987; Krzanowski and Lai, 1988; Vendramin *et al.*, 2010; Arbelaitz *et al.*, 2013). The main criteria that are used by the validity indices concern the quality of clusters expressed in terms of compactness and separation of the resulting information granules (Davies and Bouldin, 1979). The compactness can be quantified by the sum of distances between the data belonging to the same cluster, whereas the separation is expressed by determining a sum of distances between the prototypes of the clusters. The lower the sum of the distances in the compactness measure and the higher the differences between the prototypes, the better the solution. For instance, the Xie-Beni (XB) index is reflective of this characterization of these properties by taking into account the following ratio:

$$V = \frac{\sum_{i=1}^{c} \sum_{k=1}^{N} u_{ik}^m \|x_k - v_i\|^2}{N \min_{\substack{i,j \\ i \neq j}} \|v_i - v_j\|} \tag{8.28}$$

Note that the nominator is a descriptor of the compactness while the denominator captures the separation aspect. The lower the value of V, the better the quality of the clusters. The index shown later is developed along the same line of thought with an exception that in the nominator one has the distance of the prototypes from the total mean $\tilde{v}$ of the entire dataset:

$$V = \frac{\sum_{i=1}^{c} \sum_{k=1}^{N} u_{ik}^m \|x_k - v_i\|^2 + \frac{1}{c} \sum_{i=1}^{c} \|v_i - \tilde{v}\|^2}{N \min_{\substack{i,j \\ i \neq j}} \|v_i - v_j\|} \tag{8.29}$$

Some other alternative is expressed in the following form:

$$V = \sum_{i=1}^{c} \sum_{k=1}^{N} u_{ik}^m \|x_k - v_i\|^2 + \frac{1}{c} \sum_{i=1}^{c} \|v_i - \tilde{v}\|^2 \tag{8.30}$$

As noted there are a number of other alternatives; however they quantify the criteria of compactness and separation in different ways.

Classification Error

The classification error is one of the obvious alternatives with this regard: we anticipate that clusters are homogeneous with respect to classes of patterns. In an ideal situation, a cluster should be composed of patterns belonging only to a single class. The more heterogeneous the cluster is, the lower is its quality in terms of the classification error criterion. The larger the number of clusters is, the higher their likelihood becomes homogeneous. Evidently, this implies that the number of clusters should not be lower than the number of classes. If the topology of the data forming individual classes is more complex, we envision that the number of required classes to keep them homogeneous has to be higher in comparison with situations where the geometry of classes is quite simple (typically spherical geometry of highly disjoint classes, which is easily captured through the Euclidean distance function).

Reconstruction Error

Clusters tend to reveal an inherent structure in the data. To link this aspect of the clusters with the quality of the clustering result, we determine reconstruction abilities of the clusters. Given a structure in the data described by a collection of prototypes v_1, v_2, ..., v_c, we first represent any data x with their help. For the k-means, this representation returns a Boolean vector $[0\ 0....1\ 0...0]$ where a nonzero entry of the vector indicates the prototype, which is the closest to x. For the FCM, the representation of x comes as a vector of membership grades $u(x) = [u_1\ u_2\ ...\ u_c]$ computed in a "usual" way. This step is referred to as a granulation, namely, a process of representing x in terms of information granules (clusters). In the sequel, a degranulation phase returns a reconstructed datum x expressed as

$$\hat{x} = \frac{\sum_{i=1}^{c} u_i^m(x)v_i}{\sum_{i=1}^{c} u_i^m(x)} \tag{8.31}$$

The previous formula results directly from the minimization of the following performance index with the weighted Euclidean distance used originally in the FCM algorithm and the minimization with respect to the reconstructed data while the reconstruction error is expressed as the following sum:

$$V = \sum_{k=1}^{N} ||x_k - \hat{x}_k||^2 \tag{8.32}$$

(with $||.||^2$ standing for the same weighted Euclidean distance as being used in the clustering algorithm).

8.6 INFORMATION GRANULES AND INTERPRETATION OF CLUSTERING RESULTS

8.6.1 Formation of Granular Descriptors of Numeric Prototypes

Numeric prototypes produced by FCM and k-means are initial descriptors of data. It is rather surprising to anticipate that a single numeric vector (prototype) can fully describe a collection of numeric data. To make this description more comprehensive and reflective of the diversity of the data, we form granular prototypes by taking advantage of the principle of justifiable granularity. Starting from the existing numeric prototype v_i, we determine the coverage and specificity subsequently forming a granular prototype V_i by maximizing the product of these two criteria.

Proceeding with more details, the coverage is expressed in a usual way as follows.

For the k-means clustering, we have

$$\text{cov}(V_i, \rho_i) = \frac{1}{N_i}\text{card}\left\{x_k \in X_i \mid \|x_k - v_i\|^2 \leq \rho_i^2\right\} \tag{8.33}$$

where X_i is a set of data belonging to the ith cluster and N_i stands for the number of data in X_i.

For the FCM algorithm, in computing the coverage instead of counting, we carry out an accumulation of the membership degrees:

$$\text{cov}(V_i, \rho_i) = \frac{\sum_{x_k : \|x_k - v_i\|^2 \leq \rho_i^2} u_{ik}}{\sum_{k=1}^{N} u_{ik}} \tag{8.34}$$

In both expressions shown earlier, the distance used implies the geometry of the developed information granule. For instance, in the case of the Euclidean distance, the granules are of hyper-ellipsoidal shapes described in the form

$$\frac{(x_1 - v_{i1})^2}{\sigma_1^2} + \frac{(x_2 - v_{i2})^2}{\sigma_2^2} + \cdots + \frac{(x_n - v_{in})^2}{\sigma_n^2} \leq \rho_i^2 \tag{8.35}$$

In case of the Chebyshev distance, one has rectangular shapes of the granules:

$$\max\left[\frac{|x_1 - v_{i1}|}{\sigma_1}, \frac{|x_2 - v_{i2}|}{\sigma_2}, \ldots, \frac{|x_n - v_{in}|}{\sigma_n}\right] \leq \rho_i \tag{8.36}$$

The specificity can be expressed as a linearly decreasing function of ρ_i, $\text{sp}(V_i, \rho_i) = 1 - \rho_i$ assuming that the values of ρ_i are in the unit interval. The optimal granular prototype V_i is determined by maximizing the product of the coverage and specificity:

$$\rho_{i,\max} = \arg\text{Max}_{\rho_i \in [0, 1]}\left[\text{cov}(V_i, \rho_i) \cdot \text{sp}(V_i, \rho_i)\right] \tag{8.37}$$

In the buildup of information granules along with the incorporation of the excitatory data (viz., belonging to the clusters of interest), we also include inhibitory data (viz., those belonging to clusters different from the one being of interest). We introduce the coverage implied by the data belonging to the cluster of interest and the others not belonging to this cluster.

Proceeding from the k-means, we compute the following expressions:

$$\text{cov}(V_i, \rho_i)^+ = \frac{1}{N_i}\text{card}\left\{x_k \in X_i \mid \|x_k - v_i\|^2 \leq \rho_i^2\right\}$$

$$\text{cov}(V_i, \rho_i)^- = \frac{1}{N_i}\text{card}\left\{x_k \notin X_i \mid \|x_k - v_i\|^2 \leq \rho_i^2\right\} \tag{8.38}$$

The coverage expression involving the excitatory and inhibitory data comes as follows:

$$\text{cov}(V_i, \rho_i) = \max\left[0, \text{cov}(V_i, \rho_i)^+ - \text{cov}(V_i, \rho_i)^-\right] \tag{8.39}$$

As implied by the principle of justifiable granularity, optimal information granule is determined through the maximization of the product of the coverage and specificity.

A related approach is concerned with a concise description of information granules in the form of so-called shadowed sets (Pedrycz, 2009).

8.6.2 Granularity of Data and Their Incorporation in the FCM Algorithm

The FCM algorithm discussed so far processed numeric data. An interesting and practically viable suite of applications concerns with granular data (information granules). Clustering information granules opens a new avenue in the realm of clustering. Interestingly, the clustering algorithms we discussed so far are applicable here. The critical issue is to position information granules in a new features space so that the nature of information granules can be captured. In general, as information granules are more advanced than numeric entities, it is apparent that they require more parameters to characterize granules. This entails that the features space in which information granules are described becomes of higher dimensionality. Obviously, there are a variety of approaches of forming features spaces. With this regard, two main categories of forming information granules are considered:

a. *Parametric.* Here it is assumed that information granules to be clustered are represented in a certain parametric format, say, they are described by triangular fuzzy sets, intervals, Gaussian membership functions, etc. In each case there are several parameters associated with the membership functions, and the clustering is carried out in the space of the parameters (Hathaway *et al.*, 1996). Then the resulting prototypes are also described by information granules having the same parametric form as the information granules being clustered. Obviously, the dimensionality of the space of the parameters in which the clustering takes place is usually higher than the original space. For instance, in the case of triangular membership functions, the new space is R^{3n} (given the original space is R^n).

b. *Nonparametric.* There is no particular form of the information granules, so some characteristic descriptors of information granules used to capture the nature of these granules (Pedrycz *et al.*, 1998) are formed and used to carry out clustering.

8.7 HIERARCHICAL CLUSTERING

In contrast to the FCM and k-means clustering, where a structure is revealed for some predetermined number of clusters, hierarchical clustering is about building clusters in an iterative fashion.

Hierarchical clustering aims at building clusters in a hierarchical manner by successively merging individual data into clusters or successively splitting collections

Distance

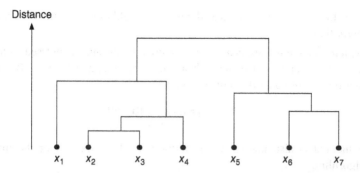

Figure 8.2 Example of a dendrogram formed through hierarchical clustering; in the case of agglomerative bottom-up clustering, starting from the number of clusters being equal to the number of data in X, larger groups are formed by aggregating the two clusters that are the closest to each other.

of data into smaller groups (subsets). In the first case, one starts with the number of clusters set to the number of data, and the merging of the closest data points is realized step by step. In the successive divisions, the starting point is a single cluster (the entire dataset X), which is then successively split into smaller clusters. The results of clustering are displayed in the form of a dendrogram. An illustrative dendrogram is visualized in Figure 8.2.

On the vertical scale shown are the values of distances obtained between the consecutively formed clusters. In the formation of the clusters, as illustrated earlier, a distance between clusters plays a pivotal role. The agglomerative method works as follows. At the first step, as we start with the individual data (each data point forms an individual cluster). Here the distances between the clusters are computed in a straightforward manner by invoking any distance discussed earlier, say, the Euclidean one. Here the distance between x_2 and x_3 is the smallest one, and these two data are merged, forming a cluster $\{x_2, x_3\}$. At the next step, the distances are computed between all other data $x_1, x_4, x_5, \ldots, x_7$ and $\{x_2, x_3\}$. The distance between x_4 and $\{x_2, x_3\}$ is the lowest, and in virtue of this, we build a three-element cluster $\{x_2, x_3, x_4\}$. Next x_6 and x_7 are merged, forming another cluster. At this stage, one has three clusters: $\{x_2, x_3, x_4\}$, $\{x_1\}$, and $\{x_5, x_6, x_7\}$. Finally, two clusters, $\{x_1, x_2, x_3, x_4\}$ and $\{x_5, x_6, x_7\}$, are merged.

When forming the clusters, it becomes apparent that the concept of distance needs further clarification as it applies now to a single data and a collection of data. Here the construct is not unique and several alternatives can be exercised. In particular, the following three options are commonly considered. Depending on them, the resulting clustering is referred to as a single linkage, complete linkage, and average linkage. Consider two clusters already constructed, say, X_i and X_j. We have the following ways of determining the distance between the clusters (Duda *et al.*, 2001):

Single-linkage hierarchical clustering uses the distance expressed as the minimum of distances between the pairs of data belonging to X_i and X_j:

$$||X_i - X_j|| = \min_{\substack{x \in X_i \\ y \in X_j}} ||x - y|| \qquad (8.40)$$

This method is also referred to as the nearest neighbor algorithm or the minimum algorithm.

Complete-linkage hierarchical clustering uses the distance expressed as the maximum of distances between the data belonging to X_1 and X_2 (viz., the most distant points are taken into consideration):

$$||X_i - X_j|| = \max_{\substack{x \in X_i \\ y \in X_j}} ||x - y|| \tag{8.41}$$

The method is also known as the farthest neighbor method or the maximum algorithm.

Average-linkage hierarchical clustering uses the distance expressed as the average of distances determined between the individual data located in X_1 and X_2:

$$||X_i - X_j|| = \frac{1}{\operatorname{card}(X_i)\operatorname{card}(X_j)} \sum_{x \in X_i} \sum_{y \in X_j} ||x - y|| \tag{8.42}$$

These three alternatives are illustrated in Figure 8.3.

The choice of the number of clusters can be determined by observing a level at which the dendrogram could be "cut" in a visibly convincing way, that is, a level when moving from c to $c + 1$ clusters results in a significant changes in terms of the distances reported between the clusters. For instance (see Figure 8.2), the visible number of clusters is 3 as there is significant increase in the values of distances when moving to 2 clusters (as this requires an involvement of the significantly higher value of the distance as already used in the aggregation of clusters realized so far).

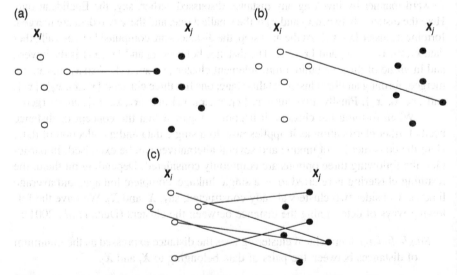

Figure 8.3 Computing distances in single-linkage (a), complete-linkage (b), and average-linkage (c) variants of hierarchical clustering.

Each cluster can be further characterized in the form of a single information granule by applying the principle of justifiable granularity to the elements comprising X_i. In terms of further more insightful data analysis, there are two points supported by such information granularity-based pursuits.

First, the number of clusters could be determined by optimizing the overall sum of the performance indices used in the principle of justifiable granularity applied to the individual clusters at a certain level of cutting the dendrogram. Consider that we have selected c clusters. Running the principle of justifiable granularity (whose values are maximized), we have $Q_1, Q_2, ..., Q_c$. The overall performance index associated with this family of clusters is determined as the average $(Q_1 + Q_2 + \cdots + Q_c)/c$. Second, the principle of justifiable granularity is helpful in identifying potential granular outliers, namely, the data that do not fall within the realm of the union of the information granules.

The results of the tasks highlighted earlier are helpful in identifying the semantically sound information granules and this way support further processing focused on classification or prediction.

8.7.1   Description of Results of Clustering Through Information Granules

Here we engage the principle of justifiable granularity by augmenting the representation abilities of numeric prototypes, building granular prototypes or forming granular prototypes in cases when prototypes are not constructed explicitly (such as in hierarchical clustering).

Formation of Granular Prototypes on a Basis of Hierarchical Clustering
Hierarchical clustering does not return prototypes (as in the case of the FCM algorithm); however on a basis of the data falling forming a cluster, a numeric representative and its granular counterpart can be constructed.

8.8   INFORMATION GRANULES IN PRIVACY PROBLEM: A CONCEPT OF MICROAGGREGATION

Data privacy is concerned with efficient ways of concealing complete data, namely, avoiding data disclosure. Among the existing methods, microaggregation is about forming small clusters of data aimed at no disclosing the individual data. The data are replaced by the prototypes of the clusters, meaning that the data are not disclosed. Higher levels of privacy are achieved for a smaller number of clusters. Higher information loss (resulting in higher values of the reconstruction error) comes with the smaller number of clusters. There is a need to set up a certain sound compromise.

Clustering results in information granules and the granules (clusters) deliver a required level of abstraction through which the data are avoided from being disclosed.

The method proposed by Torra (2017) is an example of a direct usage of the FCM clustering. It consists of the following main steps:

1. Running the FCM algorithm for given number of clusters (c) and a certain value of the fuzzification coefficient m_1 (>1).

2. For each datum x in the dataset of interest X, determine the membership values using the well-known formula of computing the entries of the partition matrix for some other value of the fuzzification coefficient m_2 (>1), namely,

$$u_i(x) = \frac{1}{\sum_{j=1}^{c}\left(\frac{\|x-v_i\|}{\|x-v_j\|}\right)^{2/(m_2-1)}} \qquad (8.43)$$

3. For each x generate a random number ξ drawn from the uniform distribution over the unit interval and choose the cluster i_0 (prototype) so that the following relationship is satisfied:

$$\sum_{i=1}^{i_0} u_i(x) < \xi < \sum_{i=i_0}^{c} u_i(x) \qquad (8.44)$$

Typically, the number of clusters is selected in a way that each cluster should consist of at least k data; the dependency between k and c is expressed in the following form:

$$k < \mathrm{card}(X)/c < 2k \qquad (8.45)$$

8.9 DEVELOPMENT OF INFORMATION GRANULES OF HIGHER TYPE

Information granules of higher type emerge in classes of applications when dealing with situations when data are distributed over time and/or space. A representative example in this domain comes under the rubric of hotspot identification and description. Figure 8.4 illustrates a situation wherein when starting with a collection of individual data (forming a so-called data level) $D_1, D_2, \ldots, D_s$, one realizes clustering (FCM) and then on their basis constructs information granules as discussed through the maximization of (8.34) producing for each dataset a collection of granular hotspots (lower IG level). In the sequel, a collection of information granules built as a result of aggregation is formed (upper IG level), and owing to the elevated level of abstraction, those naturally emerge as information granules of higher type. In more detail, the following processing phases are realized:

Data level. The datasets present at each data site $x_k(I)$, $I = 1, 2, \ldots, s$ are processed individually. The FCM algorithm applied to each dataset returns prototypes $v_i(I)$ and partition matrix $U(I)$.

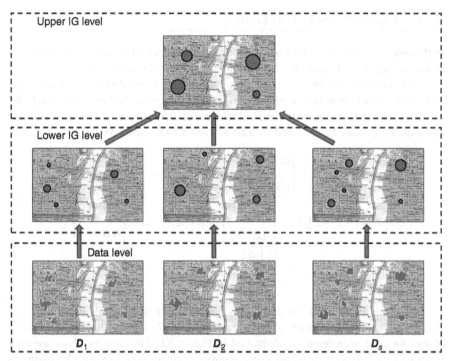

Figure 8.4   From data to hotspots and information granules of higher type.

Lower level of information granules. The numeric prototypes are augmented by structural information coming from the partition matrix, yielding granular prototypes as comprehensive descriptors of the structure in the data. They are formed by applying the principle of justifiable granularity so that the radius of the granule built around the numeric prototype is expressed in the form $\rho_i(I) = \arg\max_\rho[\text{cov}*\text{sp}]$. The resulting information granule is described as a triple $V_i(I) = (v_i(I), \rho_i(I)), \mu_i(I)$ with $\mu_i(I)$ standing for the associated optimized value of the criterion used in the principle of justifiable granularity.

Upper level of information granules. Here the information granules formed at the lower level are clustered, and a collection of representatives is being formed. First, note that the available information granules are represented in a three-dimensional space of the coordinates of the prototypes $v_i(I)$ (two variables) and a length of the radius $\rho_i(I)$ of the granule. Once the clustering has been completed, the results are returned as a family of prototypes and radii, namely, $\tilde{v}_j, \tilde{\rho}_j, j = 1, 2, \ldots, c$. Applying the principle of justifiable granularity, we transform them into information granules of type 2 in which the prototypes are information granules $\tilde{v}_j$ (rather than single numeric entities). Also granular radii $\tilde{R}_j$ are developed. Both these descriptors are information granules spanned over $\tilde{v}_j, \tilde{\rho}_j$ constructed so far.

8.10 EXPERIMENTAL STUDIES

In this section, we present selected illustrative examples to illustrate the performance of clustering methods and offer an interpretation of obtained results.

We consider a collection of two-dimensional synthetic data forming a mixture of patterns drawn from five normal (Gaussian) distributions, 200 data from each distribution. The parameters of the individual normal distributions (mean vectors v and covariance matrices Σ) are listed as follows (see also Figure 8.5):

$$v_1 = [2 \ \ 5], \Sigma_1 = \begin{bmatrix} 0.2 & 0.0 \\ 0.0 & 0.6 \end{bmatrix} \quad v_2 = [2 \ \ 5], \Sigma_2 = \begin{bmatrix} 1.0 & 0.5 \\ 0.5 & 2.0 \end{bmatrix}$$

$$v_3 = [-3 \ \ -3], \Sigma_1 = \begin{bmatrix} 0.6 & 0.0 \\ 0.0 & 0.7 \end{bmatrix} \quad v_4 = [2 \ \ 5], \Sigma_4 = \begin{bmatrix} 1.0 & 0.0 \\ 0.0 & 0.2 \end{bmatrix}$$

$$v_5 = [2 \ \ 5], \Sigma_5 = \begin{bmatrix} 0.2 & 1.0 \\ 1.0 & 1.0 \end{bmatrix}$$

We run FCM with the commonly used fuzzification coefficient $m = 2$ and change the number of clusters c varying it from 2 to 7. The results shown in terms of the obtained prototypes are displayed in Figure 8.6. Different symbols are used to indicate data belonging to the corresponding clusters (allocation to the cluster is done on a basis of the maximum membership grades).

The quality of the structure revealed by the clustering method is quantified in terms of the obtained reconstruction criterion V (see (8.30)); the values produced for the corresponding clusters are collected in Table 8.1. Furthermore it is shown that by

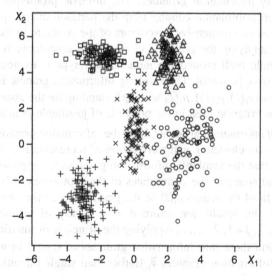

Figure 8.5 Synthetic data coming as a mixture of normal distributions; various symbols denote data coming from the individual distributions.

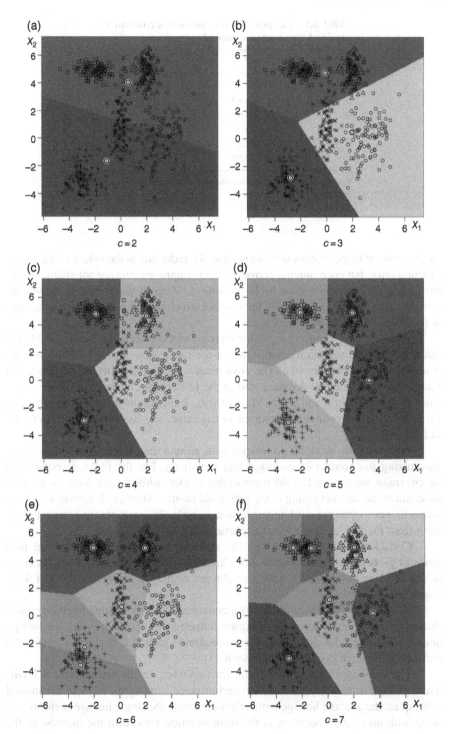

Figure 8.6 Results of fuzzy clustering; included are prototypes superimposed over the existing data.

TABLE 8.1 Values of the reconstruction criterion V obtained for selected values of c

c	V	$V(c + 1)/V(c)$
2	1065.29	
3	541.80	**0.51**
4	264.95	**0.49**
5	150.47	**0.57**
6	144.37	0.96
7	129.54	0.90

The highest values of reduction of the reconstruction error are shown in boldface.

moving toward higher values of c, we achieve the reduction in the values of the reconstruction error, but once moving above $c = 5$, the improvements are not visible (as the ratio achieves the values close to 1, showing no drop in the values of V over the increase of c). This coincides with the original structure in the data having a mixture of five Gaussian probability functions.

As could have been expected, the increase in the values of c leads to the lower values of the reconstruction error; however the major changes (error reduction) are observed for lower values of c. The error reduction for the values of c exceeding 5 becomes quite limited as demonstrated in the last column of the table showing the ratios of the reconstruction errors $V(c + 1)/V(c)$. Running the k-means clustering on the same data, the results are reported in the same way as before (see Figure 8.7 and Table 8.2).

Now we form granular prototypes built around their numeric counterparts by maximizing the product of coverage and specificity. For the FCM, in computing the coverage, we accumulate the membership grades, while for the k-means we take the count of the data belonging to the individual cluster. Another alternative exercised here is to incorporate the inhibitory information by looking at a negative impact stemming from the data assigned to other clusters.

Considering these alternatives, the resulting prototypes are visualized in a series of plots displayed in Figure 8.8. Granular prototypes generated on the basis of the FCM and k-means clustering results are displayed in Figures 8.9 and 8.10, respectively.

We report the results produced by hierarchical clustering. Applying hierarchical clustering, the three modes of clustering are considered (single, complete, and average linkage). Selecting the cuts across the dendrogram yielding 2, 5, and 8 clusters, the obtained groups are displayed in Figure 8.11.

The associated values of the reconstruction error are covered in Table 8.3. Different linkage strategies yield different performance. Among the three of the studied techniques, the average linkage works the best, while the single linkage performs the worst with no visible reduction in the reconstruction error with the increase of the number of the clusters.

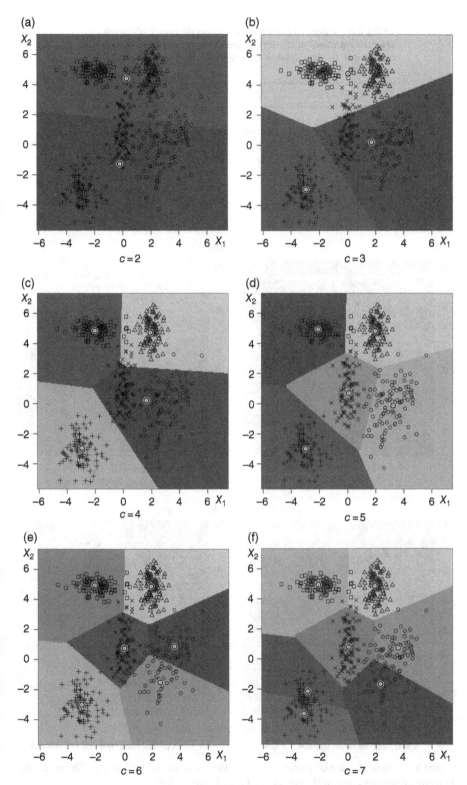

Figure 8.7   Reconstruction results obtained for the *k*-means clustering; the plots include the obtained prototypes superimposed over the data.

TABLE 8.2 Values of the reconstruction criterion V obtained for selected numbers of clusters c; the drop in the values of $V(c)$ are included in the last column with the higher values of drop marked in boldface

c	V	$V(c + 1)/V(c)$
2	1208.07	
3	629.04	**0.52**
4	361.53	**0.57**
5	198.66	**0.55**
6	187.79	0.94
7	149.26	0.80

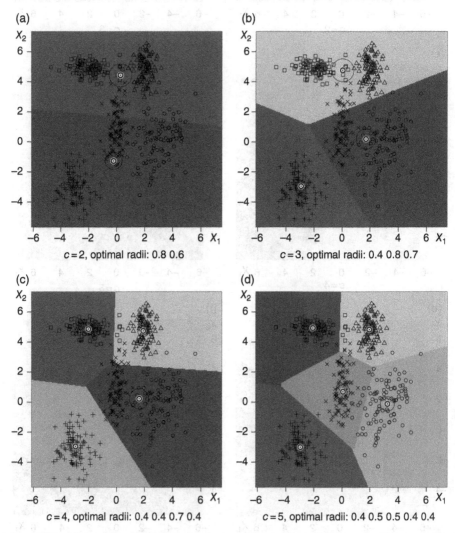

Figure 8.8 Plots of granular prototypes for prototypes produced by k-means. Shown are the numeric prototypes along with their radii (shown in dark color).

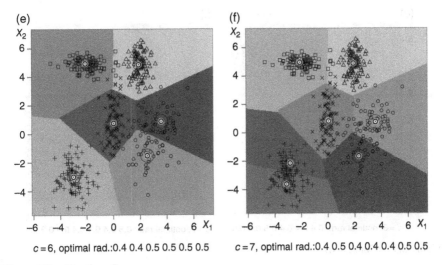

(e) $c=6$, optimal rad.:0.4 0.4 0.5 0.5 0.5 0.5

(f) $c=7$, optimal rad.:0.4 0.5 0.4 0.4 0.4 0.5 0.5

Figure 8.8 (Continued)

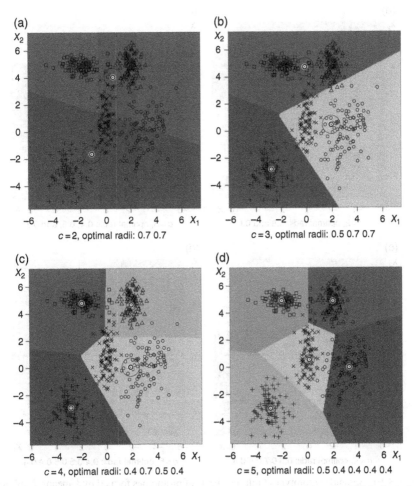

(a) $c=2$, optimal radii: 0.7 0.7

(b) $c=3$, optimal radii: 0.5 0.7 0.7

(c) $c=4$, optimal radii: 0.4 0.7 0.5 0.4

(d) $c=5$, optimal radii: 0.5 0.4 0.4 0.4 0.4

Figure 8.9 Granular prototypes constructed for results produced by the FCM clustering; shown are the numeric centers and their radii (shown in dark color).

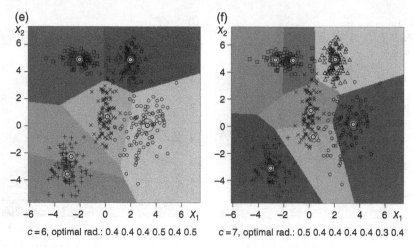

c = 6, optimal rad.: 0.4 0.4 0.4 0.5 0.4 0.5   c = 7, optimal rad.: 0.5 0.4 0.4 0.4 0.4 0.3 0.4

Figure 8.9   (Continued)

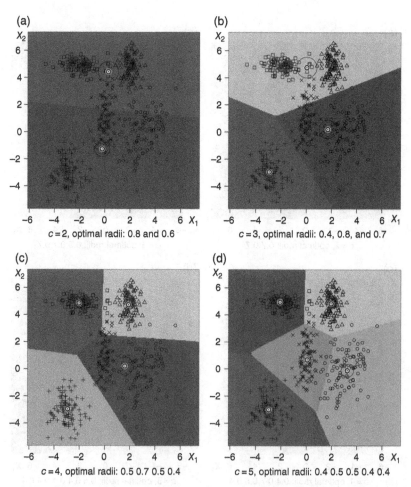

c = 2, optimal radii: 0.8 and 0.6   c = 3, optimal radii: 0.4, 0.8, and 0.7

c = 4, optimal radii: 0.5 0.7 0.5 0.4   c = 5, optimal radii: 0.4 0.5 0.5 0.4 0.4

Figure 8.10   Granular prototypes built on a basis of prototypes formed by k-means. Inhibitory information is included when forming information granules, centers, and radii are shown in dark color.

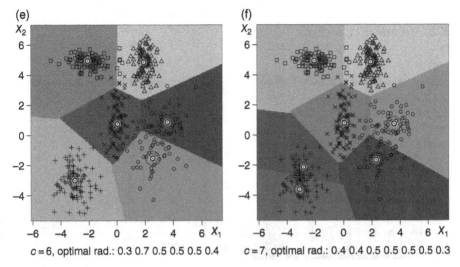

(e) X_2 ... $c = 6$, optimal rad.: 0.3 0.7 0.5 0.5 0.5 0.4

(f) X_2 ... $c = 7$, optimal rad.: 0.4 0.4 0.5 0.5 0.5 0.5 0.3

Figure 8.10 (Continued)

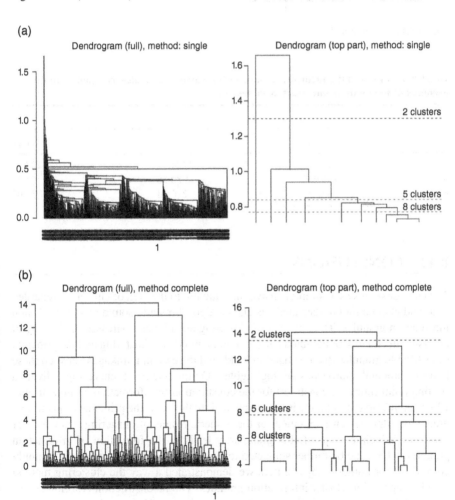

Figure 8.11 Dendrograms obtained for hierarchical clustering for $c = 2$, 5, and 8: (a) single linkage, (b) complete linkage, and (c) average linkage.

(c)

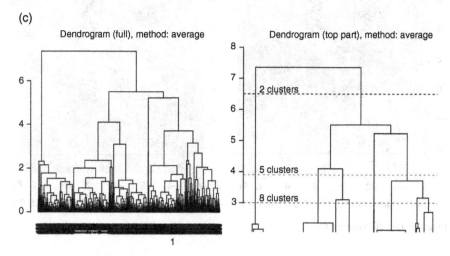

Figure 8.11 (Continued)

TABLE 8.3 Values of the reconstruction error V obtained for single-, complete-, and average-linkage method and selected values of c

c	V		
	Single	Complete	Average
2	1989.79	1224.99	1128.50
5	1961.32	280.28	363.28
8	1938.35	216.48	166.07

8.11 CONCLUSIONS

The chapter serves as a comprehensive introduction to the area of clustering regarded as a fundamental tool to data analysis and an algorithmically sound vehicle to design information granules. The diversity of clustering techniques is stressed, and a way of assessing the quality of obtained clusters is elaborated on. Clustering and the principle of justifiable granularity come hand in hand, and the essential linkages between these two fundamental approaches are highlighted. On the one hand, clustering delivers a starting point (numeric construct) for the construction of information granule around which an information granule is being spanned. In this way, the principle of justifiable granularity delivers an enrichment of the numeric results of clustering. On the other hand, the principle of justifiable granularity leads to the formation of information granules, which can be further subjected to clustering. While clustering is commonly considered for numeric objects, here we demonstrated a way the clustering method could be applied to cluster information granules. The reconstruction criterion serves as a useful way to quantify the reconstruction abilities of information granules.

REFERENCES

O. Arbelaitz, O. Arbelaitz, I. Gurrutxaga, J. Muguerza, J. M. Pérez, and I. Perona, An extensive comparative study of cluster validity indices, *Pattern Recognition* 46(1), 2013, 243–256.

J. Bezdek, *Pattern Recognition with Fuzzy Objective Function Algorithms*, New York, Plenum Press, 1981.

D. L. Davies and D. W. Bouldin, A clustering separation measure, *IEEE Transactions on Pattern Analysis and Machine Intelligence* 1, 1979, 224–227.

R. O. Duda, P. E. Hart, and D. G. Stork, *Pattern Classification*, 2nd ed., New York, John Wiley & Sons, Inc., 2001.

J. C. Dunn, Well separated clusters and optimal fuzzy partitions, *Journal of Cybernetics* 4, 1974, 95–104.

R. J. Hathaway, J. Bezdek, and W. Pedrycz, A parametric model for fusing heterogeneous fuzzy data, *IEEE Transactions on Fuzzy Systems* 4, 1996, 270–281.

A. K. Jain, Data clustering: 50 years beyond K-means, *Pattern Recognition Letters* 31(8), 2010, 651–666.

A. K. Jain and R. C. Dubes, *Algorithms for Clustering Data*, Upper Saddle River, NJ, Prentice-Hall, 1988.

W. J. Krzanowski and Y. T. Lai, A criterion for determining the number of groups in a data set using sum of squares clustering, *Biometrics* 1, 1988, 23–34.

V. Loia, W. Pedrycz, and S. Senatore, Semantic web content analysis: A study in proximity-based collaborative clustering, *IEEE Transactions on Fuzzy Systems* 15(6), 2007, 1294–1312.

G. W. Milligan and M. Cooper, An examination of procedures for determining the number of clusters in a dataset, *Psychometrika* 50(2), 1985, 159–179.

W. Pedrycz, *Knowledge-Based Clustering: From Data to Information Granules*, Hoboken, NJ, John Wiley & Sons, Inc., 2005.

W. Pedrycz, From fuzzy sets to shadowed sets: Interpretation and computing, *International Journal of Intelligence Systems* 24(1), 2009, 48–61.

W. Pedrycz, J. Bezdek, R. J. Hathaway, and W. Rogers, Two nonparametric models for fusing heterogeneous fuzzy data, *IEEE Transactions on Fuzzy Systems* 6, 1998, 411–425.

W. Pedrycz and P. Rai, A multifaceted perspective at data analysis: A study in collaborative intelligent agents systems, *IEEE Transactions on Systems, Man, and Cybernetics. Part B, Cybernetics* 39(4), 2009, 834–844.

W. Pedrycz and J. Waletzky, Fuzzy clustering with partial supervision, *IEEE Transactions on Systems, Man, and Cybernetics* 5, 1997, 787–795.

P. Rousseeuw, Silhouettes: A graphical aid to the interpretation and validation of cluster analysis, *Journal of Computational and Applied Mathematics* 20, 1987, 53–65.

V. Torra, Fuzzy microaggregation for the transparency principle, *Journal of Applied Logic* 23, 2017, 70–80.

L. Vendramin, R. J. Campello, and E. R. Hruschka, Relative clustering validity criteria: A comparative overview, *Statistical Analysis and Data Mining* 3(4), 2010, 209–235.

CHAPTER 9

QUALITY OF DATA: IMPUTATION AND DATA BALANCING

In this chapter, we discuss important and commonly encountered problems of incomplete and imbalanced data and present ways of alleviating them through mechanisms of data imputation and data balancing. Along with an elaboration on different imputation algorithms, a particular attention is focused on the role of information granularity in the quantification of the quality of data both when coping with incompleteness and a lack of balance of data.

9.1 DATA IMPUTATION: UNDERLYING CONCEPTS AND KEY PROBLEMS

In real-world problems, quite often data are incomplete, meaning that some values of their features (attributes) are missing. Such situations occur in various scenarios where the data are collected, say, in survey data, observational studies, and even controlled experiments. Further processing of data in prediction or classification problems requires their completeness. Fixing the dataset by reconstructing missing is referred to as *imputation*. The overall process is referred to as data imputation or missing value imputation (MVI). The most common application of imputation is with statistical survey data (Rubin, 1987). One can encounter imputation problems in many other areas, say, DNA microarray analysis (Troyanskaya *et al.*, 2001). There are numerous ways of realizing data imputation, and the quality of the results depends upon the number of missing data and a reason behind the data being missed.

When dealing with incomplete data, in addition to imputation, another alternative is to consider a so-called a complete-case analysis. Here data with missing attributes (features) are simply excluded, and any further processing uses the remaining complete data. The approach is impractical in many cases because of at least two reasons: (i) introduction of potential bias. When the eliminated data differ systematically from the observed complete data, the ensuing models are subject to some shifts, and

Pattern Recognition: A Quality of Data Perspective, First Edition. Władysław Homenda and Witold Pedrycz.

(ii) in case of a large number of variables, there might be a few complete data left as most of the data would be discarded.

Before discussing the main approaches used in imputation, it is instructive to provide a detailed insight into the nature of the problem and the essence of the phenomenon behind missing data.

Consider a single variable, Y, whose values could be missing and a single covariate Z whose values are fully available. Let M be a Boolean variable, which takes the value 1 if Y is missing and 0 if Y is observed. The essence of the mechanism causing occurrence of missing data is captured by the relationship between M, Y, and Z. The following types of the phenomenon of missing data (missingness) are encountered (Rubin, 1987; Little and Rubin, 2002).

Missing Completely at Random (MCAR)
In this scenario, $P(M = 1|Y, Z) = P(M = 1)$ meaning that the missing value does not depend upon Y and Z.

Missing at Random (MAR)
Here $P(M = 1|Y, Z) = P(M = 1|Z)$, which states that the missing value is conditional on Z. In both *MCAR* and *MAR*, the value of Y does not depend on the unobserved values of Y.

Missing Not at Random (MNAR)
Here the probability Y is missing depends upon the missing value itself, that is, depends on Y.

Deterministic approaches to imputation consider any missing value as a problem of prediction, with the objective to reconstruct the most likely value for the ones that are missing.

9.2 SELECTED CATEGORIES OF IMPUTATION METHODS

There are a large number of imputation techniques. Several of them are commonly encountered in a variety of studies, owing to the simplicity, low computational overhead, effectiveness, and other factors of the imputation algorithm. Here we recall some of the commonly encountered alternatives (refer to Figure 9.1).

Statistics-Based Imputation
The crux of this approach is to replace the missing value of a given attribute by the probabilistic representative of this attribute. For instance, one takes a simple statistical representative such as a mean or median (see Figure 9.1a). The intuition behind this class of methods is obvious: if the value is missing, we consider a statistical representative of the available values of the attribute under consideration. The method is simple to implement. There are several drawbacks, however. All the missing values are replaced by the same value, say, the mean or median. The variance of this variable is underestimated. As only a single variable (attribute) is involved in the imputation, this method distorts relationships existing among the variables by lowering estimates of the correlation between the variables to 0.

(a)   (b)

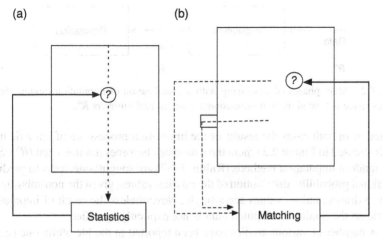

Figure 9.1   Illustration of the principle of the selected imputation methods: (a) statistics-based imputation and (b) hot deck imputation.

Random Imputation
To avoid bias present in the previous methods, here the missing value is replaced by some random value coming from the distribution of the attribute under consideration. The limitations of the variance/correlation reduction become somewhat alleviated.

Hot Deck Imputation
Here the missing value is replaced by the value encountered in the data whose distance to the data under consideration is the smallest one (the matching index is the highest) (see Figure 9.1b). The bias present in the previous model is then reduced as the imputed values could vary from case to case. The matching measure, which is used to compare closeness of two data, has to be provided in advance (Reilly, 1993; Andridge and Little, 2010).

Regression Prediction
This category of imputation mechanism falls under the rubric of model-based imputation. The imputed value is determined by forming a regression model between the variable whose values are to be imputed and some other variable(s) present in the data (see Figure 9.1). Here both one-dimensional regression and multivariable regression models could be sought. This imputation method is more advanced by taking into several variables in the determination of the missing value; however the form (structure) of the regression model has to be specified. Along with commonly used regression models, other nonlinear relationships could be considered such as neural networks or rule-based models.

An overall flow of processing indicating a position of the imputation phase along with the ensuing general framework of prediction or classification is displayed in Figure 9.2.

It is worth noting that in spite of the visible diversity, once the imputation has been completed, one cannot tell apart the originally available data and those being

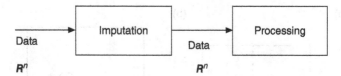

Figure 9.2   Main phases of processing with a visualization of imputation module; here the features space is treated as an n-dimensional space of real numbers $\boldsymbol{R}^n$.

imputed as in both cases the results of the imputation process are of numeric nature. This is stressed in Figure 9.2 where the data space becomes not impacted ($\boldsymbol{R}^n$). Sometimes random imputation methods (Rubin, 1987) are sought: one seeks to predict the conditional probability distribution of the missing values, given the non-missing ones, and then draws random values accordingly. Nevertheless the result of imputation is numeric so the quality of imputed data is not explicitly specified.

A number of various studies have been reported in the literature; one can refer here to Liu and Brown (2013), Branden and Verboven (2009), and Di Nuovo (2011).

We fix notation to be used in the subsequent methods. The data points form a collection of N data defined in the n-dimensional space of real numbers. They are denoted as vectors $\boldsymbol{x}_1, \boldsymbol{x}_2, ..., \boldsymbol{x}_k, ..., \boldsymbol{x}_N$ where $\boldsymbol{x}_k \in \boldsymbol{R}^n$, $k = 1, 2, ..., N$. We also introduce a Boolean matrix, $\boldsymbol{B} = [b_{kj}]$, $k = 1, 2, ..., N$; $j = 1, 2, ..., n$, to capture information about the missing data. In this matrix the kjth entry is set to 0, that is, $b_{kj} = 0$ if the jth variable of the kth data point is missing. Otherwise for the available data the corresponding entry of the matrix is set to 1.

9.3   IMPUTATION WITH THE USE OF INFORMATION GRANULES

As pointed out, the format of the data being imputed should reflect this fact and be made distinct from those data that were originally available. Those original data are numeric. To make a distinction from the numeric data and at the same time, the result of imputation gives rise to the data of different (lower) quality. The fact that the data is imputed and its quality becomes lower manifests as the result of imputation coming as an information granule formalized in one of the existing ways (intervals, fuzzy sets, probability functions, etc.)

The overall processing scheme positioning granular imputation as one of the functional module in the entire scheme of pattern recognition is displayed in Figure 9.3.

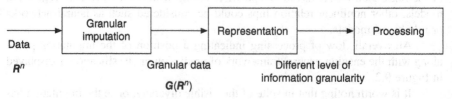

Figure 9.3   Process of granular imputation yielding a new features space.

The notable aspect is the augmentation of the features space; while the original space in which the data a positioned is R^n, after imputation, we encounter an augmented features space of higher dimensionality where we accommodate the parameters of information granules that are built as a result of imputation. For instance, when accepting information granules in the form of intervals, we require including the bounds of the intervals, and this doubles the dimensionality of the space of numeric data to R^{2n}. In case of triangular fuzzy numbers, we require the features space R^{3n} so that the parameters of the fuzzy sets (bounds and the modal value) are represented. Then the data become heterogeneous; the numeric ones are positioned in this new space by admitting the same values of the bounds of the intervals or making the three parameters of fuzzy numbers identical whereas those being imputed have the values of these parameters different.

9.4 GRANULAR IMPUTATION WITH THE PRINCIPLE OF JUSTIFIABLE GRANULARITY

The method exhibits some resemblance to the hot deck technique; however it offers substantial generalization in comparison with the existing methods outlined earlier by augmenting the results of imputation by information granularity and generating information granules in place of missing numeric values. The two-phase process of data imputation consists of the following steps:

1. Invoking a certain method of imputing numeric data (here any technique discussed earlier can be considered)
2. Building information granules on the basis of the numeric imputations realized during the first phase

We elaborate on details by considering a mean imputation method dealing separately with each column (variable). In what follows we determine two statistics (mean and variance) for the individual variables computed on a basis of available data, namely,

$$m_l = \frac{\sum_{k=1}^{N} x_{kl} b_{kl}}{\sum_{k=1}^{N} b_{kl}} \tag{9.1}$$

$$\sigma_l^2 = \frac{\sum_{k=1}^{N} (x_{kl} - m_l)^2 b_{kl}}{\sum_{k=1}^{N} b_{kl}} \tag{9.2}$$

where $l = 1, 2, \ldots, n$. Then, as in some standard method, the average is sought as an imputed value for the missing values in the lth column; note it is computed on a basis of available data.

During the second design phase, we build information granules around the numeric imputed values. It is worth stressing that the method used in the first phase takes into account only a single variable and all possible relationships among other variables are not considered at all. When moving with the second phase, we introduce some improvement of the generic imputation method by looking at possible dependencies between the data x for which we do imputation and any other data vector, say, z. We determine the weighted Euclidean distance between these two vectors as follows:

$$\rho(x, z) = \sum_{l=1}^{n} \frac{(x_l - z_l)^2}{\sigma_l^2} \tag{9.3}$$

where σ_l is a standard deviation of the lth variable. The calculations completed earlier are realized for the coordinates of the vectors for which the values of both entries (x_l and z_l) are available (viz., the corresponding entries of the Boolean vectors are equal to 1).

For the given x we normalize the values of these distances across all data, thus arriving at the following expression:

$$\rho'(x, z_k) = \frac{\rho(x, z_k) - \min_l \rho(x, z_l)}{\max_l \rho(x, z_l) - \min_l \rho(x, z_l)} \tag{9.4}$$

where $x \neq z_k$, $k = 1, 2, \ldots, N$. Subsequently we construct the weight associated with x expressing an extent to which x can be associated with z. In other words, we can use z to determine eventual missing coordinates of x:

$$w(x, z) = 1 - \rho'(x, z) \tag{9.5}$$

The higher the value of this weight, the more visible the association between x and z becomes and the more essential contribution of z to the imputation of the missing entries of x.

Having all these prerequisites in place, we outline an overall procedure of data imputation, leading to a granular result of this imputation. We consider the jth variable for which a numeric imputed value is m_j (this could be a median, mean, or any sound numeric representative). Around this numeric m_j we form a granular imputed value by invoking the principle of justifiable granularity in which weighted data are considered. The data used for the formation of the information granule come in the form of the pairs (data, weight) with the weights determined using (9.5). Thus we have

$$x_{1j}, w(x, x_1), x_{2j}, w(x, x_2), \ldots, x_{Nj}, w(x, x_M) \tag{9.6}$$

where x is a data vector for which the jth variable is subject to imputation $M < N$ as only take the data, which are complete with respect to the jth variable.

The bounds of the imputed information granule are obtained by maximizing the performance index Q used in the principle of justifiable granularity (refer to Chapter 7). In the case of intervals and triangular fuzzy numbers, the optimization process is realized for the lower and upper bounds individually. The returned normalized performance index describes the quality of the imputed information granule. For the lth imputed interval data, the associated quality ξ_l is expressed in the form $\xi_l = (Q_l)/(\max_{i \in I} Q_i)$ where I is the set of imputed data for the jth variable. Once completed, the imputed result comes in the form of the interval and its weight, namely, $([x_l^-, x_l^+], \xi_l)$. In the case of the triangular fuzzy sets, this yields the triple of the parameters of the membership function associated with the quality of the granule, namely,

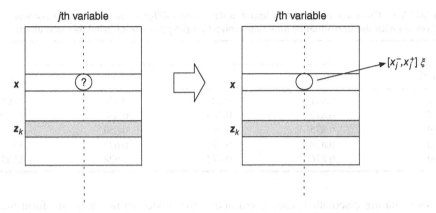

Figure 9.4   Granular imputation realized with the principle of justifiable granularity.

$\left(\left(x_l^-, m_l, x_l^+ \right), \xi_l \right)$. The underlying processing realizing this imputation mechanism is illustrated in Figure 9.4.

The main features of the method include (i) involvement of all variables in the imputation process and (ii) formation of information granule instead of a single numeric imputation result. These two features highlight the advantages of the developed approach.

To illustrate the performance of this imputation method, we consider synthetic data composed of 1000 eight-dimensional data. Each variable is governed by a normal distribution of some mean value and standard deviation. In detail, we have $m_1 = -5$, $\sigma_1 = 2$, $m_2 = 11$, $\sigma_2 = 3$, $m_3 = -6$, $\sigma_3 = 1$, $m_4 = 0$, $\sigma_4 = 1$.

We randomly eliminate $p\%$ of data and then run the method, leading to the formation of the intervals of imputed data. Then the results are quantified in terms of the achieved coverage where the coverage is determined by comparing the produced intervals with the original data, which were randomly deleted. The results displayed for different values of p are presented in Table 9.1.

It is noticeable, as it could have been expected, that with the increase in the percentage of the missing values, the coverage drops or fluctuates; however the differences are quite low.

From the functional perspective, the new features space that is formed through the imputation process is illustrated in Figure 9.4. As we are concerned with interval information granules, the dimensionality of the original space is doubled in terms of its dimensionality as for each original variable there are two bounds of the interval. The quality of the imputed result is quantified in terms of ξ_l (as presented earlier). Two points have to be emphasized. First, there could be several variables missing for the same data. In this situation, we can compute the average of ξ_l associated with the missing entries. For the kth data denote this average by $\bar{\xi}(k)\mathbf{x}(k)$. This serves as an indicator of the performance of the data point $\mathbf{x}(k)$, which can be used in the building a classification or prediction model F by quantifying the relevance of the individual data. For the complete data points, the weights are set to 1, while the weights coming with the imputed data are lower than 1, and thus

TABLE 9.1 Coverage produced by imputed data versus different values of p; the parameter β comes with the performance index maximized by the principle of justifiable granularity

	β			
p (%)	0.5	1	2	3
1	0.871	0.802	0.629	0.561
5	0.872	0.785	0.638	0.550
10	0.875	0.786	0.633	0.546
20	0.870	0.778	0.625	0.559
50	0.876	0.773	0.628	0.555

these data are discounted when constructing the model. In this way, we form the following categories of data:

Complete data $(x(k), \text{target}(k))\ k \in K$

Imputed data $(x(k), \text{target}(k), \bar{\xi}_k)\ k \in L$

The performance index is reflective of this situation and comes in the following sum composed of the two components:

$$Q = \sum_{k \in K} (M(x(k)) - \text{target}(k))^2 + \sum_{k \in L} \bar{\xi}_k \left(M(x(k)) - \text{target}(k) \right)^2 \qquad (9.7)$$

Here M stands for the constructed model; we show this as illustrated in Figure 9.5. Overall, when using new data x, the processing is realized in two phases:

1. **Imputation**, if required. The variables whose values are not available are imputed using the algorithm used earlier. The result comes in the form of some information granule, say, an interval or a fuzzy set.

2. **Determining** the output of the model, say, the prediction one. Here the inputs of the model are either numeric or granular.

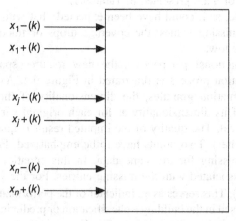

Figure 9.5 Building augmented features space resulting from data imputation to be used for models of classification and prediction.

9.5 GRANULAR IMPUTATION WITH FUZZY CLUSTERING

Fuzzy clustering, say, fuzzy c-means (FCM) (Bezdek, 1981), is a generic vehicle to reveal a structure in data. In light of this observation, once the clustering has been completed for all available complete data, we can impute the missing values. In the context of data imputation, there are some benefits: (i) all variables and relationships among them are taken into account when building representatives (prototypes) of the data so that (or and hence) these prototypes can be effectively used in the imputation method, and (ii) clustering builds some general dependencies, and we are not confined to the detailed functional relationships (that might be difficult to verify), which are behind some sophisticated imputation techniques.

In what follows, we briefly review the method as it is considered in the setting of incomplete data. The objective function shown later and guiding a process of formation of clusters is expressed as a sum of distances between data and prototypes where $v_1, v_2, \ldots, v_c$ are the prototypes of the clusters and $U = [u_{ik}]$ is a partition matrix, while $m, m > 1$, is a fuzzification coefficient impacting a geometry of the membership functions (entries of the partition matrix):

$$Q = \sum_{i=1}^{c} \sum_{k=1}^{N} u_{ik}^m \|x_k - v_i\|_B^2 \tag{9.8}$$

The distance $\|.\|$ standing in the previous formula is a weighted Euclidean distance function expressed as

$$\|x_k - v_i\|_B^2 = \sum_{j}^{n} \frac{\left(x_{kj} - v_{ij}\right)^2}{\sigma_j^2} b_{kj} \tag{9.9}$$

where σ_j is a standard deviation of the jth variable. Note that the Boolean matrix B used in the previous calculations emphasizes that only available data are involved in the computations.

The FCM method returns a collection of prototypes and the partition matrix, which are determined in an iterative fashion using the following formulas:

$$u_{ik} = \frac{1}{\sum_{j=1}^{c} \left(\frac{\|x_k - v_i\|_B}{\|x_k - v_j\|_B}\right)^{2/(m-1)}}$$

$$v_{ij} = \frac{\sum_{k=1}^{N} u_{ik}^m x_{kj} b_{kj}}{\sum_{k=1}^{N} u_{ik}^m b_{kj}} \tag{9.10}$$

where $i = 1, 2, \ldots, c; j = 1, 2, \ldots, n; k = 1, 2, \ldots, N$.

Note that these are modified expressions used originally in the FCM; here in the computations we eliminated the missing entries (using the Boolean values of B).

Now a certain input x whose some values are missing becomes reconstructed, namely, the missing values are imputed. Let us associate with x a Boolean vector b whose values set to 0 indicate that the corresponding entry (variable) is missing. The missing values are imputed as a result of a two-phase process:

1. Determination of membership grades computed on a basis of the prototypes and using only available inputs of x:

$$\tilde{u}_i = \frac{1}{\sum_{j=1}^{c} \left(\frac{\|x_k - v_i\|_B}{\|x_k - v_j\|_B} \right)^{2/(m-1)}} \qquad (9.11)$$

2. Computing the missing entries of x coming as a result of a reconstruction process:

$$\tilde{x}_j = \frac{\sum_{i=1}^{c} \tilde{u}_i^m v_{ij}}{\sum_{i=1}^{c} \tilde{u}_i^m} \qquad (9.12)$$

where the indices j are those for which the entries of b are equal to 0.

As noted, fuzzy clustering yields a nonzero reconstruction error. To compensate for it and quantify the quality of reconstruction, one admits that the results of reconstruction are information granules instead of numeric results. In the simplest case the granules are intervals. Consider complete data (where no imputation was completed). For the jth variable, the reconstruction returns $\hat{x}_{kj}$ that is usually different from x_{kj}. The differences $e_{kj} = x_{kj} - \hat{x}_{kj}$ are collected, and for each variable $j = 1, 2, \ldots, n; k = 1, 2, \ldots, N$. Based on this dataset of differences, we form an information granule E_j with the bounds $\left[e_j^-, e_j^+ \right]$ using the principle of justifiable granularity. Typically the modal value of errors is 0, which makes the error negative and positive. By taking the result of the reconstruction, we build an interval $X_{kj} = x_{kj} \oplus E_j$ where the sum is produced following the calculus of intervals. This yields

$$X_{kj} = x_{kj} \oplus E_j = \left[x_{kj} + e_j^-, x_{kj} + e_j^+ \right] \qquad (9.13)$$

When the imputation is completed for some new data x, we determine the numeric result of imputation and then make all its coordinates granular by building the corresponding intervals as in (9.13).

It is worth stressing that the information granule E_j is the same for any value of x_j, which results in the information granule of the same specificity. The mapping $e_j = e_j(\hat{x}_j)$ can be made more refined by admitting that this mapping is described through a collection of c rules:

$$-\text{if } \hat{x}_j \text{ is } A_i \text{ then } \varepsilon = \left[\varepsilon_i^-, \varepsilon_i^+ \right] \qquad (9.14)$$

where $A_1, A_2, \ldots, A_c$ are fuzzy sets defined in the space of imputed values. The rules associate the predicted value with the interval of error.

9.6 DATA IMPUTATION IN SYSTEM MODELING

Interestingly, the granular nature of imputed results can be encountered when dealing with incomplete linguistic data, namely, the data whose values are information granules (say, fuzzy sets), and some of the entries are missing (expressed by fuzzy sets such as unknown). Following the principle outlined in the study, the imputed results are information of higher type than the granular data one originally started with (say, fuzzy sets of type 2). The role of granular data resulting from the imputation process can be used and exploited in the ensuing granular models (such as classifiers or predictors). The results of the models are now granular and this way effectively quantify the performance of the model by making it more in rapport with reality and reflecting the fact that it has been constructed in the presence of incomplete (and then imputed) data, implying a varying level of information granularity of the classification and prediction results depending upon the region of the features space and the quality of data present there.

There are several alternatives that are to be treated in the development of the model (when estimating its parameters). The performance index has to be modified to cope with granular data. In what follows, we use capital letters to denote the imputed data. Denote the model by M, so for any input x, the model returns $y = M(x)$.

Distinguished are four alternatives depending on the nature of the imputed data:

1. (x_k, target_k)—no imputation; there are original numeric data and subsequently there is no change to the typical performance index used in the construction of numeric models. It could be a commonly encountered RMSE index, say,

$$\text{RMSE} = \sqrt{\frac{1}{M_1} \sum_{k=1}^{M_1} (M(x_k) - \text{target}_k)^2} \tag{9.15}$$

   where M_1 is the number of data for which no imputation was applied.

2. (X_k, target_k) imputation of the input data (concerning one or several input variables). The granular input X_k implies that the result of the model is granular as well, $Y_k = M(X_k)$. In this regard, one evaluates an extent to which target is included (contained) in Y_k, say, $\text{incl}(\text{target}_k, Y_k)$. The measure of inclusion can be specialized depending upon the nature of information granule of the imputed data. In the case of sets (intervals), the concept is binary and relates to the inclusion predicate

$$\text{incl}(\text{target}_k, Y_k) = \begin{cases} 1 & \text{if } \text{target}_k \in Y_k \\ 0 & \text{otherwise} \end{cases} \tag{9.16}$$

   In the case of fuzzy sets (Y_k being described by some membership function), the inclusion predicate returns a degree of membership of target_k in Y_k, namely, $Y_k(\text{target}_k)$.

3. (x_k, Target_k) imputation realized for the output data. The output of the model is numeric, $y_k = M(x_k)$, and we compare it with the granular target, Target_k. As before we use the same way of evaluation of the quality of the model as discussed in the previous scenario.

4. $(X_k, Target_k)$ imputation involves both the input and output variables. The output of the model is granular Y_k, which has to be compared with $Target_k$. Here we use the degree of matching $\xi(Y_k, Target_k)$.

Overall, the performance index comprises two components: the one that is computed on a basis of numeric results produced by the model (scenario 1) and another one that involves inclusion or matching two information granules (scenarios 2–4). More formally, we can structure the global performance index as follows:

$$Q = \sqrt{\frac{1}{M_1} \sum_{k=1}^{M_1} (M(x_k) - target_k)^2}$$

$$+ \beta \left\{ \frac{1}{M_2} \sum_{k=1}^{M_2} (1 - incl(target_k, Y_k)) + \frac{1}{M_3} \sum_{k=1}^{M_3} (1 - \xi(target_k, Y_k)) \right\} \tag{9.17}$$

where β is a certain weight factor, which helps striking a sound balance between the two components of the performance index involving original data and those being subject to imputation. M_2 denotes the number of data concerning the second and the third scenario, while M_3 is the number of cases dealing with the fourth scenario. The parameters of the model are optimized so that the performance index Q becomes minimized.

9.7 IMBALANCED DATA AND THEIR GRANULAR CHARACTERIZATION

Imbalanced data create a challenge to a variety of classification methods. The results produced in the presence of imbalanced data are always worsened by the nature of the data. To improve the quality, various ways of making the balanced are considered. One of the fundamental indicators with imbalanced data classification is the imbalance ratio (viz., the ratio of the number of data located in the minority class to the total number of data).

To formulate the problem in a systematic way, we consider a two-class problem where the data belonging to class ω_1 are dominant so the ratio of the data in class ω_1 to those belonging to class ω_2 is significantly higher than 1. In other words, patterns in class ω_2 are underrepresented. The balancing of the data concerns undersampling of the data in class ω_1. In contrast to the commonly encountered undersampling methods whose results are just numeric data forming a selected subset of patterns used further for classifier design, we form a subset of granular data—information granules where the facet of information granularity associated with them compensates for the use of the subset of the dominant class rather than all the data.

To fix the notation, the data are located in the n-dimensional space R^n and those belonging to the dominant class ω_1 consists of N of patterns, whereas the ones in the underrepresented class ω_2 have M patterns where $M \ll N$. The set of all patterns is denoted by X.

9.7.1 Main Approaches to Data Balancing: An Overview

As the data balancing techniques exhibit a great deal of diversity, it is convenient to organize them into several main categories, in particular sampling methods, algorithmic methods, and cost-sensitive methods.

Sampling Methods

The sampling approaches are employed to address the skewed class distribution problem caused by the imbalanced nature of data at data level. The intent is to modify the original imbalanced dataset to provide a balanced distribution (He and Garcia, 2009; Galar *et al.*, 2012). Sampling approaches, which are applied prior to any learning, can be split into three main groups, namely, undersampling, oversampling, and hybrid methods, which combine both sampling methods (Alibeigi *et al.*, 2012). The undersampling approach removes data from the majority class to adjust the class distribution, while the oversampling approach generates new data belonging to the minority class.

Random undersampling and oversampling are the two commonly used techniques for balancing dataset. A number of other undersampling and oversampling techniques have been developed. A single-sided selection (OSS) method, which undersamples representative subset from the majority class, has been discussed in Kubat and Matwin (2000). The condensed nearest neighbor (CNN) rule and Tomek Links integration method were discussed in Batista *et al.* (2004). A synthetic minority oversampling technique (SMOTE) was proposed in Chawla *et al.* (2012), in which the oversampling is achieved through creating synthetic minority class examples. An oversampling method based on k-means clustering and the SMOTE method was proposed in Santos *et al.* (2015) to construct a representative dataset for predicting the survival of hepatocellular carcinoma patients. A method, called majority weighted minority oversampling technique (MWMOTE), was presented in Barua *et al.* (2014). The MWMOTE method first identifies the hard-to-learn minority class data, assigns relative weights to the selected samples, and then generates synthetic samples using clustering. In He *et al.* (2008), an adaptive synthetic (ADASYN) sampling approach was discussed.

These sampling approaches exhibit several shortcomings. Because of lack of information about the structure of the training data, random undersampling tends to remove certain important instances, while the random oversampling can potentially lead to overfitting problems. Sampling methods are straightforward. One advantage of sampling methods is that they are easily applicable to various available classification frameworks without the need to change the underlying learning strategies (Sun *et al.*, 2009).

Algorithmic Level Methods

The algorithm-based approaches are aimed to solve the imbalanced problem through creating new classification algorithms or modifying existing ones. For example, in Wu and Chang (2003), the class-boundary-alignment algorithm was proposed to augment support vector machines (SVMs) to deal with imbalanced data. Through the combination of an integrated sampling technique with an ensemble of SVMs, SVMs with asymmetric misclassification costs achieved better prediction performance (Wang and Japkowicz, 2010).

Algorithm-level approaches are usually problem specific, which means they are applicable to particular types of problems. To find a solution to the problem at hand at the algorithmic level, one has to articulate the essence of the specific learning algorithms and accommodate the knowledge about the application domain.

Cost-Sensitive Learning Methods

Cost-sensitive learning methods aim at increasing classification accuracy by specifying unequal misclassification costs between classes, instead of handling all misclassification errors equally. It is apparent that the cost related with misclassification of minority class is usually higher than that with the majority class (Sun *et al.*, 2007). A wrapper that discovers the amount of resampling for an imbalanced dataset with optimizing some evaluation functions such as cost, cost curves, and the cost-dependent *f*-measure has been proposed in Chawla *et al.* (2008). The problem of optimal learning and decision-making when different misclassification errors incur different penalties has been discussed in Elkan (2001). Cost-sensitive methods involve a cost matrix, which defines the penalty associated with misclassification errors. Some specific knowledge and extra learning overhead become necessary to set up the cost matrix (Sun *et al.*, 2009).

9.7.2 Granular Representation of Oversampled Data

The crux of the method is to combine the undersampling mechanism with the formation of information granules built around the selected patterns (data). The corresponding information granule is built in a way to reflect the nature of the data around the selected pattern.

Consider that a certain percentage p out of N data has been selected randomly, resulting in total of P data, $P = pN$. Denote one of the selected patterns by z. Around it we form a set-based granule $\Omega(z)$ embracing the data coming from both the classes:

$$\Omega(z, \rho) = \text{card}\{x \in X \,|\, \|x-z\| < \rho\} \tag{9.18}$$

where ρ assumes values in [0,1]. The geometry of information granule is implied by the type of the distance ||.|| used in (9.18). The value of ρ is optimized in such a way so that the product of coverage and specificity becomes maximized. In other words, we have

$$\rho_{\text{opt}} = \arg\text{Max}_\rho [\text{card}\{x \in X \,|\, \|x-z\| < \rho\} * (1-\rho)] \tag{9.19}$$

Depending upon a location of z, we can visualize three typical situations that are encountered depending upon a location of z versus the data coming from the minority class (see Figure 9.6).

Repeating the process for all P data, we end up with a subset of data coming from the overrepresented (majority) class $x(1)$, $x(2)$, ..., $x(P)$. With each of them comes the number of data embraced by the information granule formed as discussed earlier. Denote these numbers by $N(1)$, $N(2)$, ..., $N(P)$. Obviously, the smallest value of $N(.)$ is set to 1. We normalize these counts into the [1, 2] interval following a linear transformation (scaling):

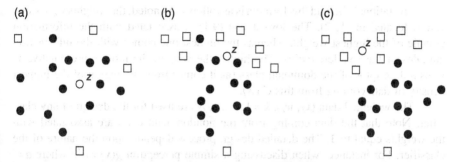

Figure 9.6   Location of z and the resulting information granule (circles, majority class). (a) z surrounded by data coming from the majority class, (b) z located in the region exhibiting a mixture of data coming from the majority and minority class, and (c) z located in the region occupied by data coming from the minority class.

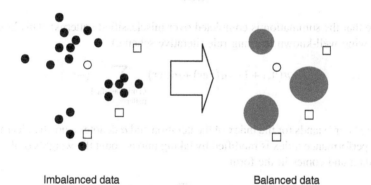

Imbalanced data   Balanced data

Figure 9.7   From a set of imbalanced data to their balanced counterpart involving granular data.

$$w(k) = \frac{N(k) - \min_k N(k)}{\max_k N(k) - \min_k N(k)} + 1 \qquad (9.20)$$

The weight values over 1 indicate that the data from the majority class are weighted higher than those coming from the minority class; however the effect is not overwhelming as the weighting happens in the quite limited range of [1, 2]. Obviously, the effect coming from the majority class could be enhanced by bringing another positive scaling factor α with the values greater than 1, αw_k.

Figure 9.7 illustrates that the balance between the numbers of data belonging to two different classes is achieved by forming a small number of granular data coming from the dominant class.

Having the radius ρ_k of the data (pattern) x_k coming from the dominant class, we convert the individual radii of the granular data into the corresponding weights w_k as follows:

$$w_k = \frac{\rho_k - \rho_{\min}}{\rho_{\max} - \rho_{\min}} + 1 \qquad (9.21)$$

The rationale behind this formula is as follows. As noted, the weights w_k assume values located in [1, 2]. The lowest one (9.1) is associated with the information granule of the highest weight, whereas the highest one comes with the information granules of the smallest radius. The higher the weight, the more representative is x_k as a descriptor of the dominant class (as it comes from the region of the highest density of data coming from this class).

The weighted data (x_k, w_k), $k = 1, 2, \ldots, N$ are used for the design of any classifier. Note that the data coming from the nondominant class are associated with the weights equal to 1. The detailed design process depends upon the nature of the classifier. For instance, when discussing a simple perceptron $g(x) = a^T z$ where $a = [a_0\ b]$ is a vector of parameters and $z = [1\ x]$ is an extended feature vector, its original performance index Q is expressed as

$$Q(a) = \sum_{\substack{k-\text{misclassified} \\ \text{patterns}}} \left(-a^T z_k\right) \tag{9.22}$$

(note that the summation is completed over misclassified patterns). This leads to the following well-known learning rule (iterative scheme)

$$a(\text{iter} + 1) = a(\text{iter}) + \alpha(\text{iter}) \sum_{\substack{k:\text{misclassified} \\ \text{patterns}}} \left(-z_k\right) \tag{9.23}$$

where "iter" stands for the index of the iteration and α denotes a positive learning rate. The performance index is modified by taking into account the weights of the individual data and comes in the form

$$Q(a) = \sum_{\substack{k-\text{misclassified} \\ \text{patterns}}} \left(-a^T z_k\right) w_k \tag{9.24}$$

If a classifier is a neural network, the minimized performance index is a weighted sum of the squared errors:

$$Q = \sum_{k=1}^{N} w_k \left(NN(x_k) - t_k\right)^2 \tag{9.25}$$

where $N(x_k)$ is an output of the network and t_k is the corresponding required target (say, class label) of the kth pattern.

In case of the k-NN classifier, the classification rule involves counts of weights associated with the closest patterns. For instance, if $k = 5$ and the closest patterns are

Class ω_1 (dominant class) 1.8 1.7

Class ω_2 (minority class) 1 1 1

then the pattern is assigned to class ω_1 because $3.5 > 3$.

9.8 CONCLUSIONS

This chapter tackles the crucial issues of the data quality. The incomplete data are commonly encountered in numerous applications. Imbalanced data present in a variety of tasks dealing with anomaly detection of various characters (say, fraud detection, malicious behavior on the network, etc.). Imputation and balancing are the two categories of essential activities aimed to enhance the quality. The underlying ideas inherently involve information granularity. Information granules resulting as a result of imputation and balancing are useful in flagging the quality of the results of these processes and quantifying their performance. The principle of justifiable granularity arises here as a sound algorithmic framework.

REFERENCES

M. Alibeigi, S. Hashemi, and A. Hamzeh, DBFS: An effective density based feature selection scheme for small sample size and high dimensional imbalanced data sets, *Data and Knowledge Engineering* 81–82, 2012, 67–103.

R. Andridge and R. Little, A review of hot deck imputation for survey non-response, *International Statistical Review* 78(1), 2010, 40–64.

S. Barua, M. M. Islam, X. Yao, and K. Muras, MWMOTE—majority weighted minority over-sampling technique for imbalanced data set learning, *IEEE Transactions on Knowledge and Data Engineering* 26(2), 2014, 405–425.

G. E. A. P. A. Batista, R. C. Prati, and M. C. Monard, A study of the behavior of several methods for balancing machine learning training data, *ACM SIGKDD Explorations Newsletter* 6(1), 2004, 20–29.

J. C. Bezdek, *Pattern Recognition with Fuzzy Objective Function Algorithms*, New York, Plenum Press, 1981.

K. V. Branden and S. Verboven, Robust data imputation, *Computational Biology and Chemistry* 33(1), 2009, 7–13.

N. V. Chawla, K. W. Bowyer, L. O. Hall, and W. P. Kegelmeyer, SMOTE: Synthetic minority over-sampling technique, *Journal of Artificial Intelligence Research* 16(1), 2012, 321–357.

N. V. Chawla, D. A. Cieslak, L. O. Hall, and A. Joshi, Automatically countering imbalance and its empirical relationship to cost, *Data Mining and Knowledge Discovery* 17(2), 2008, 225–252.

A. G. Di Nuovo, Missing data analysis with Fuzzy C-Means: A study of its application in a psychological scenario, *Expert Systems with Applications* 38(6), 2011, 6793–6797.

C. Elkan, The foundations of cost-sensitive learning. In: *Proceedings of 17th International Joint Conference on Artificial Intelligence*, Seattle, WA, August 4–10, 2001, 973–978.

M. Galar, A. Fernandez, E. Barrenechea, H. Bustince, and F. Herrera, A review on ensembles for the class imbalance problem: Bagging-, boosting-, and hybrid-based approaches, *IEEE Transactions on Systems, Man, and Cybernetics Part C: Applications and Reviews* 42(4), 2012, 463–484.

H. He, Y. Bai, E. A. Garcia, and S. Li, ADASYN: Adaptive synthetic sampling approach for imbalanced learning. In: *Proceedings of IEEE International Joint Conference on Neural Networks*, IEEE, Hong Kong, China, 2008, 1322–1328.

H. B. He and E. A. Garcia, Learning from imbalanced data, *IEEE Transactions on Knowledge and Data Engineering* 21(9), 2009, 1263–1284.

M. Kubat and S. Matwin, Addressing the curse of imbalanced training sets: One-sided selection. In: *Proceedings of 7th International Conference on Machine Learning*, Stanford, CA, June 29–July 2, 2000, 179–186.

R. J. A. Little and D. B. Rubin, *Statistical Analysis with Missing Data*, 2nd ed., Hoboken, NJ, Wiley-Interscience, 2002.

Y. Liu and S. D. Brown, Comparison of five iterative imputation methods for multivariate classification, *Chemometrics and Intelligent Laboratory Systems* 120, 2013, 106–115.

M. Reilly, Data-analysis using hot deck multiple imputation. *Statistician* 42(3), 1993, 307–313.

D. B. Rubin, *Multiple Imputation for Non response in Surveys*, New York, John Wiley & Sons, Inc., 1987.

M. S. Santos, P. H. Abreu, P. J. García-Laencin, A. Simão, and A. Carvalho, A new cluster-based oversampling method for improving survival prediction of hepatocellular carcinoma patients, *Journal of Biomedical Informatics* 58, 2015, 49–59.

Y. M. Sun, M. S. Kamel, A. K. C. Wong, and Y. Wang, Cost sensitive boosting for classification of imbalanced data, *Pattern Recognition* 40(12), 2007, 3358–3378.

Y. Sun, A. K. C. Wong, and M. S. Kamel, Classification of imbalanced data: A review, *International Journal of Pattern Recognition and Artificial Intelligence* 23(4), 2009, 687–719.

O. Troyanskaya, M. Cantor, G. Sherlock, P. Brown, T. Hastie, R. Tibshirani, D. Botstein, and R. B. Altman, Missing value estimation methods for DNA microarrays, *Bioinformatics* 17(6), 2001, 520–525.

B. X. Wang and N. Japkowicz, Boosting support vector machines for imbalanced data sets, *Knowledge and Information Systems* 25(1), 2010, 1–20.

G. Wu and E. Y. Chang, Class-boundary alignment for imbalanced dataset learning. In: *Proceedings of International Conference on Machine Learning 2003 Workshop on Learning from Imbalanced Data Sets II*, ICML, Washington, DC, 2003.

INDEX

Pattern Recognition: A Quality of Data Perspective, First Edition. Władysław Homenda
and Witold Pedrycz.
© 2018 John Wiley & Sons, Inc. Published 2018 by John Wiley & Sons, Inc.